# Objective Tests in O Level Chemistry

# Objective Tests in O Level Chemistry

**Derek Stebbens**

*Head of Science, Westminster School*

**Heinemann Educational Books**

Heinemann Educational Books Ltd
22 Bedford Square, London WC1B 3HH
LONDON EDINBURGH MELBOURNE AUCKLAND
HONG KONG SINGAPORE KUALA LUMPUR
NEW DELHI IBADAN NAIROBI JOHANNESBURG
EXETER(NH) KINGSTON PORT OF SPAIN

First published 1982

**British Library Cataloguing in Publication Data**

Stebbens, Derek
    Objective tests in O Level Chemistry
    1. Chemistry—Examinations, questions, etc.
    I. Title
    540'.76        QD42

    ISBN 0-435-64844-6

Typeset by Mid-County Press, London SW15
and printed in Great Britain
by Spottiswoode Ballantyne Ltd., Colchester

# Preface

During the last decade, multiple-choice questions have become an important part of examinations at O level. At first, attempts were made to keep these questions 'secure' and all papers had to be returned to the Examining Boards. The problems of maintaining complete security, pressure from teachers and growing expertise in maintaining a flow of acceptable, tested questions have all combined to produce a growing openness in this area of examining. An increasing number of Boards allow schools to retain the multiple-choice papers and in some cases it is possible to purchase copies of recent papers.

The production of a multiple-choice examination paper is likely to be a lengthy and expensive operation. It first involves the gathering together, from a writing team, of a number of items considerably in excess of that required for the final paper. A surplus of items is necessary because of the rejection of many at later stages in the process of compiling the paper. These raw items will be seen by a number of experienced teachers who prepare the items for pretesting, frequently completely rewriting them in the process. The items that are still being considered at this point are made up into a series of multiple-choice papers. Each paper is tested under examination conditions on a large number of pupils. These pupils are selected to make up a population as near identical as possible to that expected for the real examination in which the questions are to be used.

Computer analysis of the performance of each item and the pre-test as a whole gives information about such factors as the difficulty and discriminating power of individual items. This information is used to decide which items could go forward to be considered for inclusion in the final paper. The paper will be built up according to a specification which will be different for each Examining Board. The basic framework will be decided by the total number of questions and the number of different types of items to be used. Working within this framework, there will be scope for variety in the order of the item-type, the range of difficulty of items accepted and the overall difficulty of the paper. There is also the opportunity for considerable variety from Board to Board in the testing of chemical knowledge and understanding This will be related to the syllabus on which each examination is based, the importance with which each area of the syllabus is viewed, and the weight given to the different abilities which an examination of this type can test.

This book is made up of questions that have successfully passed through a selection process similar to that outlined here. The questions used have appeared in the O level multiple-choice papers of London chemistry and Nuffield chemistry examinations. With great generosity, the University of London University Entrance and School Examinations Council has made available over 2200 multiple-choice items from past papers. Items have been selected from this 'bank' to compose 14 tests of various areas of O level chemistry. The tests are intended for use at some time after a topic has been completed. Each test contains sufficient items for a selection to be made of those that are suitable for the course being followed.

This book has two unusual features. The first is the use of genuine questions from past papers which have been pre-tested, analysed, accepted and used. The second unusual feature is the discussion of the answers to these questions. All questions are adequately discussed, some at considerable length. This has produced a book where the space occupied by the answers is comparable with the space occupied by the questions. The aim is to help the student develop in general chemical understanding while, at the same time, gaining in confidence and competence at dealing with questions of this type. This approach should be valuable for students in schools or colleges, those working on their own and those who would benefit from guidance over particular difficulties.

DEREK STEBBENS

# Contents

# Introduction

Five types of multiple-choice question are used in the tests in this book. Most tests use all types of question. In every test, the order in which the types appear is the same. The order is:

1 classification sets
2 multiple-completion
3 multiple-choice
4 assertion/reason
5 situation sets

Examples of each type of question will be given to help you if you are not already familiar with the directions for answering the question, the appearance of the question and the steps involved in arriving at the correct answer. The correct answer is always one letter chosen from A, B, C, D and E and is called the 'key'.

## 1 Classification sets

A classification set begins with the presentation of five similar headings. There is scope for considerable variety in the headings which might, for example, be five different numbers, five different diagrams of apparatus, five different graphs, five different chemicals or five different reactions. A set of questions is based on these five headings and, in each case, you must decide which one of the five headings is the best answer. In the complete set, each heading may appear as the correct answer once, more than once or not at all.

Questions 1–5 concern the following techniques:

    A distillation

    B filtration

    C crystallization

    D chromatography

    E electrolysis

    Select, from A to E, the technique you would use to separate

1 hydrated copper(II) sulphate from its aqueous solution

2 water from a salt solution

3 unused zinc from the result of reacting excess zinc with dilute sulphuric acid

4 the constituents of the colouring matter obtained from grass

5 nitrogen from air

### Question 1
You will probably have carried out this separation or one very similar to it. An aqueous solution of the wanted substance is concentrated by evaporation, frequently using heat from a Bunsen flame to carefully boil the solution in an evaporating basin. When enough water has been evaporated (how do you recognize this point?) the solution is left to cool. Crystals of the substance separate out and these are washed with a little distilled water and then dried. The technique used here could be called 'evaporation' or 'crystallization'. The second term appears in the question and the key is **C**.

### Question 2
This is rather similar to the situation in question 1. Here, however, it is the water that is required rather than the dissolved solid. Distillation (key **A**) is the technique, the water vapour evaporated from the solution being cooled, condensed and collected.

### Question 3
It is likely that you have reacted a metal, such as zinc, with a dilute acid. Any metal left over remains as an insoluble solid in an aqueous solution. The technique of filtration (key **B**) is used to separate the solid from such a mixture.

### Question 4
The separation of the various constituents of the colouring matter in ink, flower petals, sweets or grass is another experiment that you probably will have carried out. When the colouring matter is soluble, the technique of chromatography (key **D**) is often used to separate and identify the substances responsible for the colour.

**Question 5**

In the laboratory, nitrogen can be separated from the other components of air by a technique that could be called 'chemical reaction'. Water vapour, carbon dioxide and oxygen are each removed by a selective reaction and nitrogen is the main component of the remaining gas. What other gases does this gas contain?

No technique resembling 'chemical reaction' is listed in the question. Only one of the five techniques involves changing materials in a chemical way — can you spot which this is? In answering question 5, it is necessary to recognize that the term 'distillation' is very close to the method of 'fractional distillation' which is used on a large scale to separate air into its various components, including nitrogen. This is discussed further in question 7. The correct key is **A**. Note that **A** has been the correct key twice in this set of questions while **E** has not appeared at all.

## 2 Multiple-completion

A statement is made and this is followed by four responses, any one or more than one of which may follow correctly from and complete the original statement. When it has been decided which responses are correct, the key is chosen by using the following code:

A If only **1**, **2** and **3** are correct

B if only **1** and **3** are correct

C if only **2** and **4** are correct

D if only **4** is correct

E if some other response, or combination of responses, of those given is correct

| Directions summarized for question 6 | | | | |
|---|---|---|---|---|
| **A** 1, 2, 3 only correct | **B** 1, 3 only correct | **C** 2, 4 only correct | **D** 4 only correct | **E** Some other response or combination of responses is correct |

**Question 6**

Which substances decrease in mass when heated in air over a Bunsen burner?

1 Copper(II) carbonate

2 Magnesium oxide

3 Hydrated calcium chloride

4 Copper metal

Copper(II) carbonate loses carbon dioxide gas on heating and leaves a black solid, copper(II) oxide. As a gas is lost, substance **1** will decrease in mass on heating. You may have heated malachite, a similar compound of copper, which also decomposes to copper(II) oxide, decreasing in mass in the process.

Magnesium oxide is the product of heating magnesium in oxygen. Further heating in air or oxygen will produce no change in composition and no change in mass.

Hydrated calcium chloride contains water of crystallization, $CaCl_2.6H_2O$. The action of heat will drive off the water of crystallization and the mass will decrease. It is possible that heating will drive off something else as well. Heat a little of the compound in an ignition tube. Test the vapour that leaves the tube with blue cobalt chloride paper and with moist indicator paper. What are you testing for here? When the remaining solid is cool, add a few drops of indicator solution. From your results, decide what, apart from water, is lost on heating, and what is left behind.

Copper is a metal of low reactivity but when heated strongly in air it combines with oxygen to form copper(II) oxide. The metal becomes covered with a thin layer of black oxide. As the copper gains oxygen, it will increase in mass on heating.

The question is concerned with a decrease in mass on heating and this only happens with copper(II) carbonate and hydrated calcium chloride, **1** and **3**. Having decided that **1** and **3** are correct, you should refer to the code. Do not worry about the code before this point. If you can mark your book or examination paper, you might find it helpful to make a pencil tick or cross against the number of each response as soon as you have decided whether or not it is correct. In question 6 your marks should be $\checkmark \times \checkmark \times$. When you refer to the code you will see that the correct key is **B**.

When you meet this type of question in the tests, the directions inside the box will be printed before the group of questions using this code. If you want more information, you should refer back to this part of the introduction.

## 3 Multiple-choice

In a question of this type, a statement is made, or a question is asked, and this is followed by five responses. You are required to select the **one** that is most correct, most complete or follows most closely from the original statement. You are choosing the **best** response of those given.

### Question 7

The constituents of liquid air are listed below, together with their boiling points:

| | |
|---|---|
| Argon | $-186°C$ |
| Helium | $-269°C$ |
| Krypton | $-153°C$ |
| Nitrogen | $-196°C$ |
| Neon | $-246°C$ |
| Oxygen | $-183°C$ |
| Xenon | $-108°C$ |

On fractionation, liquid air can be separated into two fractions, one of which is mainly oxygen and the other mainly nitrogen.

Which of the following groups of substances would you expect to find with oxygen in the oxygen fraction?

**A** Helium, neon and xenon

**B** Argon, neon and krypton

**C** Neon, helium and argon

**D** Xenon, krypton and argon

**E** Krypton, neon and helium

If you look at the boiling points of the two main constituents of liquid air, nitrogen and oxygen, you will see that nitrogen has the lower boiling point. Oxygen has the higher (less negative) boiling point. One would not expect the oxygen fraction to contain either of the two components with the lowest boiling points, helium and neon. **A**, **C** and **E** contain both of these components and cannot be the correct key. **B** contains neon and is unlikely to be the correct key — the oxygen fraction is more likely to contain nitrogen than it is to contain neon. The gases in key **D** all have boiling points higher than nitrogen and this must be the correct key.

The gases in this question have been listed alphabetically. If you find this causes difficulties, it might help if you arranged the gases in order of boiling point.

| | Boiling point in °C |
|---|---|
| Helium | $-269$ |
| Neon | $-246$ |
| Nitrogen | $-196$ |
| Argon | $-186$ |
| Oxygen | $-183$ |
| Krypton | $-153$ |
| Xenon | $-108$ |

If three gases, other than nitrogen, were to be found in the oxygen fraction, these gases would most likely be argon, krypton and xenon (key **D**).

If the nitrogen fraction also contained three other gases, what would you expect these three gases to be?

## 4 Assertion/reason

These questions consist of two statements. The first statement is numbered **1** and the second statement is numbered **2**.

### Question 8

1 Anhydrous copper(II) sulphate is used to test for water.

2 Anhydrous copper(II) sulphate gets hot when water is added to it.

Before selecting the correct key, you have to answer two, or sometimes three, questions:

(1) Is the first statement, on its own, a true or a false statement? In question 8, the first statement is true. Anhydrous copper(II) sulphate is used to test for water.

(2) Is the second statement, on its own, a true or a false statement? In question 8, the second statement is also true. Anhydrous copper(II) sulphate does get hot when water is added to it.

If one of the statements is false, the correct key will be **C**, **D** or **E** (see the code which follows). If both statements are true, you have to answer the third question.

(3) Is the second statement a correct explanation of the first statement? In question 8, the answer is 'no'. Anhydrous copper(II) sulphate becomes hot when substances other than water are added to it. One test for the *presence* of water in a liquid is to see whether anhydrous copper(II) sulphate turns to the blue colour of hydrated copper(II) sulphate — not to see whether anhydrous copper(II) sulphate becomes hot.

You might like to carry out an experiment in which dry ammonia is passed over cold, anhydrous copper(II) sulphate in a combustion tube. Why is it advisable to set up the experiment in a fume cupboard? You will find that the solid becomes very hot even though water is not being added to it. There will also be a variety of colour changes ending with a far larger volume than at the start of an intensely purple solid. If you know about 'moles', you could weigh the solid before and after to see if you have made a compound of formula $CuSO_4.5NH_3$. To obtain this result it is necessary to make certain that the solid you start with is freshly made anhydrous copper(II) sulphate.

The code for deciding the correct key is shown below. When assertion/reason questions appear in the tests in this book, they will be preceded by the directions summarized in the box. If you require more information, you should refer back to this part of the introduction.

Decide whether the **first statement** is true or false. Decide whether the **second statement** is true or false. Then choose

A if both statements are true and the second statement is **A CORRECT EXPLANATION** of the first statement

B if both statements are true but the second statement is **NOT A CORRECT EXPLANATION** of the first statement

C if the first statement is true, but the second statement is false

D if the first statement is false, but the second statement is true

E if both statements are false

| Directions summarized for question 8 | | |
|---|---|---|
| First statement | Second statement | |
| A True | True | Second statement is a correct explanation of the first |
| B True | True | Second statement is NOT a correct explanation of the first |
| C True | False | |
| D False | True | |
| E False | False | |

In question 8, both statements are true but the second is not a correct explanation of the first. The key is **B**.

## 5 Situation sets

A situation set consists of two or more multiple-choice questions (type 3) based on an experiment, an industrial process, a table of information, a piece of apparatus or some other common theme. Each question contains five responses and you choose the best one of those given as described for question 7 earlier.

Tests 15 and 16 at the end of the book contain questions on all topics. The answers are as follows:

**Test 15 Answers**

| | | | | | |
|---|---|---|---|---|---|
| 1 C | 6 E | 11 C | 16 B | 21 C | 26 D |
| 2 E | 7 C | 12 A | 17 C | 22 B | 27 B |
| 3 A | 8 B | 13 C | 18 A | 23 B | 28 C |
| 4 D | 9 D | 14 B | 19 A | 24 C | 29 E |
| 5 B | 10 D | 15 B | 20 D | 25 C | 30 D |

**Test 16 answers**

| | | | | | |
|---|---|---|---|---|---|
| 1 A | 6 C | 11 E | 16 D | 21 C | 26 A |
| 2 B | 7 D | 12 A | 17 B | 22 A | |
| 3 B | 8 A | 13 A | 18 A | 23 C | |
| 4 E | 9 B | 14 D | 19 B | 24 C | |
| 5 A | 10 E | 15 D | 20 A | 25 D | |

# Test 1  The Periodic Table

Questions **1–3** concern the diagram below which shows the Periodic Table divided into five sections labelled **A, B, C, D** and **E**.

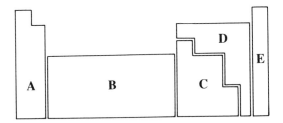

Into which section would the element described in each of the following questions be most likely to fall?

1  A metal which floats on water

2  A metal which forms coloured, hydrated ions of different charges

3  A coloured gas

## Questions 4–6
In the incomplete Periodic Table below, the letters **A** to **E** do not represent the true symbols of the elements concerned.

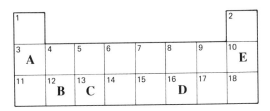

In the questions which follow choose, from the letters **A** to **E**, the one element which fits most closely the description given.

4  The element which would be expected to react most rapidly with water

5  The element which is least likely to form compounds with other elements

6  The element which is likely to combine most readily with sodium

Questions **7–9** concern the following families of elements:

    **A** Alkali metals (e.g. sodium)

    **B** Alkaline earth metals (e.g. magnesium)

    **C** Halogens (e.g. chlorine)

    **D** Noble gases (e.g. argon)

    **E** Transition metals (e.g. copper)

Choose from **A** to **E** the family which

7  has electronic configurations with seven electrons in the outer shell

8  forms ions $X^{2+}$ only (where X is the symbol of the element)

9  has molecules which each contain two atoms at room temperature

| Directions summarized for questions 10 to 16 | | | | |
|---|---|---|---|---|
| **A** 1, 2, 3 only correct | **B** 1, 3 only correct | **C** 2, 4 only correct | **D** 4 only correct | **E** Some other response or combination of responses is correct |

10  A vertical group in the Periodic Table contains

    1  elements with the same atomic mass

    2  elements with the same atomic number

    3  either metals or non-metals but not both

    4  elements with similar properties

11  Which of the following sets of elements consist of members of only one group of the Periodic Table?

    1  Calcium, iron, magnesium

    2  Lithium, sodium, potassium

    3  Oxygen, nitrogen, sulphur

    4  Fluorine, chlorine, iodine

12 Rubidium is an element in the alkali metal family. Rubidium will probably

1 react vigorously with water

2 form coloured salts

3 colour a Bunsen flame

4 form salts soluble in water

13 Some properties of the halogens are that they

1 form ions with a single negative charge

2 react vigorously with sodium

3 each form a gaseous compound with hydrogen

4 are all gaseous at room temperature

14 The element arsenic (As) occurs, with phosphorus, in Group 5 of the Periodic Table. Arsenic would be expected to form chlorides of formula

1 $AsCl_2$      3 $AsCl_4$

2 $AsCl_3$      4 $AsCl_5$

15 Manganese, iron, nickel, cobalt and copper are metals which

1 react slowly if at all with water

2 are ductile

3 form coloured salts

4 have low densities

16 An element $Z$ reacts violently with water to form a hydroxide ZOH which is soluble in water. The element $Z$ will probably

1 be a metal

2 form an ionic chloride

3 form a gaseous oxide

4 form colourless compounds

Directions for questions 17 to 22. Each of the questions for incomplete statements in this section is followed by five suggested answers. Select the best answer in each case.

17 Six elements have atomic numbers 2, 4, 6, 8, 10 and 12, respectively. The two LEAST reactive elements would be those with atomic numbers

A 2 and 8      D 4 and 10

B 2 and 10      E 6 and 12

C 2 and 12

18 Francium (Fr) is a member of the alkali metal family of elements (Li, Na, K, Rb, Cs, Fr). Francium would be expected to

A combine with hydrogen to form a colourless gas at room temperature of formula HFr

B form a carbonate, $Fr_2CO_3$, which is coloured

C react with chlorine to form a chloride of formula FrCl

D form an oxide $Fr_2O$ which dissolves in water to give an acidic solution

E be displaced from an aqueous solution of francium sulphate, $Fr_2SO_4$, by iron filings

19 Which of the following best describes the given oxides?

|   | Calcium oxide | Carbon dioxide | Copper(II) oxide | Sulphur trioxide |
|---|---|---|---|---|
| A | Basic | Neutral | Basic | Acidic |
| B | Acidic | Acidic | Neutral | Basic |
| C | Basic | Neutral | Neutral | Acidic |
| D | Basic | Acidic | Basic | Acidic |
| E | Acidic | Basic | Acidic | Basic |

20 Radium, symbol Ra, is in Group 2 and astatine, symbol At, is in Group 7 of the Periodic Table. The formula of the compound formed between radium and astatine would be expected to be

A RaAt      D $Ra_2At_7$

B $RaAt_2$      E $Ra_7At_2$

C $Ra_2At$

21 An element $E$ forms an ionic compound with lithium. It also forms a compound with hydrogen whose aqueous solution is strongly acidic. $E$ is most likely to be in the same group of the Periodic Table as

A chlorine      D nitrogen

B sodium      E magnesium

C aluminium

**22** Which reagent would produce a visible reaction when added to aqueous sodium bromide?

**A** aqueous barium chloride

**B** aqueous iodine

**C** aqueous potassium iodide

**D** chlorine water

**E** dilute hydrochloric acid

| Directions summarized for questions 23 to 27 | | |
|---|---|---|
| First statement | Second statement | |
| **A** True | True | Second statement is a correct explanation of the first |
| **B** True | True | Second statement is NOT a correct explanation of the first |
| **C** True | False | |
| **D** False | True | |
| **E** False | False | |

**23**

1 Chlorine is liberated when iodine is added to aqueous potassium chloride.

2 Chlorine is above iodine in the halogen group of the Periodic Table.

**24**

1 Silver nitrate solution may be used to distinguish bromides from iodides in aqueous solution.

2 Silver bromide is a deeper yellow than silver iodide.

**25**

1 When phosphorus is burnt in oxygen and the resulting product dissolved in water, the solution has a pH less than 7.

2 Non-metallic elements frequently form alkaline oxides.

**26**

1 Carbon dioxide and silicon dioxide have similar *physical* properties.

2 Carbon and silicon are in the same group of the Periodic Table.

**27**

1 The noble gases are placed in the same group of the Periodic Table.

2 The noble gases all have the same atomic number.

**Questions 28–30**

The apparatus shown was used to demonstrate that in the series

    Chlorine
    Bromine
    Iodine

one halogen will displace a lower halogen from one of its salts.

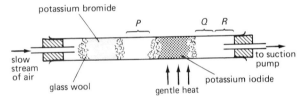

A slow, steady stream of air was drawn through the tube and the crystals of potassium iodide were gently warmed. A *small amount* of chlorine gas was introduced into the air stream and after a little while coloured substances were seen in the parts of the tube labelled *P*, *Q* and *R*.

| Region | *P* | *Q* | *R* |
|---|---|---|---|
| *Observation* | orange-brown vapour | violet vapour | black solid |

**28** What are the names of the substances which were seen in the labelled parts of the tube?

| | *P* | *Q* | *R* |
|---|---|---|---|
| **A** | Chlorine | Bromine | Iodine |
| **B** | Chlorine | Iodine | Bromine |
| **C** | Iodine | Bromine | Chlorine |
| **D** | Bromine | Bromine | Iodine |
| **E** | Bromine | Iodine | Iodine |

29 If, instead of chlorine, a small amount of bromine
vapour had been introduced into the stream of air,
what observations would have been made in the
regions *P*, *Q* and *R*?

| | P | Q | R |
|---|---|---|---|
| **A** | Green gas | Violet vapour | Black solid |
| **B** | Orange-brown vapour | Violet vapour | Black solid |
| **C** | Violet vapour | Violet vapour | Black solid |
| **D** | None | Orange-brown vapour | Orange-brown vapour |
| **E** | Orange-brown vapour | Orange-brown vapour | Orange-brown vapour |

**30** Suppose that the tube was reversed and that the
air flowed through the potassium iodide before
the potassium bromide.

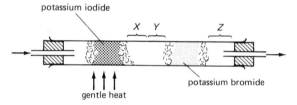

If a small amount of chlorine were introduced into
the air stream, what would be seen, after a short
time, in the regions *X*, *Y* and *Z*?

| | X | Y | Z |
|---|---|---|---|
| **A** | Green gas | Green gas | Orange-brown vapour |
| **B** | Nothing | Nothing | Orange-brown vapour |
| **C** | Violet vapour | Violet vapour | Violet vapour |
| **D** | Violet vapour | Nothing | Orange-brown vapour |
| **E** | Violet vapour | Black crystals | Nothing |

# Test 1 | Answers

1  You should recognize the five areas that appear in this Periodic Table. **A** encloses the metallic elements of Groups 1 and 2. These are the most reactive metals in the Periodic Table. Their atoms are relatively large and, in the case of Group 1, not packed as closely as possible. This gives the metals, particularly those of Group 1, low densities. In the case of lithium and sodium, the additional contribution of low mass for the atoms gives these two elements densities below that of water. The metals that float on water are to be found in **A** which is the key to this question.

Can you explain why potassium, with a density greater than that of water, also floats on water? Which element in **A** has properties completely different from all the others?

2  **B** represents the transition elements, all of which are metallic. They are composed of relatively small, massive and usually closely packed atoms. All of these factors can contribute to the high melting point, high boiling point, high density, and strength of most elements in area **B**. Another property usually shown by these elements is the formation of two or more different positively charged ions which are normally coloured when in aqueous solution or in hydrated crystals. **B** is the correct key to this question.

3  **A**, **B** and **C** are metallic elements. (There is one exception — can you spot it?) The approximate dividing line separating metals from non-metals can be seen on this table running between **C** and **D**.

**D** and **E** are non-metallic elements, frequently with a structure composed of small molecules. The smallest of these molecules contain only one or two atoms and most of such elements are gaseous at room temperature and pressure. The elements in **E** are the noble gases, all of which are unreactive, monatomic and colourless. **D** contains some coloured gases and is the key in this question. How many gaseous elements can you find? How many of these are coloured? Can you locate the gaseous element that is not in **D** or **E**?

4  The elements that react readily with water are the halogens and the most reactive metals, those of Groups 1 and 2. There is no halogen shown in the

table so choice has to be made between **A**, which is lithium, and **B**, which is magnesium. Reactivity among the metals increases on going down each group, with Group 1 being more reactive than Group 2. You will probably have carried out reactions between metals and water and know that lithium beats magnesium in reactivity here. **A** is the correct key. Decision would have been very difficult if **B** had been one position lower in Group 2, calcium.

5  The most unreactive elements are the noble gases. One of these, **E**, is shown in the table and this is the correct key here.

6  One pattern in the Periodic Table is the way in which non-metals (those on the right of the table) can combine with each other, while metals (those on the left) only combine with elements from the right. Almost all compounds are found to be one of two types, non-metal/non-metal or metal/non-metal. This question is about combination with sodium, a metal. Sodium is most likely to combine with a non-metal and the only non-metal shown in the table, apart from the noble gas **E**, is **D**. **D** must be the correct key. What will be the name and formula of the compound formed between sodium and **D**? (Hint — what is the name and formula of the compound formed between sodium and the element above **D** in the Periodic Table?)

7  One of the few perfect patterns in the Periodic Table concerns the number of electrons in the outer shells of atoms of non-transition elements. The number runs from 1 for every element of Group 1 to 8 for the elements of Group 8 (Group 0). The number of outer electrons is the same as the Group number and is the underlying reason for the repetition of chemical and physical properties of elements which gives the Periodic Table its observed form. The element in this question has 7 electrons in its outer shell, so it will be in Group 7 — a halogen, key **C**.

What is the number of electrons in the outer shell of helium atoms? Is the pattern as perfect as suggested above?

8  Another almost perfect pattern found in the Periodic Table concerns the 'bonding power' of non-transition elements. This might be the charge

on an ion, if one is formed, or the number of bonds used in non-ionic bonding. Another term used to describe the value of this property is 'valency'.

The maximum valency of an element is the same as the Group number. For example, the maximum (and only) valency of Group 1 elements is 1. The maximum valency of Group 4 elements, such as carbon, is 4 as in $CCl_4$. The maximum valency of Group 7 elements is 7 as in $IF_7$ or $IO_4^-$. The maximum valency is the same as, and arises from, the number of electrons in the outer shell of the atom.

The valency can be 2 less than the maximum (if this number is possible) or 2 less again or even 2 less again. The values for the valencies of selected elements from Groups 1 to 8 (0) are shown below.

In question 8, X has a valency of 2 so it could be in Groups 2, 4, 6 or 8. But another pattern about valency is that positive ions are formed by

10  Only the fourth statement is correct and the key is **D**. **1** and **2** are completely incorrect. There is rather more room for debate with **3**. The Groups with which you are probably most familiar — 1, 2, 7 and 8 — contain all metals or all non-metals. But the Groups in the middle of the Periodic Table show a change from non-metal at the top to metal at the bottom. Perhaps the best group to show this is carbon, silicon, germanium, tin and lead where the first two elements are non-metallic and the final two are metallic. This change is associated with crossing the dividing line between metals and non-metals as pointed out in the answer to question 3.

11  **1** and **2** are sets of metallic elements but only **2** contains metals of only one Group. **3** and **4** are both sets of non-metallic elements but here only **4** contains members of one Group alone. **2** and **4** are correct, key **C**.

| Group number | Element | Valencies and possible compounds | | | | |
|---|---|---|---|---|---|---|
| 1 | Na | 1  $Na^+Cl^-$ | | | | |
| 2 | Mg | 2  $Mg^{2+}2Cl^-$ | | | | |
| 3 | Tl | 3  $Tl^{3+}3Cl^-$ | 1  $Tl^+Cl^-$ | | | |
| 4 | Pb | 4  $PbCl_4$ | 2  $Pb^{2+}2Cl^-$ | | | |
| 5 | P/Bi | 5  $PCl_5$ | 3  $PCl_3$ | 1  $Bi^+Cl^-$ | | |
| 6 | S | 6  $SF_6$ | 4  $SF_4$ | 2  $SCl_2$ | | |
| 7 | I | 7  $IF_7$ | 5  $IF_5$ | 3  $ICl_3$ | 1  ICl | |
| 8(0) | Xe | 8  $XeF_8$ | 6  $XeF_6$ | 4  $XeF_4$ | 2  $XeF_2$ | |

metallic elements, that is those to the left and bottom of the Periodic Table. In the list above, the only 2+ ions come from Groups 2 and 4.

Looking at the description of element X, it can be seen at once that it is not a transition element. What word in the question tells you this? X must be a metal of Group 2 or Group 4 but Group 4 does not feature among the families listed in the question. Group 2, the alkaline earth metals, does appear in the question. X must be from this Group and the key is **B**.

9  This element forms small molecules and must come from the non-metallic side of the Periodic Table. If you have answered the extra question at the end of the answer to question 3 you will know, if you did not already, that the halogens all contain 2 atoms in their molecules. The key to this question is **C**.

12  Rubidium is like sodium and potassium, both of which react vigorously with water (**1** correct), and produce colour when introduced as compound or element into a Bunsen flame (**3** correct). Another general property of this group is that all the compounds are soluble in water. It is very difficult to find any examples of really insoluble Group 1 salts. **4** is also correct.

Group 1 salts are only coloured if, like potassium permanganate or sodium chromate, they also contain a transition element. (Potassium permanganate and sodium chromate may be called potassium manganate(VII) and sodium chromate(VI) on the course you are following.) **2** is incorrect. **1**, **3** and **4** are correct and the key is **E**.

13  The halogens (non-metals) all react vigorously with sodium (metal) to form ionic compounds $Na^+F^-$, $Na^+Cl^-$, $Na^+Br^-$ and $Na^+I^-$ in which

the halogen has a single negative charge. **1** and **2** are correct.

Of the halogens, only fluorine and chlorine are gaseous at room temperature. Both bromine and iodine give off noticeable vapour at room temperature but the first would be regarded as a liquid and the second as a solid. **4** is incorrect. The halogens do, however, all form gaseous compounds with hydrogen of formula HF, HCl, HBr and HI. **3** is correct which, together with **1** and **2** correct gives a key of **A**.

14 You could answer this question if you knew that phosphorus forms chlorides of formulae $PCl_5$ and $PCl_3$. As arsenic is in the same Group it would be expected to form chlorides of similar formulae, $AsCl_5$ and $AsCl_3$. This makes **2** and **4** correct and the key is **C**.

You could also answer the question from the Group number rather than from the resemblance to phosphorus. Elements of Group 5 can show a valency of the Group number or 2 less or 2 less again (see the answer to question 8). This gives possible valencies of 5, 3 and 1. Arsenic appears to be showing valencies of 2 in $AsCl_2$, 3 in $AsCl_3$, 4 in $AsCl_4$ and 5 in $AsCl_5$. Of these, only the valencies of 3 and 5 are expected and, once again, the key is **C**.

15 The four metals are all transition elements. General properties of such elements include relatively low reactivity (**1** correct), general metallic properties such as being ductile (**2** correct), the formation of coloured salts (**3** correct) and the possession of relatively high densities (**4** incorrect as stated). The correct key is **A**.

16 The only elements that react readily with water are the reactive metallic elements of Groups 1 and 2 and the halogens. Of the halogens, only fluorine would be said to react 'violently' with water and here the product is HF, not FOH as the question requires.

$Z$ must be a metallic element of Group 1 or 2 (**1** correct). Being a member of such a group, the chloride will be an ionic solid (**2** correct) and its oxide will also be an ionic solid (**3** incorrect). Compounds of the element will be colourless unless it is combined with a transition element (**4** correct). **1, 2** and **4** are correct and the key is **E**. To obtain this answer it turns out to be unnecessary to place the element in its exact Group but do you know to what Group $Z$ must belong?

17 You would first look to see if there are any noble gases here, and there are two. Elements 2 and 10

are such elements and the key is **B**. If you are uncertain about this, check in the Periodic Table used for questions 4 to 6. Which noble gases are these?

18 **A** is a description of a halogen rather than of a alkali metal. In **B**, the formula of the carbonate is correct but it will not be coloured. **C** is correct, partly because it does not say too much! In **D**, the oxide will have the formula given and will dissolve in water but the solution will be alkaline. Being a very reactive element, francium will not be displaced by iron and **E** is incorrect. The only correct statement is **C**.

19 Oxides of metals such as calcium and copper would be expected to be basic. Oxides of non-metals such as carbon and sulphur would be expected to be acidic. This is stated correctly in key **D**.

20 Group 2 elements form ions with two positive charges, $Ra^{2+}$.

Halogens, when combining with metals, form ions with one negative charge, $At^-$.

The compound will be between $Ra^{2+}$ and $At^-$. To balance the charges in order to make the compound neutral, the formula will have to be $Ra^{2+}2At^-$ which can also be written $RaAt_2$. This is key **B**.

21 Formation of a compound with the metal lithium shows that $E$ must be a non-metal. **B**, **C** and **E** cannot be correct keys. The compound of $E$ with hydrogen would be strongly acidic if $E$ is chlorine (HCl) but alkaline if $E$ is nitrogen ($NH_3$). $E$ must be chlorine, key **A**.

22 Reactions of aqueous sodium bromide would include formation of a precipitate with silver ions (not included in this question) and the displacement of bromine by a more reactive halogen. This would happen with chlorine and **D** is the correct key.

23 Chlorine is above iodine in the halogen group (second statement true) and because of this chlorine will liberate iodine when added to aqueous potassium iodide. This is the opposite of the first statement which is false. The key is **D**.

24 Aqueous silver nitrate is frequently used to distinguish between bromides and iodides (first statement true). The reactions produce precipitates of silver bromide and silver iodide. These precipitates can be identified because silver bromide is pale yellow while silver iodide is a

darker yellow. This is the opposite of the second statement which is false. The key is **C**.

25 Phosphorus burns in oxygen to give phosphorus oxides which, being formed by a non-metal, are acidic (second statement false). The product dissolves in water to give a solution of pH less than 7 (first statement true). The key is **C**.

26 Carbon dioxide is a gas while silicon dioxide is a solid with a high melting point and high boiling point. They do not have similar physical properties (first statement false). Carbon and silicon are in the same group in the Periodic Table (second statement true). The key is **D**.

27 It is true that the noble gases are placed in the same group but this is not because they have the same atomic number. Every element has its own unique atomic number. The noble gases are placed in the same group because of their similar properties and electronic configuration. The first statement is true, the second false and the key is **C**.

28 As chlorine passes over the potassium bromide it would be expected to displace bromine. The orange-brown vapour in region *P* would be bromine. The key must be **D** or **E**.

The chlorine will now be combined in potassium chloride, mixed with unchanged potassium bromide. Bromine vapour will move on and displace iodine from potassium iodide. This will form violet iodine vapour in the warm region *Q* and black, solid iodine in the cooler region *R*. The correct key is **E**.

29 If chlorine was replaced by bromine vapour, there would be no displacement reaction between bromine and potassium bromide and bromine vapour would pass on into region *P*. From there, the bromine vapour would produce the same changes as if chlorine had been used. The sequence of observations in *P*, *Q* and *R* would be the same as if chlorine had been used. These are listed below the diagram at the start of this set of questions and they also appear in key **B** which is the correct key.

30 Chlorine would displace iodine from the potassium iodide. The iodine would first appear in the warm part of the tube, *X*, as violet vapour. The key must be **C**, **D** or **E**. The violet vapour would condense to a black, crystalline solid in the colder region *Y* and nothing would pass over into region *Z*. The correct key must be **E**. If the potassium bromide were warmed, would you expect orange-brown vapour to be observed at *Z*?

1  In an 'oil film' experiment for estimating the size of molecules, it is assumed that the

1  oil does not mix with water

2  molecules are touching throughout the film

3  film of oil is a monomolecular layer

4  molecules of the oil are spherical

2  When a drop of liquid bromine is placed in the bottom of a gas jar which is then covered with a lid and left for some time, the brown colour of bromine vapour spreads through the whole of the gas jar. This happens because

1  the forces between the molecules in liquid bromine are weak

2  bromine is not so dense as oxygen and nitrogen

3  bromine molecules are in continuous motion

4  molecules of bromine are diatomic

3  1 mole of atoms of ANY element

1  contains the same number of atoms

2  has the same mass

3  contains $L$ atoms, where $L$ is the Avogadro constant

4  occupies the same volume at constant temperature and pressure

4  One mole of atoms of sulphur has a mass twice that of one mole of atoms of oxygen. It follows that

1  1 atom of sulphur has a mass twice that of 1 atom of oxygen

2  20 g of sulphur and 10 g of oxygen will occupy the same VOLUME at room temperature and pressure

3  20 g of sulphur and 10 g of oxygen will each contain the same number of atoms

4  20 g of sulphur will combine with 10 g of oxygen to form sulphur dioxide $SO_2$

5  In the experimental determination of the formula of zirconium oxide, it is necessary to know the

1  masses of zirconium and oxygen which combine together

2  densities of zirconium and oxygen

3  relative atomic masses of zirconium and oxygen

4  groups in the Periodic Table in which zirconium and oxygen are to be found

6  The density of a gas suggests that it has a relative molecular mass of 28. Which of the following gases could it be? (C = 12, H = 1).

1  Propene ($C_3H_6$)

2  Ethene ($C_2H_4$)

3  A mixture of equal volumes of ethane ($C_2H_6$) and propane ($C_3H_8$)

4  A mixture of equal volumes of methane ($CH_4$) and propyne ($C_3H_4$)

Directions for questions 7 to 20. Each of the questions or incomplete statements in this section is followed by five suggested answers. Select the best answer in each case.

7  Aqueous solutions of silver nitrate and potassium chromate react together instantly on mixing and give a red precipitate. If single crystals of these two compounds are very carefully introduced so that they rest on opposite sides of the bottom of a beaker filled with distilled water, where would you expect the first sign of a precipitate to appear?

**A** On the surface of the silver nitrate crystal

**B** On the surface of the potassium chromate crystal

**C** On the surfaces of both crystals

**D** In the water between the two crystals

**E** In the water above the two crystals

8  A student attempted to determine the formula of magnesium oxide by burning some magnesium in air and obtained these results:

Mass of magnesium ribbon                    $= 0.24$ g

Mass of product after heating              $= 0.15$ g

Mass of product after re-heating          $= 0.17$ g

(Relative atomic masses: $Mg = 24$, $O = 16$)

It can be concluded from these results that

**A** 1 mole of magnesium atoms combines with 1 mole of oxygen atoms

**B** 1 mole of magnesium atoms combines with 2 moles of oxygen atoms

**C** 2 moles of magnesium atoms combine with 1 mole of oxygen atoms

**D** magnesium oxide escaped as smoke during the experiment and the results should be ignored

**E** the heating should be repeated until a constant mass is recorded

9  How many moles of atoms are present in 81 g of aluminium? ($Al = 27$)

**A** $\dfrac{1}{3}$

**B** 1

**C** 3

**D** 9

**E** 27

10  The element silicon has a relative atomic mass of 28. This indicates that

**A** each atom of silicon weighs 28 g

**B** the density of silicon is 28 g $cm^{-3}$

**C** atoms of silicon weigh 28 times as much as hydrogen atoms

**D** atoms of silicon all contain 28 protons

**E** atoms of silicon weigh 28 times as much as hydrogen molecules

11  Silicon ($Si = 28$) forms a compound with chlorine ($Cl = 35.5$) in which 5.6 g of silicon is combined with 21.3 g of chlorine. The simplest possible formula for this compound is

**A** $SiCl$

**B** $SiCl_3$

**C** $SiCl_4$

**D** $Si_5Cl_9$

**E** $Si_6Cl_2$

12  In 11.0 g of a compound of phosphorus and sulphur it is found that 4.8 g of sulphur is combined with the phosphorus. What is the formula of this sulphide of phosphorus? ($P = 31$, $S = 32$)

**A** $P_2S_3$

**B** $P_3S_3$

**C** $P_3S_2$

**D** $P_2S_4$

**E** $P_4S_3$

13  The relative atomic masses of zinc and iodine are 65 and 127 respectively. From this statement, it may be deduced that

**A** the radii of the atoms of zinc and iodine are in the ratio 65:127

**B** 65 g of zinc occupies the same volume as 127 g of iodine

**C** 65 atoms of zinc weigh the same as 127 atoms of iodine

**D** 1 g of zinc contains 65 atoms and 1 g of iodine contains 127 atoms

**E** 65 g of zinc contains the same number of atoms as 127 g of iodine

14  32 g of an element X combine with 48 g of oxygen. If $X = 32$ and $O = 16$, the simplest formula of the oxide formed is

**A** $X_3O$

**B** $X_3O_2$

**C** $X_3O_4$

**D** $X_2O_3$

**E** $XO_3$

15 What mass of oxygen contains the same number of oxygen atoms as there are hydrogen atoms in 1 g of hydrogen? ($H = 1$, $O = 16$)

A  8 g               D  24 g

B  12 g              E  32 g

C  16 g

16 How many moles of oxygen atoms contain the same number of atoms as 160 grams of bromine? ($O = 16$, $Br = 80$)

A  2                 D  32

B  8                 E  80

C  16

17 Analysis of a certain oxide shows that 0.5 mole of atoms of X combines with 4 grams of oxygen ($O = 16$). It follows that the simplest formula for this oxide is

A  XO                D  $X_2O$

B  $XO_2$            E  $X_4O_5$

C  $XO_3$

18 When 0.024 g of graphite is burned in oxygen, 48 $cm^2$ of carbon dioxide are formed. What will be the volume of carbon dioxide formed when 0.024 g of diamond is burned in oxygen?
(The volumes are measured under the same conditions of temperature and pressure.)

A  12 $cm^3$         D  96 $cm^3$

B  24 $cm^3$         E  None of the above

C  48 $cm^3$

19 The relative molecular mass of sulphur vapour at 450°C is 256. The molecular formula of sulphur vapour at this temperature would be ($S = 32$)

A  S                 D  $S_6$

B  $S_2$             E  $S_8$

C  $S_4$

20 The simplest formula of a hydrocarbon is $CH_2$ and its relative molecular mass is 112 ($H = 1$, $C = 12$). Its molecular formula is

A  $C_2H_4$          D  $C_8H_{16}$

B  $C_4H_8$          E  $C_{10}H_{20}$

C  $C_6H_{12}$

| Directions summarized for questions 21 to 25 | | |
|---|---|---|
| First statement | Second statement | |
| A  True | True | Second statement is a correct explanation of the first |
| B  True | True | Second statement is NOT a correct explanation of the first |
| C  True | False | |
| D  False | True | |
| E  False | False | |

21
1 If liquid bromine is placed at the bottom of a gas jar, bromine vapour does NOT diffuse through the air in the jar.

2 When liquid bromine is placed at the bottom of a gas jar, bromine molecules are prevented from diffusing through the air by collisions with oxygen and nitrogen molecules.

22
1 When a jar of air is inverted over a jar of carbon dioxide, some carbon dioxide passes into the upper jar.

2 Carbon dioxide is less dense than air.

23
1 Ammonia gas (relative molecular mass $= 17$) diffuses more rapidly than hydrogen chloride (relative molecular mass $= 36.5$), when the temperatures and pressures are identical.

2 The molecule of ammonia contains 4 atoms and the molecule of hydrogen chloride contains only 2 atoms.

**24**

  **1** One mole of hydrogen atoms has the same mass as one mole of carbon atoms.

  **2** One mole of hydrogen atoms contains the same number of atoms as one mole of carbon atoms.

**25**

  **1** The volume of 1 g of gaseous carbon dioxide is greater than the volume of 1 g of solid carbon dioxide.

  **2** 1 g of gaseous carbon dioxide contains more molecules than 1 g of solid carbon dioxide.

**27** A suitable supply of hydrogen for this experiment could best be obtained by

  **A** adding powdered copper to dilute sulphuric acid

  **B** adding sodium to water

  **C** reacting granulated zinc with dilute hydrochloric acid

  **D** reacting magnesium ribbon with steam

  **E** electrolysing dilute hydrochloric acid

| *Results of pupils* | 1 | 2 | 3 | 4 | 5 |
|---|---|---|---|---|---|
| Mass of boat (g) | 32.0 | 30.0 | 35.0 | 20.0 | 28.0 |
| Mass of boat + metal oxide (g) | 40.0 | 38.0 | 39.0 | 28.0 | 42.4 |
| Mass of boat + metal, after reduction (g) | 38.4 | 36.7 | 38.2 | 36.4 | 40.8 |

**Questions 26–30**

In an experiment to determine the formula of an oxide of a metal X, a known mass of the oxide was placed in a previously weighed porcelain boat. Hydrogen was then passed over the boat in a reduction tube and the boat heated until the reaction was complete.

  The boat, which now contained the metal X, was allowed to cool with hydrogen still passing over it. When cold, the boat was re-weighed.

  Five pupils, 1 to 5, performed this experiment and their results are shown above.

**26** The metal produced by reduction was allowed to cool in the stream of hydrogen because

  **A** this ensures that the metal cools quickly

  **B** this sweeps out the water vapour formed

  **C** the excess hydrogen prevents the formation of hydrogen peroxide, $H_2O_2$

  **D** the presence of hydrogen prevents any possible oxidation of the metal

  **E** the reduction of the last trace of metal oxide takes place most quickly at low temperatures

**28** The best method of ensuring that the oxide had been completely converted to metal would be to

  **A** observe when the last trace of the oxide colour had disappeared

  **B** pass a rapid stream of hydrogen over the porcelain boat

  **C** continue heating the porcelain boat in the hydrogen for about 30 minutes

  **D** pass hydrogen until no further water vapour could be detected in the hydrogen escaping from the apparatus

  **E** heat the porcelain boat in the hydrogen until the mass of its contents became constant

**29** Which TWO of the pupils obtained results ,in agreement with one another?

  **A** 1 and 2          **D** 2 and 5

  **B** 1 and 3          **E** 4 and 5

  **C** 2 and 4

**30** The relative atomic masses of X and O are 64 and 16, respectively.

  What was the most likely formula of the metal oxide?

  **A** XO          **D** $X_2O_3$

  **B** $XO_2$          **E** $X_3O_4$

  **C** $X_2O$

# Test 2 | Answers

1 The 'oil film' or 'oil drop' experiment is the only method available in a normal school laboratory for obtaining information about the size of molecules. You may have worked with the variation in which a 'drop of a drop' of undiluted oil is placed on a water surface. Or you may have used a normal size drop of liquid composed of a very dilute solution of the oil in a volatile solvent.

Whichever variation you used, the oil can spread into a thin, circular layer on the water surface. Measurement of the radius of the circle gives information about the volume of oil involved, using the formula $\pi r^2 h$. In this formula, $r$ is the radius of the circle of oil and $h$ is the thickness of the layer. Earlier measurements will have already given a value for the volume of oil used, $V$, so

$$V = \pi r^2 h \quad \text{and} \quad h = \frac{V}{\pi r^2}$$

The only unknown in the equation is $h$ which can therefore be calculated. The value of $h$ gives the size of the molecules in one direction, that of the film thickness, if the film is assumed to be one molecule thick. **3** is one of the assumptions that has to be made. In equating $V$ and $\pi r^2 h$, it is assumed that no oil has been lost by a process such as mixing with the water. **1** is an assumption that has to be made, as is **2**, because if the molecules were not touching, $V$ and $\pi r^2 h$ would not be the same.

The experiment gives a value for the size of molecules in one direction. It is not assumed that the molecules are spherical and **4** is incorrect. You probably know that the molecules are more like matchsticks, with the experiment giving information only about the length of the match. **1**, **2** and **3** are assumptions that have to be made, and the key is **A**.

2 In the experiment described, the molecules in liquid bromine are fairly closely packed. The forces between the molecules in liquid bromine are relatively weak (**1** correct) and the molecules readily become vapour. The molecules vaporize because they are in continuous motion, with many molecules possessing sufficient energy to escape from the liquid. Once in the vapour, the continuous motion of the molecules allows them to fill the whole of the gas jar (**3** correct).

The molecules in bromine vapour are moving at about 150 metres per second. Can you see why it takes some time for the brown colour to spread through the whole of the gas jar? See the answer to question 21 for the solution.

The spread of bromine due to the motion of its molecules, called diffusion, occurs despite bromine being more dense than oxygen and nitrogen (**2** incorrect). Molecules of bromine are diatomic, which is another way of saying that they contain two atoms per molecule, but this is not a reason for the diffusion. **4** is incorrect. Only **1** and **3** are correct and the key is **B**.

3 One mole of atoms of every element contains the same number of atoms, this number being called the Avogadro constant. **1** and **3** are correct.

The mass of one mole of atoms varies from element to element and **2** is not correct. It is only in special cases that one mole of atoms of different elements occupy the same volume and **4** is also incorrect. Only **1** and **3** are correct and the key is **B**.

Can you find values for the volumes of one mole of atoms of oxygen, nitrogen, argon and neon? Is there a pattern about your answers which helps to make further predictions about the volume of one mole of atoms for any other elements?

4 Sulphur is a solid while oxygen is a gas, so the volume of 10 g of oxygen will be far greater than the volume of 20 g of sulphur. **2** is incorrect. If you have solved the problem at the end of question 3, you may realize that **2** would only be correct if sulphur and oxygen were both gases containing the same number of atoms in their molecules. It has already been pointed out that they are not both gases. Can you discover whether they have the same number of atoms in their molecules?

Only **1** and **3** follow from the information given and the key is **B**.

Because **3** is correct, **4** must be wrong. 20 g of sulphur and 10 g of oxygen do contain the same number of atoms. If these masses were to combine, the formula of the oxide formed would be SO or $S_2O_2$ or $S_xO_x$. What mass of oxygen is needed to combine with 20 g of sulphur to form sulphur dioxide, $SO_2$?

**5** The experimental determination of the formula of zirconium oxide would involve the following steps.

(a) Finding the masses of zirconium and oxygen which combine. **1** is correct.

(b) Using the relative atomic masses of zirconium and oxygen to obtain the masses of one mole of each element. This would be followed by dividing the masses in (a) by the mass of one mole to find the ratio by moles in which the two elements combine. **3** is correct.

(c) Scaling up the ratio from (b) to obtain the simplest whole-number ratio of moles of zirconium and oxygen. This would give the simplest possible formula called the 'empirical formula'.

**2** and **4** are not involved so the key is **B**.

**6** The relative molecular masses of the gases in the question are

**1** $C_3H_6$ $\quad 3 \times 12 + 6 = 42$

**2** $C_2H_4$ $\quad 2 \times 12 + 4 = 28$

**3** $C_2H_6$ $\quad 2 \times 12 + 6 = 30$
$\quad\;\; C_3H_8$ $\quad 3 \times 12 + 8 = 44$
$\quad\qquad\qquad$ average $= 37$

**4** $CH_4$ $\qquad 12 + 4 = 16$
$\quad\;\; C_3H_4$ $\quad 3 \times 12 + 4 = 40$
$\quad\qquad\qquad$ average $= 28$

**2** and **4** are correct and the key is **B**.

Do not be worried by the way key **B** is repeated so many times. This is unlikely to happen in a real examination. These questions were selected from different examination papers to test and help you on various points. It is just coincidence that **B** has arisen so often.

**7** It is unlikely that you will have met this experiment. You must imagine what is happening. The single crystals will begin to dissolve and will become surrounded by their aqueous solutions. The particles in these aqueous solutions will diffuse through the water in all directions. Somewhere in the water between the two crystals, those particles diffusing towards each other will meet and produce a red precipitate. **D** is the correct key.

This experiment is similar to one you may have carried out in which ammonia gas and hydrogen chloride gas diffuse towards each other in a long, sealed tube. A white deposit of ammonium chloride is formed where the diffusing gases meet.

Do you think that the red precipitate in question 7 will be formed mid-way between the crystals?

**8** When magnesium is heated in air, it should increase in mass as it combines with gases in the air to form magnesium oxide and some magnesium nitride. In the problem here, the mass does not increase in the expected way so some difficulty must have arisen in the experiment. This is explained by **D** which is the correct key.

**9** The mass of one mole of atoms of aluminium is 27 g. In 81 g of aluminium there are $\dfrac{81 \text{ g}}{27 \text{ g/mole}} = 3$ moles. This is key **C**.

**10** 'Silicon has a relative atomic mass of 28' means that atoms of silicon have a mass of 28 compared with whatever is taken as the standard of unit atomic mass. The present standard is that the isotope of carbon, $^{12}C$, is taken as having a mass of 12 units. This choice does not appear in question 10. You have to select the 'best' answer from those available, which is key **C**. This describes the original standard of relative atomic mass. Do you know which element was chosen to replace hydrogen as the original standard? Was it carbon or was there another standard before the present one?

**11** The numbers of moles of silicon and chlorine in the compound are

Silicon $\qquad \dfrac{5.6 \text{ g}}{28 \text{ g/mole}} = 0.2$ mole

Chlorine $\qquad \dfrac{21.3 \text{ g}}{35.5 \text{ g/mole}} = 0.6$ mole

The ratio of silicon to chlorine is 0.2:0.6 or 1:3.

The simplest possible formula is $SiCl_3$ and the key is **B**. You may be puzzled by the answer, expecting the formula to be $SiCl_4$. In this problem, silicon is forming the compound $Si_2Cl_6$. The simplest possible formula of this compound, giving only the ratio of silicon to chlorine, is $SiCl_3$. This formula is called the 'empirical formula'.

**12** This is a slightly more difficult variation of question 11. The total mass of phosphorus and sulphur is 11.0 g, from which the mass of sulphur has to be subtracted to find the mass of phosphorus, 6.2 g.

The numbers of moles of phosphorus and sulphur in the compound are

Phosphorus $\dfrac{6.2 \text{ g}}{31 \text{ g/mole}} = 0.2$ mole

Sulphur $\dfrac{4.8 \text{ g}}{32 \text{ g/mole}} = 0.15$ mole

The ratio of phosphorus to sulphur is $0.2{:}0.15 = 20{:}15 = 4{:}3$.

    The empirical formula of the compound is $P_4S_3$ which makes key **E** the best choice. Can you see why the word 'best' had to be used in the previous sentence rather than the word 'correct'?

**13** The statement in the question gives no information about the radii of atoms (key **A**) or about the volumes of known masses of elements (key **B**). Key **C** is nearer the mark but suggests, incorrectly, that zinc atoms are heavier than iodine atoms. Would **C** be correct if it were changed to '127 atoms of zinc weigh the same as 65 atoms of iodine'?

    **D** could not be correct as it suggests, as does **C**, that zinc atoms are heavier than iodine atoms. It would also be impossible to state the number of atoms in a given mass unless the Avogadro constant was also provided. To what extent would **D** be better if it were changed to '1 g of zinc contains 127 atoms and 1 g of iodine contains 65 atoms'?

    Finally, **E** is reached which is a correct deduction from the statement at the beginning of the question.

**14** This is similar to question 11 but is made a little more difficult because it deals with an unknown element, X.

    The numbers of moles of X and oxygen in the masses that combine are

X $\dfrac{32 \text{ g}}{32 \text{ g/mole}} = 1$ mole

Oxygen $\dfrac{48 \text{ g}}{16 \text{ g/mole}} = 3$ moles

The ratio of X to oxygen is 1:3 which gives the simplest formula as $XO_3$, key **E**.

**15** There is one mole of hydrogen atoms in 1 g of hydrogen. What mass of oxygen contains this same number of oxygen atoms? Answer: one mole of oxygen atoms, which is 16 g. This is key **C**.

**16** 160 g of bromine contains $\dfrac{160 \text{ g}}{80 \text{ g/mole}} = 2$ moles of bromine atoms. Therefore 2 moles of oxygen atoms contain the same number of atoms as 160 g of bromine and the correct key is **A**.

**17** This is a variation of question 14 with the moles of X given rather than the mass of X.

4 g of oxygen contain $\dfrac{4 \text{ g}}{16 \text{ g/mole}} = 0.25$ mole of oxygen atoms.

The information given shows that 0.5 mole of atoms of X combine with 0.25 mole of oxygen atoms. This is a ratio of 2 atoms of X to 1 atom of oxygen which gives a simplest formula of $X_2O$, key **D**.

**18** Have you been caught by this one? Graphite and diamond are different crystalline forms of the same element, carbon. Equal masses of graphite and diamond will contain equal numbers of carbon atoms and will produce equal volumes of carbon dioxide on burning in oxygen. If 0.024 g of graphite produce 48 cm³ of carbon dioxide, this same volume of carbon dioxide will be formed on burning 0.024 g of diamond. The correct key is **C**.

**19** The information shows that 1 mole of sulphur molecules has a mass of 256 g. 1 mole of sulphur atoms has a mass of 32 g.

In 1 mole of sulphur molecules there will be $\dfrac{256 \text{ g}}{32 \text{ g/mole}}$ of atoms $= 8$ moles of atoms.

This will be the case if each molecule of sulphur contained 8 atoms which is key **E**. This is the answer to question 19 but is it the only solution to the problem? What, for example, would be the relative molecular mass of sulphur vapour if it was composed of equal numbers of molecules of formula $S_6$ and $S_{10}$?

**20** There are various ways to solve this problem. One is to use the relative atomic masses given for carbon and hydrogen in order to calculate the relative molecular mass of each compound, stopping when you come to one with a value of 112.

**A** $C_2H_4$  $RMM = 2 \times 12 + 4 = 28$

**B** $C_4H_8$  $RMM = 4 \times 12 + 8 = 56$

**C** $C_6H_{12}$  $RMM = 6 \times 12 + 12 = 84$

**D** $C_8H_{16}$  $RMM = 8 \times 12 + 16 = 112$

    The correct key is **D**. **E** in this question has a relative molecular mass of 140, but what would you do if **E** were $C_9H_4$ with a relative molecular mass of 112? There is such a compound possible. Its structure is

$CH{\equiv}C{-}C{\equiv}C{-}C{\equiv}C{-}C{\equiv}C{-}CH_3$

The quicker, and more certain, way to solve the problem is to use the information about the simplest formula. As this is $CH_2$, the relative molecular mass of the simplest formula is 14. How many times does 14 go into 112, the relative molecular mass of the real molecule? Answer: 8. So there must be eight $CH_2$ units in the molecule which is $C_8H_{16}$, key **D**.

21 This is a similar situation to that met in question 2. Bromine vapour leaves the liquid and does diffuse through the air. The first statement is false.

Bromine molecules do collide with oxygen and nitrogen molecules but this does not prevent the bromine from diffusing. The second statement is also false and the key is **E**.

The numerous collisions with other molecules force each bromine molecule to follow a very tortuous path called 'the drunkard's walk'. This is why it takes some time for the bromine colour to spread even though the bromine molecules are moving at about 150 metres per second.

22 Carbon dioxide will move from the lower gas jar to the upper one by diffusion. The first statement is true. Diffusion allows this mixing to occur despite carbon dioxide being more dense than air. The second statement is false and the key is **C**.

23 You have probably carried out the experiment in which ammonia gas and hydrogen chloride gas diffuse towards each other from opposite ends of a long, sealed tube. A white deposit of ammonium chloride is formed where the two gases meet. This deposit is first observed nearer the hydrogen chloride end of the tube because ammonia diffuses more rapidly. The first statement is true.

The formulae of the molecules are $NH_3$ and HCl so the second statement is also true but is not a correct explanation of the more rapid diffusion. The key is **B**.

You would not be expected to know a detailed explanation of the real reason for the ammonia diffusing more rapidly than the hydrogen chloride. If you look at the first statement you will see that ammonia molecules have a lower mass than hydrogen chloride molecules. Ammonia molecules move faster to compensate for their lower mass.

24 You are not expected to remember values for the masses of one mole of atoms for various elements but you should be aware that the first statement is false. One mole of atoms of any element contains the same number of atoms, so the second statement is true and the key is **D**.

25 The first statement is true. When a solid or liquid turns to a gas, there is a very large increase in volume. In the solid and liquid states the molecules are in close contact. On vaporization, the molecules become about 10 diameters apart, this giving an approximate volume increase of 1000 times.

1 g (a fixed mass) of gaseous and solid carbon dioxide contains the same number of molecules. The increase in volume is due to the greater separation of molecules when gaseous rather than to any difference in the number of molecules. The second statement is false and the key is **C**.

26 You will probably have carried out an experiment similar to the one described here. If it was a class experiment, you probably used methane, with perhaps a little alcohol vapour, in place of hydrogen. If the hydrogen, or other reducing gas, is turned off before the metal is cool, air might diffuse back into the reduction tube and cause oxidation of the metal. This is covered by key **D**.

27 The very best way to obtain the hydrogen would be from a cylinder of the gas. This is not among the five methods listed so you must look for the best available. This is the one outlined in **C** which will give hydrogen at a reasonably controlled rate in relative safety provided certain precautions are observed. If one is worried about the danger of explosion, **A** is the best method as it gives no hydrogen at all! All the other methods will give hydrogen. Which do you think is the most dangerous of those suggested?

28 Heating to constant mass, key **E**, is a frequently used way of checking that oxidation or reduction of a solid is complete. This method has already been met in this test in question 8. The correct key would have been **E** in that question if the three masses had been, say 0.24 g, 0.35 g and 0.37 g.

29 If it is assumed that reduction is complete in all experiments, the masses of metal and oxygen in each sample can be found by subtraction.

| | 1 | 2 | 3 | 5 |
|---|---|---|---|---|
| Mass of metal/g | 6.4 | 6.7 | 3.2 | 12.8 |
| Mass of oxygen/g | 1.6 | 1.3 | 0.8 | 1.6 |

Looking at these results, it will be seen that pupil 1 obtained exactly twice both values obtained by pupil 3. These results are in agreement and the key is **B**. Which pupil had probably not passed hydrogen over the heated metal for nearly long enough? What is the reason for omitting from the table above the mass of metal and the mass of oxygen for pupil 4?

**30** The results of pupils 1 and 3 are in agreement and both scale up (by multiplication by 10 and 20) to 64 g of metal combining with 16 g of oxygen. From the given relative atomic masses, this is a ratio of one mole to one mole. The formula is XO and the key is **A**.

**Questions 1–5** are concerned with the following particles:

**A** A metal atom

**B** A non-metal atom

**C** A noble gas atom

**D** A negative ion

**E** A positive ion

You may wish to use the part of the Periodic Table which is shown below.

| H | | | | | | | He |
|---|---|---|---|---|---|---|---|
| Li | Be | B | C | N | O | F | Ne |
| Na | Mg | Al | Si | P | S | Cl | Ar |

Select, from **A** to **E**, the particle most clearly related to the numbered structures below.

**1** A particle containing 1 proton, 0 neutrons and 0 electrons

**2** A particle containing 3 protons, 4 neutrons and 3 electrons

**3** A particle containing 8 protons, 9 neutrons and 10 electrons

**4** A particle containing 17 protons, 18 neutrons and 17 electrons

**5** A particle containing 11 protons, 12 neutrons and 10 electrons

## Questions 6–8

The following are the main particles with which chemists are concerned:

**A** Protons          **D** Ions

**B** Neutrons        **E** Atoms

**C** Electrons

Select the particle which best completes each of the following sentences.

**6** The mass of an atom is mainly due to the total number of protons and ... which it contains.

**7** The Periodic Table contains families of elements the outer shells of whose atoms contain the same number of ...

**8** Chlorine has two types of atoms with different atomic weights (relative atomic masses), but these atoms have the same number of ... in their nuclei.

| Directions summarized for qeustions 9 to 18 | | | | |
|---|---|---|---|---|
| **A** 1, 2, 3 only correct | **B** 1, 3 only correct | **C** 2, 4 only correct | **D** 4 only correct | **E** Some other response or combination of responses is correct |

**9** The atomic number of an element is numerically equal to the

  **1** number of neutrons in the nucleus of an atom of the element

  **2** number of protons in the nucleus of an atom of the element

  **3** relative atomic mass of the element

  **4** number of electrons surrounding the nucleus in a neutral atom of the element

**10** From the charge on a metal ion ONLY, it is possible to tell

  **1** how many protons are present in the ion

  **2** how many electrons are present in the ion

  **3** the reactivity of the metal atom from which it was formed

  **4** how many electrons have been lost by the atom from which it was formed

**11** The atomic numbers of the elements F, Na, Mg and K are 9, 11, 12 and 19, respectively. For which of the following ions is the electronic structure 2,8?

1 $F^-$          3 $Mg^{2+}$

2 $Na^+$         4 $K^+$

**12** Which of the following statements about electronic configuration are true?

1 Atoms of the noble gases contain an even number of electrons

2 When an alkali metal atom loses one electron the resultant ion has the electronic configuration of the nearest halogen atom

3 When a halogen atom gains one electron, the resultant ion has the electronic configuration of the nearest noble gas

4 The atomic number of an element is equal to the number of electrons present in a neutral atom of the element

**13** An element with the symbol Y has the electronic configuration 2.8.6. This element forms

1 a compound with hydrogen of formula $H_2Y$

2 an ion $Y^{2-}$

3 a compound with calcium of formula CaY

4 a compound with chlorine of formula $Y_6Cl$

**14** Isotopes of a given element have the same

1 atomic number

2 number of neutrons in their nuclei

3 atomic mass

4 number of electrons in their atoms

**15** The half-life of a radioactive element is

1 the time taken for the activity of a specimen of the element to decay to half its initial value

2 unchanged when the element reacts to form a compound

3 independent of the quantity of the element involved

4 shorter at high temperatures than at room temperature

**16** When a radioactive atom loses an α-particle

1 changes take place in the nucleus of the atom

2 the new atom formed is never radioactive

3 its atomic number decreases by two units

4 its mass number remains unchanged

**17** When an atom of a radioactive element emits a β-particle, the new atom formed

1 has one more proton in its nucleus

2 has the same mass number as the original atom

3 may be radioactive or may be stable

4 is an isotope of the original element

**18** Francium is a radioactive member of the alkali metal group. It has a half life of 20 minutes and its atoms emit β-particles when they decay. It follows that

1 the decay product must be a noble gas (He, Ne, Ar, etc.)

2 in one hour the activity of a specimen of francium will fall to one-eighth of its original value

3 there will be no measurable change in mass of the metal as it decays

4 a specimen of francium chloride will also be radioactive, with a half life of 20 minutes

---

Directions for questions 19 to 26. Each of the questions or incomplete statements in this section is followed by five suggested answers. Select the best answer in each case.

---

**19** If an element had the electronic configuration 2.8.8., it would be

A a halogen

B a transition metal

C an alkali metal

D a noble gas

E an alkaline earth metal

**20** When aluminium forms $Al^{3+}$ ions, the atom of aluminium

A  loses 3 electrons

B  loses 3 protons

C  gains 3 electrons

D  gains 3 protons

E  gains 3 protons and loses 3 electrons

**21** The element scandium has an atomic number of 21 and a relative atomic mass of 45. The scandium atom contains

|   | Protons | Neutrons | Electrons |
|---|---------|----------|-----------|
| A | 21 | 21 | 24 |
| B | 21 | 24 | 21 |
| C | 21 | 24 | 24 |
| D | 24 | 21 | 21 |
| E | 24 | 21 | 24 |

**22** The uncharged sodium atom, which contains 11 protons and 12 neutrons in its nucleus, is represented by $^{23}_{11}Na$. If the uncharged calcium atom is represented by $^{40}_{20}Ca$, the number of electrons present in the doubly charged calcium ion, $Ca^{2+}$, is

A  18              D  42

B  22              E  58

C  38

**23** An ion of formula $X^-$ contains 10 electrons. If the relative atomic mass of X is 18, what is the composition of the nucleus of the ion?

A  7 protons and 11 neutrons

B  8 protons and 10 neutrons

C  9 protons and 9 neutrons

D  9 protons and 10 neutrons

E  10 protons and 8 neutrons

**24** The figures show the initial activities and the half-life periods for five specimens of different radioactive materials.

Which specimen will be the MOST active after 20 minutes?

|   | Initial activity (counts per minute) | Half-life |
|---|--------------------------------------|-----------|
| A | 15 | 3.1 days |
| B | 20 | 13.5 years |
| C | 100 | 10 minutes |
| D | 200 | 5 minutes |
| E | 2000 | 1 minute |

**25** The atomic numbers of some elements are shown below:

| Atomic number | 85 | 86 | 87 | 88 | 89 | 90 |
|---------------|----|----|----|----|----|----|
| Element | At | Rn | Fr | Ra | Ac | Th |

The emission of an $\alpha$-particle by $^{224}Ra$ results in the formation of

A  $^{220}Ra$              D  $^{228}Rn$

B  $^{228}Ra$              E  $^{222}Rn$

C  $^{220}Rn$

**26** The isotope of thorium with an atomic number of 90 and an atomic mass of 232 undergoes seven successive radioactive changes which involve the emission of five $\alpha$-particles and two $\beta$-particles.

Which of the following represents correctly the atomic number and mass number of the atom X that is formed at the end of the changes?

A  $^{225}_{85}X$              D  $^{212}_{82}X$

B  $^{222}_{85}X$              E  $^{210}_{80}X$

C  $^{220}_{82}X$

| Directions summarized for questions 27 to 30 | | |
|---|---|---|
| First statement | Second statement | |
| A  True | True | Second statement is a correct explanation of the first |
| B  True | True | Second statement is NOT a correct explanation of the first |
| C  True | False | |
| D  False | True | |
| E  False | False | |

**27**

1 Isotopes of an element have different relative atomic masses

2 Isotopes of an element contain different numbers of electrons

**28**

1 All isotopes of a given element have the same atomic number

2 All atoms of a given element contain the same number of neutrons

**29**

1 There are several isotopes of lead

2 The nucleus of the lead atom can contain different numbers of protons

**30**

1 Thorium nitrate always contains a trace of lead nitrate

2 Thorium decays to an isotope of lead

---

# Test 3 | Answers

This test contains some questions on radioactivity which may not be a topic on the course you are following. These are questions 15–18, 24–26 and 30.

1 A particle where the number of protons is not the same as the number of electrons, will be an ion. Here the presence of a proton without a neutralizing electron will give rise to a positive ion. The key is **E**. Can you see which ion this is?

2 The numbers of protons and electrons are the same so the particle will be an atom. 3 protons makes it an atom of the third element in the Periodic Table, lithium. This is a metal atom, key **A**.

3 There are more electrons than protons so this particle will be a negative ion, key **D**. Can you work out, using the Periodic Table, which element is involved and the number of negative charges on the ion?

4 This is another atom because the numbers of protons and electrons are the same. 17 protons are present so the element must come seventeenth in the Periodic Table. The element is chlorine, a non-metal, so the key is **B**.

5 There is one more proton than electrons so the particle will be a positive ion, key **E**. Can you write the symbol for this ion with its charge correctly shown?

6 The three particles in atoms with which a chemist is concerned are protons, neutrons and electrons. The masses of protons and neutrons are almost identical and, compared with these, the mass of electrons is negligible. Almost all the mass of an atom is due to the protons and neutrons it contains. The missing word is 'neutrons', key **B**.

7 One of the most perfect patterns found in the Periodic Table is the way in which the number of outer electrons in the atoms of an element is almost always the same as the element's group number. The alkali metals (Group 1), for example, have one electron in the outer shell of their atoms, while atoms of the halogens have seven outer electrons. The missing word is 'electrons', key **C**.

8 The number of protons in the nucleus of a particle is what decides which element is involved. The number of neutrons will vary from isotope to isotope. The number of electrons will be the same as the number of protons if the particle is a neutral atom, less than the number of protons if the particle is a positive ion or more than the number of protons in a negative ion. What decides that an element is, say, chlorine is the number of protons and this is the missing word. The key is **A**.

9 The atomic number is the number of protons in the nucleus of an atom of the element (**2** correct). It is also the numerical position of the element in the Periodic Table but this does not appear among the choices in the question. The total number of electrons around the nucleus of the atom is equal to the atomic number, these equal numbers of positive and negative charges making the atom neutral (**4** correct). Only **2** and **4** are correct and the key is **C**.

10 The charge on an ion is the difference between the number of protons in the nucleus and the number

of electrons surrounding the nucleus. In the case of a positive ion, the charge indicates how many electrons have been lost in forming the ion from a neutral atom where the numbers of protons and electrons are equal. **4** is the only correct statement and the key is **D**. It is not possible from this information to tell the actual numbers of protons and electrons, and it is therefore impossible to identify the element or make any statement about its reactivity.

**11** The atomic number gives the number of protons in the nucleus of each atom. This is the same as the number of electrons for the neutral atom.

F has 9 electrons and $F^-$ will have 10
Na has 11 electrons and $Na^+$ will have 10
Mg has 12 electrons and $Mg^{2+}$ will have 10
K has 19 electrons and $K^+$ will have 18

$F^-$, $Na^+$ and $Mg^{2+}$ all have 10 electrons in their ions. These will be in the electronic configuration 2,8 and the key is **A**.

**12** Atoms of noble gases do contain an even number of electrons. This number is the same as the atomic number and you probably know the atomic numbers of the noble gases near the top of their Group — He(2), Ne(10) and Ar(18). **1** is correct. There is a very quick way of predicting whether any element contains an even or odd number of electrons. All you need to know is the Group number of the element in question. Can you see what this quick way is?

Alkali metal atoms have one electron more than found in the atom of the nearest noble gas. When an alkali metal atom loses an electron, its electronic configuration becomes that of a noble gas, not that of a halogen. **2** is incorrect.

Halogens come immediately before noble gases in the Periodic Table. Halogen atoms have one less electron than atoms of the nearest noble gas. When halogen atoms gain one electron to become a negative ion, their total number of electrons, and electronic configuration, become the same as that of a noble gas. **3** is correct.

**4** is also correct as has been discussed in the answer to question 9.

**1**, **3** and **4** are correct and the key is **E**.

**13** The number of outer electrons for atoms of Y is 6. Y must be an element of Group 6 and will resemble oxygen and sulphur. It should form a hydrogen compound $H_2Y$ like $H_2O$ and $H_2S$. It should also form an ion $Y^{2-}$ similar to $O^{2-}$ in oxides and $S^{2-}$ in sulphides. It would also be expected to form a compound CaY similar to

calcium oxide, CaO, and calcium sulphide, CaS. **1**, **2** and **3** are correct and the key is **A**.

Elements of Group 6, especially if near the bottom of the group, could show a valency of 6. (See the answer to question 6 in the Periodic Table test.) It is just possible that if Y were low enough in Group 6 it might form a compound $YCl_6$ but never a compound $Y_6Cl$, where the valency of Y seems to be a fraction, 1/6.

**14** Being a 'given element' fixes the atomic number (**1** correct), the position in the Periodic Table (not listed) and the total number of electrons in a neutral atom (**4** correct). Isotopes of a given element differ in their number of neutrons and atomic mass (**2** and **3** are incorrect). **1** and **4** are correct and the key is **E**.

**15** **1** is a correct statement of what is meant by half-life. Note the careful wording of **1** to cover the possible formation of daughter products that are themselves radioactive. If this happens, the total specimen would not lose activity as quickly as expected. The specimen 'of the element' itself will, however, decay to half its initial activity in the half-life period.

There is no way at present known for changing the half-life of a radioactive element. **2** and **3** are also correct, while **4** is false, and the key is **A**. Perhaps by the time you read this answer it will have been found to be possible to alter half-lives by vast extremes of, say, pressure or magnetic field. Even if this is the case, the key will still remain at **A**.

**16** An α-particle is a helium nucleus — that is, a helium atom that has lost its two electrons. α-particles contain two protons and two neutrons. They are lost from the nucleus of a radioactive atom which is changed because of their escape (**1** correct). The new atom formed may well be radioactive (**2** incorrect). The loss of two protons does mean that the atomic number goes down by two units (**3** correct) but this loss and the loss of two neutrons means that the mass number decreases by four units. Statement **4** is incorrect. **1** and **3** are correct and the key is **B**.

**17** β-particles are electrons and they are emitted from the nucleus rather than from the shell surrounding the nucleus. A chemist could regard the electron as being produced by the change of a neutron into an electron, which is emitted, and a proton which remains. The product of the decay is a new element with one more proton in its atom and thus with an atomic number one greater than before. **1** is correct.

Electrons ($\beta$-particles) have negligible mass so the mass number is unchanged (**2** correct). The new atom may well be radioactive but does not have to be so. It depends which isotope of which element is decaying in the first step. **3** is correct.

The new atom is of an element with atomic number one greater than before and is not, therefore, an isotope of the original. **4** is incorrect. **1**, **2** and **3** are correct and the key is **A**.

**18** $\beta$-decay, as discussed in the answer to question 17, increases the atomic number by one unit. Alkali metals appear in the Periodic Table one element after the noble gases. An increase of 1 in the atomic number of an alkali metal would convert it into a Group 2 element. **1** is incorrect. Which Group 2 element is formed by this decay?

The activity of the specimen of francium (meaning the francium in the specimen) will fall to half of its original value in 20 minutes. In a further 20 minutes the activity will fall by half again to one-quarter of the original. In the final 20 minutes of the hour the activity will fall by half once more to one-eighth of the original. **2** is correct.

A $\beta$-particle has a mass of approximately 1/2000 on the atomic mass scale. Atoms of francium will have a mass in excess of 200 on the same scale. When the metal decays, its mass will change in the ratio 200 to $200-1/2000$. Such a change would be described as 'no measurable change', especially as the original mass of francium would itself be extremely small. **3** is correct but would not be so if the word 'measurable' was omitted.

The half-life of francium would not change, whatever happens to the francium and **4** is correct. **2**, **3** and **4** are correct, key **E**.

**19** Eight electrons in the outer shell indicate a noble gas, key **D**.

**20** Atoms of aluminium contain 13 protons and 13 electrons, together with some neutrons. Loss of 3 electrons (key **A**) will give aluminium atoms a charge of $+3$ as there is now an excess of 3 in the number of positive protons compared with negative electrons.

**21** An atomic number of 21 indicates 21 protons and 21 electrons in the neutral atom. This is shown only in key **B**. Does **B** contain the expected number of neutrons?

**22** Using the information in the first sentence, it can be seen that the upper number before the symbol represents the sum of the numbers of protons and neutrons (this is called the mass number). The

lower number shows the number of protons which is the same as the number of electrons. The lower number before the symbol for calcium shows that its atoms contain 20 electrons. 2 of these will have been lost to form the calcium ion, leaving 18. The key is **A**.

**23** An ion with one negative charge contains one more electron than there are protons in the nucleus. The nucleus of X must contain 9 protons. This makes the key **C** or **D**. The number of protons and neutrons must total 18 to give this value for the relative atomic mass. This makes the neutron number 9 and the key **C**.

**24** The half-lives of **A** and **B** are sufficiently long for their activities to be close to the initial values of 15 and 20 after only twenty minutes.

**C** will have passed through two half-life periods in this time, the activity reducing from 100 to 50 after ten minutes and becoming 25 after a further ten minutes.

**D** will have passed through four half-lives, its activity reducing from 200 to 100, 100 to 50, 50 to 25 and, finally, from 25 to 12.5 by the end of twenty minutes.

**E** will have passed through twenty half-lives, its activity being halved every minute from 2000 to 1000, 1000 to 500, 500 to 250, 250 to 125, 125 to 62.5, 62.5 to 31.25 and from 31.25 to less than 25 after only seven minutes. There is no need to take the calculation any further as the value has dropped below the final value of 25 for **C**. **C** is the most active after twenty minutes.

**25** $\alpha$-particles contain 2 protons and 2 neutrons. The loss of an $\alpha$-particle reduces the atomic number by 2. In this case the atomic number of 88 for Ra will become 86 and the symbol for the element will be Rn. When an $\alpha$-particle is lost, the mass number is reduced by 4. $^{224}$Ra will become $^{220}$Rn. This is key **C**.

**26** Loss of an $\alpha$-particle reduces the atomic number by 2 units and the atomic mass by 4 units. Loss of five $\alpha$-particles will reduce the atomic number from 90 to 80 and the atomic mass from 232 to 212.

Loss of a $\beta$-particle increases the atomic number by 1 unit but leaves the atomic mass unchanged. Loss of two $\beta$-particles will mean that the atomic number will go down to only 82 rather than to 80. The atomic mass will still be the 212 predicted on $\alpha$-particle loss, so the element formed at the end of the changes is $^{212}_{82}$X. This is key **D**.

**27** Isotopes of an element contain the same number of protons but differ in their number of neutrons. This gives the isotopes different relative atomic masses and the first statement is true.

The number of electrons in the neutral atom of a particular element is fixed. It is the same as the number of protons. The second statement is false and the key is **C**.

**28** The atomic number is the number of protons contained in the nucleus. This is the same for all isotopes of a given element and the first statement is true. As the number of neutrons varies from isotope to isotope, the second statement is false and the key is **C**.

**29** The first statement is true. There are several isotopes of lead, as there are for most elements. The nucleus of a fixed element, lead, must always contain the same number of protons. The second statement is false and the key is **C**.

**30** The natural radioactive decay series of thorium and similar elements, such as uranium, produce lead as the final product or decay. This means that a sample of thorium nitrate always contains traces of all the daughter products of decay, including lead. It would be reasonable to regard the lead as being in the form of lead nitrate. Both statements are true, the second being a correct explanation of the first. The key is **A**.

# Test 4 | Structure and bonding

## Questions 1–3

The elements X and Y form a substance XY (X may be the same as Y). Five different structures for the substance XY are shown below. The shaded circles represent electrons from the outer shell of X and the unshaded circles electrons from the outer shell of Y.

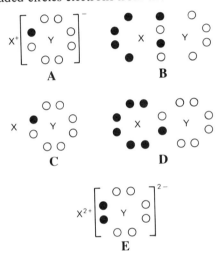

Select the structure adopted by each of the following substances (X is always the left-hand element in the formula and Y the right-hand element).

1  Bromine, $Br_2(l)$

2  Sodium bromide, NaBr(s)

3  Calcium oxide, CaO(s)

## Questions 4–7

Some possible structures for a substance are listed below:

A  A giant structure of ions

B  A giant structure of atoms of a non-metal

C  A giant structure of atoms of a metal

D  Separate, small and identical groups of atoms

E  Separate, free atoms

Select from A to E the structure which is indicated in each of the following questions.

4  A substance which conducts electricity when molten but does not conduct in the solid state

5  A hard substance which has a very high melting point and does not conduct electricity when solid or molten

6  A substance which is a good conductor of electricity when solid and when molten and is insoluble in water

7  A substance which is a gas at laboratory temperature and pressure and may be burned in air to form two new substances.

## Questions 8–10

Some ways in which atoms combine are:

A  Two atoms share one pair of electrons

B  Three atoms share two pairs of electrons

C  One atom donates one electron to another atom

D  One atom donates one electron to each of two other atoms

E  Two atoms each donate one electron to another atom

Select, from A to E, the best description of the way the atoms have combined in

8  magnesium chloride

9  sodium oxide

10  chlorine gas

25

| Directions summarized for questions 11 to 16 | | | | |
|---|---|---|---|---|
| **A**<br>1, 2, 3 only correct | **B**<br>1, 3 only correct | **C**<br>2, 4 only correct | **D**<br>4 only correct | **E**<br>Some other response or combination of responses is correct |

**11** Properties which most metals have in common include that they

1  are good conductors of electricity

2  liberate hydrogen from cold water

3  form giant structures of atoms

4  have low heats of vaporization per mole of atoms

**12** If a substance has high values for its heat of vaporization per mole, its melting point and boiling point, the substance is likely to

1  have a molecular structure

2  dissolve easily in water

3  conduct electricity when molten

4  have a giant structure

**13** Crystals can be formed

1  when a hot, saturated solution is cooled

2  when a molten metal solidifies

3  when a metal is displaced from a solution of one of its salts

4  only from liquids which conduct electricity

**14** Sulphur is composed of rings of eight atoms in

1  crystals obtained from molten sulphur

2  plastic sulphur

3  crystals obtained from a solution of sulphur in carbon disulphide

4  sulphur which has been heated until it is a thick, dark red liquid

**15** A substance melts at $-63°C$ and boils at $62°C$; its heat of vaporization is $5$ kJ $mol^{-1}$ at its boiling point. This information shows that the substance

1  is a liquid at room temperature

2  is colourless

3  has a molecular structure

4  conducts electricity

**16** In which of the following molecules would the atoms all lie in one plane?

1  $CH_4$

2  $CCl_4$

3  $NH_3$

4  $H_2O$

Directions for questions 17 to 23. Each of the questions or incomplete statements in this section is followed by five suggested answers. Select the best answer in each case.

**17** Chlorine is composed of small molecules. Evidence for this includes the fact that chlorine is

A  coloured

B  soluble in water

C  highly reactive

D  gaseous

E  an oxidizing agent

**18** A particular solid compound is a giant structure of ions. From the table below, which horizontal row represents some properties of the compound?

| | Conducts when solid | Conducts when liquid | Conducts when in aqueous solution |
|---|---|---|---|
| A | Yes | No | Yes |
| B | Yes | No | No |
| C | Yes | Yes | No |
| D | No | Yes | Yes |
| E | No | Yes | No |

19 The electrical conductivity of a metal is due to the movement of certain particles through the metal. These particles are called

A atoms

B protons

C ions

D electrons

E neutrons

20 The table below gives values for the relative atomic mass and latent heat of vaporization per gram for five elements. The element most likely to possess a giant structure at room temperature is

| | Relative atomic mass | Latent heat of vaporization in kJ per gram |
|---|---|---|
| A | 19 | 0.2 |
| B | 31 | 0.4 |
| C | 80 | 0.2 |
| D | 85 | 0.9 |
| E | 127 | 0.2 |

21 In what order (lowest first) would you expect the boiling points of sodium chloride (NaCl), phosphorus trichloride ($PCl_3$) and chlorine ($Cl_2$) to be?

A NaCl $PCl_3$ $Cl_2$

B $PCl_3$ NaCl $Cl_2$

C $PCl_3$ $Cl_2$ NaCl

D $Cl_2$ $PCl_3$ NaCl

E $Cl_2$ NaCl $PCl_3$

22 Ten polystyrene spheres are packed flat on the bench as shown

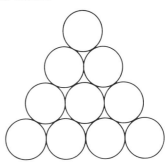

A stable, close packed second layer can be formed.

A only by 3 spheres

B only by 6 spheres

C either by 3 spheres or by 6 spheres

D only by 9 spheres

E either by 6 spheres, or by 9 spheres

23 A mixture of water and crushed ice was heated gradually, with a constant flame and efficient stirring, until the mixture boiled for a short time.
   If the temperature of the mixture was plotted against time (t), which graph would be obtained?

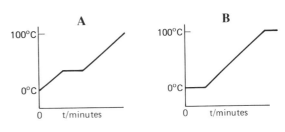

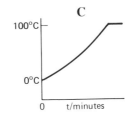

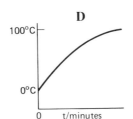

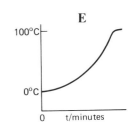

| Directions summarized for questions 24 to 27 | | |
|---|---|---|
| First statement | Second statement | |
| **A** True | True | Second statement is a correct explanation of the first |
| **B** True | True | Second statement is NOT a correct explanation of the first |
| **C** True | False | |
| **D** False | True | |
| **E** False | False | |

**24**

1 Iodine vaporises when heated gently.

2 In iodine the forces between the molecules are weak.

**25**

1 Carbon is a polymorphic (allotropic) element.

2 Carbon is able to form two oxides.

**26**

1 Most metals are good conductors of electricity.

2 Metals have a giant structure of atoms.

**27**

1 Oxygen is completely insoluble in water.

2 Oxygen has a molecular structure.

**Questions 28–30**

In an experiment to determine the solubility of a salt in water, saturated solutions of the salt were prepared at different temperatures, and known masses of the solutions were then evaporated to dryness.

The results obtained are shown in the table below.

| Temperature at which saturated solution was prepared (°C) | Mass of solution taken (g) | Mass of salt after evaporation (g) |
|---|---|---|
| 20 | 33.5 | 8.5 |
| 30 | 45.0 | 11.7 |
| 40 | 27.2 | 7.2 |

**28** The solubility of the salt in water at 20°C, expressed in g per 100 g of water, is

**A** 42.0      **D** 17.0

**B** 34.0      **E** 8.5

**C** 25.4

**29** The most likely solubility of the salt at 60°C, in g per 100 g of water, would be about

**A** 25

**B** 28

**C** 33

**D** 39

**E** 46

**30** In measuring the solubility of the salt, all of the following would be important EXCEPT

**A** making sure the solutions were saturated

**B** avoiding spitting of the solution during the evaporation

**C** evaporating at the temperature for which the solubility is required

**D** weighing the evaporating basin before adding the solution

**E** heating the evaporating basin to constant mass

1   A and E involve ionic bonding, with electrons transferred from X to Y. In B, C and D, pairs of electrons are being shared between atoms in non-ionic/covalent bonding. Bromine is composed of diatomic molecules in which the atoms are held together by one shared pair of electrons. The choice is between B, C and D. B contains two shared pairs of electrons (a double bond) while in C, element X has only one electron in its outer shell and so cannot be a bromine atom.

Perhaps you went straight to D when you answered this question. X and Y both contain the seven electrons in the outer shell which indicates that they are atoms of a halogen. Also, there is only one shared pair of electrons as would be expected for $Br_2$. The correct key is D. Which would you choose if bromine had been shown as $Br_2(s)$ or $Br_2(g)$?

2   Sodium bromide contains elements from opposite sides of the Periodic Table and will therefore be ionic. Sodium will transfer its one outer electron to the seven of the halogen, giving the metal a charge of $+1$ and the halogen a charge of $-1$. This is shown in A which is the correct key.

3   This is another compound composed of elements from the far left and far right of the Periodic Table. This compound will also be ionic but here the Group 2 element will transfer two electrons to the six which the Group 6 element already has in its outer shell. This will leave calcium with a charge of $+2$ while oxygen will have a charge of $-2$. This is shown in E which is the correct key.

B and C have not appeared in the choices above. Can you suggest substances that they might be?

4   This is a description of a compound such as sodium chloride. The ions in the giant structure of ions (key A) are too tightly held in the ionic lattice for very much conductivity to be shown. When the lattice is broken on melting, and the ions are free to move, electrical conductivity is good.

5   The hardness and high melting point indicate one of the giant structures that a substance can possess. These are listed in A, B and C. Electrical conductivity can be used to distinguish between these three possibilities. Giant structures of atoms (metallic) conduct both when solid and when molten. Giant structures of ions conduct only when molten while giant structures of atoms (non-metallic) do not conduct electricity in either state. The correct key is B.

6   The conductivity both when solid and when molten show that this is a giant structure of atoms (metallic), key C. To what extent is the information about the insolubility in water of any help?

7   Gases are composed of very small molecules so the key is either D or E. The smallest of all molecules contains only one atom and is found in the noble gases. The noble gases do not burn in air so the correct key must be D. Which one substance would D be if it had one of the following formulae: $CO_2$, $H_2$, $O_2$, $CH_4$, $CO$?

8   Magnesium chloride is composed of elements from opposite sides of the Periodic Table. It will be ionic. The Group 2 element transfers its two outer electrons per atom to two chlorine atoms. Each chlorine atom picks up one electron to become a $Cl^-$ ion. This is described in key D.

9   Once again, this is an ionic compound. The atoms of the Group 1 element transfer their one outer electron to the Group 6 element. The Group 6 element picks up two electrons in forming the $O^{2-}$ ion. Two sodium atoms are needed, each of which donates one electron to the oxygen atom. This is described in key E.

10  Chlorine is a small molecule with electron pair sharing. The halogens share one pair of electrons as shown by bromine in the first question in this test. The key is A.

11  Metals are composed of giant structures of atoms when solid and are good conductors of electricity. 1 and 3 are correct and the key is B. It is only exceptional metals that liberate hydrogen from cold water (2 incorrect). The fourth statement is completely false as the strong bonding in all giant structures gives rise to high molar heats of vaporization.

12  The high values for these three properties indicate a giant structure of some type. 4 is correct and 1 is definitely wrong. Of the giant structures, only

giant structures of ions are likely to dissolve easily in water so **2** is by no means a certain statement. **3** is not certain either, as giant structures of atoms (non-metallic) do not conduct when molten. The only statement that is 'likely to' be correct is **4** and the key is **D**.

13 You probably will have made crystals by methods **1, 2** and **3**. The 'only' in **4** makes this incorrect and the key is **A**. What examples can you think of that demonstrate that pure liquids which do not conduct electricity can produce crystals? Can you also suggest some non-conducting solutions that will deposit crystals on evaporation?

14 **1** and **3** are correct and provide answers to the two extra questions just above. Molten sulphur on first solidifying produces crystals of monoclinic sulphur which contain $S_8$ rings. Sulphur deposits as crystals of rhombic sulphur on evaporation of a solution of sulphur in carbon disulphide. These crystals also contain $S_8$ rings. Plastic sulphur and the viscous form of molten sulphur contain chains of sulphur atoms rather than rings. Only **1** and **3** are correct and the key is **B**.

15 The low values for all three properties point to a molecular structure (**3** correct). As the substance has melted before room temperature but boils above room temperature, it will be a liquid (**1** correct). The data give no information about the colour of the substance (**2** incorrect). As the substance has a molecular structure, it will not conduct electricity (**4** incorrect). Only **1** and **3** are correct, key **B**.

16 The first three molecules have a 3-dimensional shape so it is impossible for all their atoms to lie in one plane. Do you know what the general shape of these three molecules is? As a water molecule contains only three atoms, these atoms must all lie in one plane. Three points always lie in one plane (even if they lie along a line). **4** is correct and the key is **D**.

17 Small molecules have relatively weak forces acting between their molecules. This means that they are easily melted and boiled. The smallest molecules are frequently already gaseous by the time room temperature is reached. It is the gaseous nature of chlorine, key **D**, which provides evidence that chlorine is composed of small molecules.

18 Giant structures of ions have relatively strong forces holding the ionic solid together. The solid is a poor conductor of electricity so look for a 'no' in the first column of the table. The correct key must be **D** or **E**. When the giant structure of ions is broken down either by melting or dissolving in water, the substance then conducts because the ions are free to move. Look for 'yes' in the second and third columns. The correct key is **D**.

19 The electrical conductivity of a metal is due to the movement of certain mobile electrons through the structure, key **D**.

20 The strength of giant structures of all types gives them relatively high values for their molar heats of vaporization. The molar heat of vaporization is obtained from the table by multiplying the mass of one mole of atoms by the heat of vaporization. **D** has by far the highest value for the heat of vaporization and it also has the next to highest value for the mass of one mole of atoms. The product of these two high values will give **D** the highest molar heat of vaporization, so it is **D** which is most likely to possess a giant structure.

You might like to check which element **D** is, to see if this prediction is correct. You could also check that all the other four elements have a molecular structure, this being the only alternative to a giant structure. How can the table be used to identify these elements?

21 Sodium chloride possesses a giant structure of ions while the other two substances are composed of small molecules. NaCl will have by far the highest boiling point and must come last, so the key is **C** or **D**. Chlorine is the smaller of the two molecules and has the lower molar mass. Chlorine is a gas while $PCl_3$ is a liquid. Chlorine must come first in the list and the key is **D**.

22 There are nine dips in the first layer of spheres and it is into these dips that the next layer must fit if the packing is to be close. It is not possible, however, to fit a sphere into every dip in the second layer. If a sphere is put into the top dip, it obstructs the dip immediately below but two spheres can fit into the dips just below this. The two spheres obstruct the dips just below them so that there is space for only three more spheres, these going into the dips at the very bottom of the diagram. This gives a total of six spheres in the second layer.

There is another way of packing spheres closely into the second layer. This starts with one sphere in the dip below the top dip. The only other spaces that can then be occupied are the two just above the bottom three dips. This gives a total of three spheres for the second way of packing. The correct key is **C**.

The two closest packed structures are called 'hexagonal close packing' and 'cubic close packing', the second also being termed 'face centred cubic'. The packing of the close packed layers is ababab in the first type and abcabc in the second. Is it possible from the answers to the question to tell which type of packing is being constructed in each case?

**23** The mixture of water and crushed ice will be at 0°C and will remain at this temperature until all the ice has melted. This will take some time and the first portion of the graph must be a horizontal line. This is only the case in **B** which must be the correct key. When all the ice has melted, there should be a steady rise in temperature until the water reaches boiling point, 100°C. The temperature should then remain steady until all the water has boiled away. These temperature changes are also shown correctly in **B**.

**24** Iodine does vaporise when gently heated. If you look carefully at iodine in a colourless glass bottle, you will be able to see the purple vapour even at room temperature. The first statement is true.

Iodine is a molecular substance with relatively weak forces between its molecules. This is why molecular substances such as iodine vaporise readily. The second statement is true and a correct explanation of the first. The correct key is **A**.

Can you find in an earlier question the information needed to calculate the energy required to vaporise one mole of iodine molecules?

**25** Many of the elements on the border-line between metals and non-metals in the Periodic Table are able to exist in more than one crystalline form. These forms are called 'polymorphs' or 'allotropes'. Carbon is one of the elements which show this property, and is found in the forms diamond and graphite. The first statement is true.

Carbon is able to form two oxides of formulae $CO$ and $CO_2$. The second statement is also true but is not an explanation of the first. The key is **B**.

**26** Metals are good conductors of electricity, this conduction being due to the mobile electrons contained in the metallic structure. The first statement is true.

There are three types of giant structure. These are giant structure of ions, giant structure of atoms (non-metallic) and giant structure of atoms

(metallic). The second statement is true but is not the reason for metals being good conductors of electricity. The key is **B**.

**27** Nothing is completely insoluble in water. Even one of the most insoluble of substances, mercury(II) sulphide, would produce about ten ions of both $Hg^{2+}(aq)$ and $S^{2-}(aq)$ in a saturated aqueous solution with the volume of a swimming pool. Perhaps this substance would be described as 'completely insoluble' but the term could certainly not be applied to oxygen. Dissolved oxygen and nitrogen can be seen coming out of solution in the form of small bubbles when water is heated. The fairly high solubility of oxygen in water is vital for organisms that live in water.

The first statement is false but the second is true. Oxygen does have a molecular structure. The key is **D**.

**28** This substance is sufficiently soluble in water to have its solubility determined by the mass change on evaporation of a saturated solution. At 20°C, the loss in mass on evaporating the solution is 25 g. 8.5 g of the salt remain and this was originally dissolved in 25 g of water. This means that 34 g of the salt would dissolve in 100 g of water. The key is **B**.

**29** Similar calculations show that the solubility of the salt in 100 g of water is 35.1 at 30°C and 36.0 g at 40°C. The solubility at 60°C is likely to be in excess of 36.0 g, the solubility at 40°C. If the solubility continues to increase at approximately the same rate up to 60°C, a value of 46 g, key **E**, would be too large. 39 g is close to what would be expected and this must be the correct key.

**30** **A**, **B**, **D** and **E** are all essential in the experiment. The saturated solution is made up at the temperature for which the solubility is required. It would then be normal to heat this solution until boiling and then gently evaporate the water. It would not be normal to allow the water to evaporate at the temperature for which the solubility was required and **C** is the key in this question.

## Questions 1–4

Answer questions 1–4 by choosing the appropriate volume from the list **A–E**.

**A** 12 cm$^3$        **D** 48 cm$^3$

**B** 24 cm$^3$        **E** 96 cm$^3$

**C** 36 cm$^3$

All gas volumes are measured at room temperature and pressure. Assume that one mole of gas molecules occupies 24 000 cm$^3$ at room temperature and pressure.

**1** The volume of a 1.0 M solution of potassium iodide required to precipitate the lead ions completely from 48 cm$^3$ of 1.0 M lead nitrate solution according to the equation

$$Pb^{2+}(aq) + 2I^-(aq) \rightarrow PbI_2(s)$$

**2** The volume of carbon dioxide formed when excess dilute hydrochloric acid reacts with 0.002 mole of sodium carbonate, $Na_2CO_3$, according to the equation

$$Na_2CO_3(s) + 2HCl(aq) \rightarrow$$
$$2NaCl(aq) + H_2O(l) + CO_2(g)$$

**3** The volume of oxygen formed when 0.003 mole of potassium nitrate, $KNO_3$, decomposes according to the equation

$$2KNO_3(s) \rightarrow 2KNO_2(s) + O_2(g)$$

**4** The volume of hydrogen formed when 96.5 coulombs of electricity are passed through dilute hydrochloric acid HCl(aq). [1 mole of electrons (1 faraday) = 96 500 coulombs]

## Questions 5–8

In the first of a series of experiments, 50 cm$^3$ of 0.1 M sodium hydroxide was added to 50 cm$^3$ of 0.1 M hydrochloric acid. The pH of the resulting solution was 7. The solution was then evaporated to dryness and the mass of sodium chloride formed was W g.

In further separate experiments only the changes in the questions below were made to the original experiment. This pH and the mass of sodium chloride were compared with those in the original experiment.

| | pH of solution | Mass of sodium chloride on evaporation |
|---|---|---|
| **A** | 7 | W |
| **B** | 7 | Less than W |
| **C** | 7 | More than W |
| **D** | Less than 7 | W |
| **E** | Less than 7 | Less than W |

Select, from **A** to **E**, the letter which best describes the comparison with the original experiment when

**5** 25 cm$^3$ of 0.1 M sodium hydroxide was added to 25 cm$^3$ of 0.1 M hydrochloric acid

**6** 25 cm$^3$ of 0.1 M sodium hydroxide was added to 50 cm$^3$ of 0.1 M hydrochloric acid

**7** 15 cm$^3$ of 0.2 M sodium hydroxide was added to 60 cm$^3$ of 0.1 M hydrochloric acid

**8** 25 cm$^3$ of 0.2 M sodium hydroxide was added to 50 cm$^3$ of 0.1 M hydrochloric acid

| Directions summarized for questions 9 to 14 | | | | |
|---|---|---|---|---|
| **A** | **B** | **C** | **D** | **E** |
| 1, 2, 3 only correct | 1, 3 only correct | 2, 4 only correct | 4 only correct | Some other response or combination of responses is correct |

9 The reaction between dilute sulphuric acid and aqueous strontium chloride can be represented by the equation

$$H_2SO_4(aq) + SrCl_2(aq) \rightarrow SrSO_4(s) + 2HCl(aq)$$

The information *provided by this equation alone* indicates that

1 the reaction is fairly rapid at room temperature

2 a precipitate is formed when the reaction takes place

3 heat is evolved when the reaction takes place

4 2 moles of HCl can be formed from 1 mole of $H_2SO_4$

10 Phosphine ($PH_3$) and hydrogen iodide (HI) react together when mixed at room temperature according to the following equation

$$PH_3(g) + HI(g) \rightarrow PH_4I(s)$$

When 100 cm³ of phosphine is added to 80 cm³ of hydrogen iodide at room temperature and pressure

1 one mole of product is obtained

2 a solid is formed

3 the volume after mixing is 100 cm³

4 there is a 20 cm³ excess of phosphine

11 Hydrogen gas reacts with oxygen gas above 100°C to form steam according to the equation:

$$2H_2(g) + O_2(g) \rightarrow 2H_2O(g)$$

If all volumes were measured above 100°C and at 1 atmosphere pressure, you would expect

1 a decrease in the total volume after the reaction

2 that the volume of steam formed would be equal to the volume of hydrogen used

3 that, if equal volumes of hydrogen and oxygen were used to start with, one-third of the final volume would be oxygen left over

4 the mass of hydrogen reacting to be twice the mass of the oxygen reacting

12 Solutions containing 1 mole of dissolved substance include

1 1 litre of 1 M HCl

2 0.5 litre of 2 M NaOH

3 10 litres of 0.1 M NaCl

4 0.5 litres of 1 M $H_2SO_4$

13 Which of the following solutions contain the same mass of dissolved sodium chloride?

1 500 cm³ of 3 M NaCl

2 500 cm³ of 2 M NaCl

3 250 cm³ of 3 M NaCl

4 250 cm³ of 4 M NaCl

14 20 cm³ of 1.0 M aqueous sodium hydroxide will exactly neutralize

1 10 cm³ of 2.0 M nitric acid

2 20 cm³ of 1.0 M hydrochloric acid

3 40 cm³ of 0.5 M nitric acid

4 40 cm³ of 1.0 M sulphuric acid

Directions for questions 15 to 26. Each of the questions or incomplete statements in this section is followed by five suggested answers. Select the best answer in each case.

15 What mass of calcium chloride ($CaCl_2$) is required to prepare 500 $cm^3$ of a 1.0 M solution? (Ca = 40, Cl = 35.5)

A  0.50 g          D  55.5 g

B  1.00 g          E  500 g

C  37.75 g

16 How many $cm^3$ of 0.2 M hydrochloric acid would be needed to react completely with 25.0 $cm^3$ of 0.1 M sodium carbonate solution?

$$Na_2CO_3 + 2HCl \rightarrow 2NaCl + H_2O + CO_2$$

A  6.25          D  50.0

B  12.5          E  100.0

C  25.0

17 What volume of ammonia gas at room temperature and pressure would just react completely with 250 $cm^3$ of 1.0 M sulphuric acid?

$$2NH_3 + H_2SO_4 \rightarrow (NH_4)_2SO_4$$

(The molar volume of a gas at room temperature and pressure is 24 000 $cm^3$)

A  3 000 $cm^3$          D  24 000 $cm^3$

B  6 000 $cm^3$          E  48 000 $cm^3$

C  12 000 $cm^3$

18 Experiment shows that 1.2 g of magnesium reacts with 8.0 g of bromine to give 9.2 g of magnesium bromide.

How much magnesium bromide will be formed if 1.2 g of magnesium is treated with 6.0 g of bromine?

A  0.9 g          D  7.2 g

B  4.6 g          E  9.2 g

C  6.9 g

19 The element lithium reacts with water according to the equation

$$2Li + 2H_2O \rightarrow 2LiOH + H_2$$

If the molar volume is 24 litres at room temperature and atmospheric pressure (r.t.p.), the number of litres of hydrogen liberated at r.t.p. from 0.1 moles of lithium and excess water would be

A  2.4          D  0.3

B  1.2          E  0.15

C  0.6

20 What volume of oxygen is used when 20 $cm^3$ of propane ($C_3H_8$) are burnt completely in oxygen, assuming that the gas volumes are measured under the same conditions of temperature and pressure?

A  60 $cm^3$          D  120 $cm^3$

B  80 $cm^3$          E  140 $cm^3$

C  100 $cm^3$

21 When 1 mole of a hydrocarbon was burnt in excess oxygen, 2 moles of carbon dioxide and 3 moles of water were formed.

The molecular formula of the hydrocarbon would be

A  $CH_4$          D  $C_2H_6$

B  $C_2H_2$          E  $C_6H_6$

C  $C_2H_4$

22 When 15 $cm^3$ of a gaseous hydrocarbon is completely burned in oxygen, 45 $cm^3$ of carbon dioxide and 45 $cm^3$ of water vapour are obtained, all volumes being measured at the same temperature and pressure. The formula of the hydrocarbon is

A  $C_2H_6$          D  $C_3H_4$

B  $C_3H_8$          E  $C_6H_6$

C  $C_3H_6$

**23** When ammonia gas is oxidized by oxygen in the presence of a catalyst, there is a slight increase in volume. The volume of the gaseous products is reduced by 60% when the steam is removed. Which of the following equations could represent the reaction?

A $4NH_3(g) + 5O_2(g) \rightarrow 4NO(g) + 6H_2O(g)$

B $2NH_3(g) + 2O_2(g) \rightarrow N_2O(g) + 3H_2O(g)$

C $4NH_3(g) + 7O_2(g) \rightarrow 4NO_2(g) + 6H_2O(g)$

D $NH_3(g) + 2O_2(g) \rightarrow HNO_3(g) + H_2O(g)$

E $4NH_3(g) + 3O_2(g) \rightarrow 2N_2(g) + 6H_2O(g)$

**24** The equation for the reaction between silver nitrate solution and potassium chromate solution to produce insoluble silver chromate may be written as

$$2Ag^+(aq) + CrO_4^{2-}(aq) \rightarrow Ag_2CrO_4(s)$$

What is the minimum volume of 0.1 M silver nitrate ($AgNO_3$) solution required to obtain the maximum amount of precipitate from 5 cm³ of 0.2 M potassium chromate ($K_2CrO_4$) solution?

A 40 cm³      D 5 cm³

B 20 cm³      E 2.5 cm³

C 10 cm³

**25** Aqueous sodium hydroxide reacts with a certain metal chloride solution ($MCl_n$) to form a precipitate of the metal hydroxide.

$$MCl_n + nNaOH \rightarrow M(OH)_n + nNaCl$$

10 cm³ of 3.0 M NaOH were found to react exactly with 10 cm³ of 1.5 M $MCl_n$.
    The formula of the metal chloride could be

A MCl          D $M_2Cl$

B $MCl_2$        E $M_2Cl_3$

C $MCl_3$

**26** The number of moles of hydrochloric acid required to liberate all the available carbon dioxide from 1 mole of 'sodium sesquicarbonate' $Na_2CO_3.NaHCO_3.2H_2O$, is

A 1           D 4

B 2           E 5

C 3

**Questions 27–30** are concerned with an investigation of the reaction between magnesium metal and silver ions in aqueous solution.
    Various known masses of magnesium powder were added to an excess of aqueous silver nitrate in separate experiments. A precipitate of silver was formed in each case and this was filtered, washed, dried and then weighed.

**27** The masses of silver obtained from various known masses of magnesium are shown on the graph below.

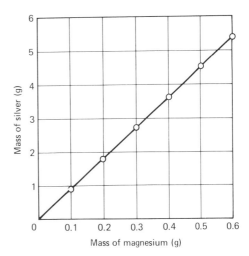

The experiment was repeated with another mass of magnesium and 5 g of silver was obtained. The number of grams of magnesium used for this experiment would have been about

A 0.55        D 4.50

B 0.60        E 5.00

C 0.65

**28** From the graph in question 27, the number of grams of silver which would be produced by the reaction of 1 mole of magnesium (Mg = 24) with excess aqueous silver nitrate is about

A 9           D 220

B 55          E 270

C 110

**29** From the experiment, the number of moles of silver $(Ag = 108)$ produced when 1 mole of magnesium $(Mg = 24)$ reacts with excess aqueous silver nitrate is nearest to

**A** 0.5         **D** 2.2

**B** 1            **E** 108

**C** 2

**30** The equation for the reaction between magnesium and aqueous silver nitrate could be written as

**A** $Mg(s) + 2Ag^+(aq) \rightarrow Mg^{2+}(aq) + 2Ag(s)$

**B** $Mg(s) + Ag^+(aq) \rightarrow Mg(s) + Ag^+(aq)$

**C** $Mg(s) + 2Ag^+(aq) \rightarrow Mg^{2+}(s) + 2Ag(aq)$

**D** $Mg(s) + Ag^+(aq) \rightarrow Mg^{2+}(aq) + Ag(s)$

**E** $2Mg(s) + Ag^+(aq) \rightarrow 2Mg^{2+}(aq) + Ag(s)$

---

# Test 5     Answers

---

**1** The chemical equation that is provided shows that 1 mole of lead ions reacts with 2 moles of iodide ions. The formulae of the compounds involved are $Pb(NO_3)_2$ and $KI$, so one mole of each compound will provide one mole of the relevant ions. This means that one mole of lead nitrate reacts with two moles of potassium iodide.

> $1000 \ cm^3$ of molar solutions contain 1 mole of dissolved substance, so

> $1000 \ cm^3$ of 1.0 M lead nitrate react with $2000 \ cm^3$ of 1.0 M potassium iodide

> $48 \ cm^3$ of 1.0 M lead nitrate react with $96 \ cm^3$ of 1.0 M potassium iodide

The correct key is **E**.

**2** The chemical equation shows that one mole of carbon dioxide is formed from every mole of sodium carbonate. 0.002 mole of carbon dioxide will be formed if 0.002 mole of sodium carbonate is used. The volume of 1 mole of gaseous molecules is given as $24\,000 \ cm^3$ which means that 0.002 mole of carbon dioxide will occupy

> $24\,000 \ cm^3/mole \times 0.002 \ mole = 48 \ cm^3$.

The correct key is **D**.

**3** You have to look at the chemical equation more closely here than in question 2.

> 2 moles of potassium nitrate produced 1 mole of oxygen molecules.

> 1 mole of potassium nitrate produces 0.5 mole of oxygen molecules.

> 0.003 mole of potassium nitrate produces

> $\dfrac{0.003 \ mole}{1 \ mole} \times 0.5$ mole of oxygen molecules

> this is 0.0015 mole of oxygen molecules

The volume of this number of moles of oxygen molecules is

> $24\,000 \ cm^3/mole \times 0.0015 \ mole = 36 \ cm^3$.

The correct key is **C**.

**4** The reaction that liberates hydrogen during electrolysis of aqueous acids is

> $2H^+(aq) + 2e^- \rightarrow H_2(g)$.

The equation shows that 2 moles of electrons ($2 \times 96\,500$ coulombs) are needed to produce 1 mole of hydrogen gas.

> $2 \times 96\,500$ coulombs give 1 mole of hydrogen molecules

> 1 coulomb gives $\dfrac{1}{2 \times 96\,500}$ mole of hydrogen molecules.

> 96.5 coulombs give $\dfrac{1}{2 \times 96\,500} \times 96.5$ mole of hydrogen molecules.

This is $\dfrac{1}{2000}$ mole of hydrogen molecules which would occupy a volume of

$\dfrac{1}{2000}$ mole $\times 24\,000 \ cm^3/mole$.

The volume is $12 \ cm^3$ which is key **A**.

**5** If you have any problems with this set of questions, it may help to write down the original quantities and the change, one below the other.

Original 50 cm³ 0.1 M NaOH    50 cm³ 0.1 M HCl

Change  25 cm³ 0.1 M NaOH    25 cm³ 0.1 M HCl

There are still equal numbers of moles of acid and alkali so a neutral solution of pH 7 will be formed. The key must be **A**, **B** or **C**. The reduction in the quantities will give a lower mass of sodium chloride on evaporation, so the key is **B**.

**6** Original 50 cm³ 0.1 M NaOH    50 cm³ 0.1 M HCl

Change  25 cm³ 0.1 M NaOH    50 cm³ 0.1 M HCl

There is insufficient NaOH to neutralize all the acid in the new experiment. The pH will be less than 7, key **D** or **E**. The quantity of sodium chloride will be determined by the NaOH used and will be half the original amount, so the key is **E**.

**7** Original 50 cm³ 0.1 M NaOH    50 cm³ 0.1 M HCl

Change  15 cm³ 0.2 M NaOH    60 cm³ 0.1 M HCl
or 30 cm³ 0.1 M NaOH

The 15 cm³ of 0.2 M NaOH is equivalent to using 30 cm³ of 0.1 M NaOH. As in question 6, there is insufficient NaOH to neutralize all the acid in 50 or 60 cm³ of 0.1 M HCl. The solution will be acidic and the quantity of sodium chloride will be lower than W. The key is **E**.

**8** Original 50 cm³ 0.1 M NaOH    50 cm³ 0.1 M HCl

Change  25 cm³ 0.2 M NaOH    50 cm³ 0.1 M HCl
or 50 cm³ 0.1 M NaOH

The 25 cm³ of 0.2 M NaOH is equivalent to using 50 cm³ of 0.1 M NaOH. The volume of solution before evaporation will be lower than in the original but the pH will remain at 7 and the mass of sodium chloride at W. The key is **A**.

**9** The chemical equation gives no information about the rate of reaction and **1** is incorrect. Had the equation been given as one involving ions

$$SO_4^{2-}(aq) + Sr^{2+}(aq) \rightarrow SrSO_4(s),$$

it would be regarded as highly likely to be rapid. This is because ionic precipitations, which involve the meeting of ions of opposite charge, are usually rapid. They are not, however, always rapid. You might like to try this reaction, using dilute sulphuric acid and an aqueous solution of a strontium salt, to see how rapidly the precipitate is formed.

The state symbol (s) attached to $SrSO_4$ shows that a precipitate is formed. **2** is correct.

Some chemical equations give information about the energy change in a reaction. The equation would then end '$\Delta H =$ value for energy change'. This equation does not contain the information needed and **3** is incorrect.

The equation says 'one mole of aqueous sulphuric acid reacts with one mole of aqueous strontium chloride to produce one mole of solid strontium sulphate and two moles of aqueous hydrochloric acid'. The information stated in **4** is provided. **2** and **4** are correct and the key is **C**.

**10** 24 000 cm³ of both gases (1 mole of each) would be required to produce 1 mole of product. **1** is incorrect. The state symbol (s) shows that the product is solid. **2** is correct.

The equation shows that 80 cm³ of $PH_3$ will react with 80 cm³ of HI. This will leave 20 cm³ of $PH_3$ and **4** is correct. The 20 cm³ of $PH_3$ is the only gas after the reaction, so **3** is incorrect. **2** and **4** are correct and the key is **C**.

**11** The chemical equation shows that 2 moles of hydrogen react with 1 mole of oxygen to form 2 moles of steam. By Avogadro's law, two moles of gas occupy twice the volume of one mole of gas. This allows moles to be replaced by 'volumes' in the chemical equation, which becomes

2 volumes of $H_2$ react with 1 volume of $O_2$ to give 2 volumes of steam.

The total volume will decrease from 3 volumes to 2 volumes so **1** is correct. 2 volumes of hydrogen will form 2 volumes of steam which means that **2** is also correct.

Statement **3** is rather more difficult to decide on. Consider the 'equal volumes' as '2 volumes'. 2 volumes of hydrogen would react with one of the two volumes of oxygen leaving the other unchanged. 2 volumes of steam will be formed. There will be a total of 3 volumes after reaction, of which 1 volume is oxygen. This means that one-third of the final volume is oxygen and statement **3** is correct.

Statement **4** would be correct if 'mass' were replaced by 'volume' both times. In its present form, **4** is incorrect. **1**, **2** and **3** are correct and the key is **A**.

**12** The reference point for solving this problem could be taken as

1 litre of 1 M solution contains 1 mole (**1** is correct).

Therefore 1 litre of 2 M solution contains 2 moles, and 0.5 litre of 2 M solution contains $2 \text{ moles} \times \dfrac{0.5 \text{ litre}}{1 \text{ litre}}$. This is 1 mole and statement 2 is correct.

In **3**, 1 litre of 0.1 M solution contains 0.1 mole. Therefore 10 litres of 0.1 M solution contain $0.1 \text{ mole} \times \dfrac{10 \text{ litre}}{1 \text{ litre}}$.

This is 1 mole and statement **3** is correct.

In **4**, 1 litre of 1 M solution contains 1 mole. Therefore 0.5 litre of 1 M solution contains $1 \text{ mole} \times \dfrac{0.5 \text{ litre}}{1 \text{ litre}}$.

This is only 0.5 mole and statement **4** is not correct.

**1, 2** and **3** are correct and the key is **A**.

**13** Working in a similar way to question 12, the numbers of moles of sodium chloride in the solutions are

**1** $3 \text{ mole} \times \dfrac{500 \text{ cm}^3}{1000 \text{ cm}^3} = 1.5 \text{ mole}$

**2** $2 \text{ mole} \times \dfrac{500 \text{ cm}^3}{1000 \text{ cm}^3} = 1 \text{ mole}$

**3** $3 \text{ mole} \times \dfrac{250 \text{ cm}^3}{1000 \text{ cm}^3} = 0.75 \text{ mole}$

**4** $4 \text{ mole} \times \dfrac{250 \text{ cm}^3}{1000 \text{ cm}^3} = 1 \text{ mole}$

**2** and **4** both contain 1 mole of sodium chloride and will therefore contain the same mass of the compound. The key is **C**.

**14** Sodium hydroxide reacts with nitric acid in the ratio 1 mole to 1 mole.

$$NaOH + HNO_3 \rightarrow NaNO_3 + H_2O$$

20 cm$^3$ of 1.0 M NaOH would neutralize 20 cm$^3$ of 1.0 M HNO$_3$ or 10 cm$^3$ of 2.0 M HNO$_3$ (**1** correct) or 40 cm$^3$ of 0.5 M HNO$_3$ (**3** correct).

Sodium hydroxide also reacts with hydrochloric acid in the ratio 1 mole to 1 mole. Therefore equal volumes of the two solutions of the same molarity would exactly neutralize each other and **2** is also correct.

There is a problem with **4** as sodium hydroxide can 'neutralize' sulphuric acid in two different ways.

*Either* $NaOH + H_2SO_4 \rightarrow NaHSO_4 + H_2O$

Here the molar ratio for reaction is one to one, so 20 cm$^3$ of 1.0 M NaOH would 'neutralize' 20 cm$^3$ of 1.0 M sulphuric acid. This does not agree with the statement in **4**.

*Or* $2NaOH + H_2SO_4 \rightarrow Na_2SO_4 + H_2O$

Here the molar ratio for reaction is 1 mole of NaOH to 0.5 mole of sulphuric acid. 20 cm$^3$ of 1.0 M NaOH would only require 10 cm$^3$ of 1.0 M sulphuric acid for 'neutralization'. This does not agree with the statement in **4**.

**1, 2** and **3** are correct and the key is **A**. Can you see why the particular figures in **4** have been selected for the incorrect statement?

**15** Work from the starting point that 1000 cm$^3$ of 1 M solution contain 1 mole.

500 cm$^3$ of 1.0 M CaCl$_2$ will contain $1 \text{ mole} \times \dfrac{500 \text{ cm}^3}{1000 \text{ cm}^3} = 0.5 \text{ mole}$.

The molar mass of calcium chloride is 111 g, so the mass of compound required will be 0.5 mole × 111 g/mole = 55.5 g. The correct key is **D**. The question is obviously concerned with a 1.0 M solution 'of calcium chloride'. What would the answer be if the question was 'What mass of calcium chloride is required to prepare 500 cm$^3$ of a 1.0 M solution of chloride ions?'? Why does your answer to this alternative question not appear among the values in question 15?

**16** The chemical equation shows that a given number of moles of sodium carbonate requires twice as many moles of hydrochloric acid. As the acid is already twice the molarity of the sodium carbonate, equal volumes of acid and carbonate will react completely. The volume of acid will be 25 cm$^3$, key **C**.

**17** 1000 cm$^3$ of 1.0 M sulphuric acid contain 1 mole. 250 cm$^3$ of 1.0 M sulphuric acid contain $1 \text{ mole} \times \dfrac{250 \text{ cm}^3}{1000 \text{ cm}^3} = 0.25 \text{ mole}$

The chemical equation given shows that twice as many moles of ammonia will be required compared with the number of moles of sulphuric acid. 0.5 mole of ammonia will be needed and this will occupy a volume of 0.5 mole × 24 000 cm$^3$/mole = 12 000 cm$^3$. The correct key is **C**.

18 The information given in the question is that 8.0 g of bromine react with 1.2 g of magnesium. This is confirmed by the mass of magnesium bromide which is the sum of the masses of the two elements. No bromine or magnesium is left over if these masses are used.

   6.0 g of bromine is 6/8 or 3/4 of the original mass of bromine. This will react with 3/4 of 1.2 g of magnesium, which is 0.9 g. The mass of magnesium bromide will be the sum of 6 g and 0.9 g, which is 6.9 g. The correct key is **C**.

19 The equation shows that 2 moles of lithium release 1 mole of hydrogen molecules. 0.1 mole of lithium would release 0.05 mole of hydrogen molecules.

   The volume of 0.05 mole of molecules is 0.05 mole × 24 litres/mole, which is 1.2 litres. The correct key is **B**.

20 The complete combustion of propane will produce carbon dioxide and water.

$$C_3H_8(g) + 5O_2(g) \rightarrow 3CO_2(g) + 4H_2O(1)$$

The chemical equation shows that 1 mole of propane requires 5 moles of oxygen. 20 cm$^3$ of oxygen will contain the same number of molecules as 20 cm$^3$ of propane (Avogadro's law). 100 cm$^3$ of oxygen will contain five times as many moles (or molecules) as 20 cm$^3$ of propane. 100 cm$^3$ is the volume of oxygen required, key **C**.

21 The formation of 2 moles of carbon dioxide shows that one mole of the hydrocarbon contains 2 moles of carbon. The formula must be $C_2H_x$ and the correct key will be **B**, **C** or **D**.

   The formation of 3 moles of water shows that one mole of the hydrocarbon provides 3 moles of hydrogen molecules or 6 moles of hydrogen atoms. The formula must be $C_2H_6$, key **D**.

22 The information given in the question shows that 1 volume of hydrocarbon reacts with oxygen to give 3 volumes of $CO_2$ and 3 volumes of $H_2O(g)$. From this, by Avogadro's law, we know that 1 molecule of hydrocarbon reacts with oxygen to give 3 molecules of $CO_2$ and 3 molecules of $H_2O$. So one molecule of the hydrocarbon must contain 3 atoms of carbon and 6 atoms of hydrogen. The formula will be $C_3H_6$, key **C**.

23 The slight increase in volume means that there must be more gaseous molecules after the reaction than before. This is only the case in **A** and **E**. If steam is removed from the gaseous products in **A**, for every 10 molecules before removal there will

be 4 after. This is a reduction in both number of molecules and volume (by Avogadro's law) of 60%. The correct key is **A**. What will be the percentage reduction in volume in **E** when the water vapour is removed?

24 The given equation shows that 2 moles of silver nitrate (2 moles of $Ag^+(aq)$) are needed for every 1 mole of potassium chromate (1 mole of $CrO_4^{2-}(aq)$).

   This means that 5 cm$^5$ of 0.2 M potassium chromate would need 10 cm$^3$ of 0.2 M silver nitrate. This silver nitrate could also be provided by using 20 cm$^3$ of 0.1 M solution. This makes **B** the correct key.

25 In the information about the volumes and molarities it can be seen that although an equal volume of NaOH is required to react with the 10 cm$^3$ of $MCl_n$, the NaOH is twice the molarity of the $MCl_n$. This means that 2 moles of NaOH are needed for every mole of $MCl_n$.

   The chemical equation states that $n$ moles of NaOH are needed for every mole of $MCl_n$. $n$ must be 2 and the key is **B**.

26 To answer this question it is necessary to know the chemical equations, or the molar ratios, for the reactions of hydrochloric acid with sodium carbonate and sodium hydrogen carbonate.

$$2HCl + Na_2CO_3 \rightarrow CO_2 + H_2O + 2NaCl$$
$$HCl + NaHCO_3 \rightarrow CO_2 + H_2O + NaCl$$

2 moles of HCl are needed for every mole of $Na_2CO_3$. 1 mole of HCl is needed for every mole of $NaHCO_3$. If a compound such as 'sodium sesquicarbonate' contains 1 mole of both sodium compounds, 3 moles of HCl will be needed. The key is **C**.

   The presence of water of crystallization does not affect the number of moles of HCl required. All this influences is the mass of compound needed to make up one mole.

   Can you see how the wording of the question has carefully excluded the other reaction of hydrochloric acid with sodium carbonate where the equation is

$$HCl + Na_2CO_3 \rightarrow NaHCO_3 + NaCl?$$

27 Find 5 g of silver on the mass of silver axis. Move from here horizontally to the right until the graph line is met. Go down vertically from this point to the mass of magnesium axis. The mass of magnesium indicated is a little less than 0.56 g. **A** must be the correct key.

**28** You do not have to be provided with a graph going up to 24 g of magnesium to answer this question! You can read off from the graph the mass of silver produced by 0.24 g of magnesium. This is 2.2 g of silver. Then scale up by a factor of 100.

    0.24 g of magnesium give 2.2 g of silver
    24 g of magnesium give 220 g of silver.

**D** is the correct key.

**29** The answer to question 28 shows that 220 g of silver is produced by 1 mole of magnesium. This mass of silver is very slightly over 2 moles of silver. It is much closer to 2 than to 2.2 so the key is **C**.

**30** The equation must show 2 moles of silver produced from 1 mole of magnesium. This is only the case in **A** and **C**, but in **C** the state symbols for the products are the wrong way round. **A** is the correct key.

**Questions 1–4** concern pairs of chemical reactions in which you have to decide whether the temperature change in the second reaction is

**A** larger than in the first

**B** smaller than in the first

**C** equal to that in the first and occurs at the same speed

**D** equal to that in the first but occurs more quickly

**E** equal to that in the first but occurs more slowly

1 First reaction: 1 g of zinc rod is dissolved in an excess of dilute hydrochloric acid.

Second reaction: 1 g of zinc filings is dissolved in the same volume of the same acid.

2 First reaction: 50 cm$^3$ of 2.0 M sodium hydroxide solution is added to 50 cm$^3$ of 2.0 M hydrochloric acid.

Second reaction: 20 cm$^3$ of 2.0 M sodium hydroxide solution is added to 20 cm$^3$ of 2.0 M hydrochloric acid.

3 First reaction: 0.01 mole of magnesium atoms is dissolved in 100 cm$^3$ of 1.0 M hydrochloric acid (a large excess).

Second reaction: 0.01 mole of magnesium atoms is dissolved in 200 cm$^3$ of 1.0 M hydrochloric acid.

4 First reaction: Enough silver nitrate solution is added to 20 cm$^3$ of 0.50 M sodium chloride solution to precipitate all the chloride ions as silver chloride.

Second reaction: The same volume of the silver nitrate solution is added to 20 cm$^3$ of 0.50 M potassium chloride solution.

**Questions 5–6** concern the following equations:

**A** $BiCl_3(aq) + H_2O(l) \rightleftharpoons BiOCl(s) + 2HCl(aq)$

**B** $C_5H_{12}(l) + 8O_2(g) \rightarrow 5CO_2(g) + 6H_2O(l)$

**C** $NaOH(aq) + HCl(aq) \rightarrow NaCl(aq) + H_2O(l)$

**D** $2KMnO_4(s) + 16HCl(aq) \rightarrow$
$2KCl(aq) + 2MnCl_2(aq) + 8H_2O(l) + 5Cl_2(g)$

**E** $C_{12}H_{26}(l) \rightarrow C_8H_{18}(l) + 2C_2H_4(g)$

Select, from **A** to **E**, a reaction in which

5 the energy change would be the same as that for the reaction

$$H^+(aq) + OH^-(aq) \rightarrow H_2O(l)$$

6 combustion occurs

| Directions summarized for questions 7 to 12 | | | | |
|---|---|---|---|---|
| **A** 1, 2, 3 only correct | **B** 1, 3 only correct | **C** 2, 4 only correct | **D** 4 only correct | **E** Some other response or combination of responses is correct |

7 The equation

$$Zn(s) + CuSO_4(aq) \rightarrow Cu(s) + ZnSO_4(aq);$$
$$\Delta H = -218 \text{ kJ mol}^{-1}$$

tells us that

1 only one molecule of copper sulphate is reacting

2 one mole of atoms of copper forms when one mole of atoms of zinc reacts

3 the reaction takes place rapidly

4 the reaction is exothermic

**8** Endothermic changes take place when

**1** water is added to anhydrous copper(II) sulphate

**2** iodine is vaporized

**3** acids are neutralized

**4** ice melts

**9** The following diagram shows the energy changes associated with one stage of the heating of water under atmospheric pressure. (H = 1, O = 16)

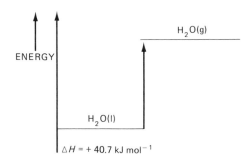

Correct statements about this system include that

**1** the conversion of $H_2O(l)$ to $H_2O(g)$ is exothermic

**2** when 18 g of steam at 100°C condense to water at 100°C, 40.7 kJ of energy are given out

**3** steam at 100°C is hotter than water at 100°C

**4** steam at 100°C contains more energy than the same mass of water at 100°C

**10** Ammonium chloride dissolves in water according to the equation:

$$NH_4Cl(s) + aq \rightarrow NH_4Cl(aq);$$
$$\Delta H = +15 \text{ kJ mol}^{-1}$$

When 0.1 mole of ammonium chloride dissolves in 50 cm³ of water

**1** some heat energy is lost to the surroundings

**2** the temperature of the water falls

**3** the concentration of the solution is 0.5 M

**4** the heat change is 1.5 kJ

**11** The heat change for the reaction shown below can be measured by mixing a solution containing calcium ions with a solution containing carbonate ions.

$$Ca^{2+}(aq) + CO_3^{2-}(aq) \rightarrow CaCO_3(s)$$

In carrying out the experiment

**1** calcium chloride and potassium carbonate solutions could be used

**2** the initial temperatures of the two solutions must be recorded

**3** the heat capacity of the mixed solutions may be taken as equal to that of an equal volume of water

**4** if equal volumes of 2 M solutions at the same temperature are mixed, the change in temperature is independent of the volumes used

**12** The neutralization of hydrochloric acid by aqueous sodium hydroxide can be represented by the following equation

$$HCl(aq) + NaOH(aq) \rightarrow NaCl(aq) + H_2O(l);$$
$$\Delta H = -56 \text{ kJ mol}^{-1}$$

This equation shows that the reaction will evolve 112 kJ when

**1** 1 mole of HCl(aq) is mixed with 2 moles of NaOH(aq)

**2** 2 moles of HCl(aq) are mixed with 2 moles of NaOH(aq)

**3** 2 moles of HCl(aq) are mixed with 1 mole of NaOH(aq)

**4** 3 moles of HCl(aq) are mixed with 2 moles of NaOH(aq)

---

Directions for questions 13 to 18. Each of the questions or incomplete statements in this section is followed by five suggested answers. Select the best answer in each case.

---

**13** Consider the equation

$$Ag^+(aq) + Cl^-(aq) \rightarrow Ag^+Cl^-(s);$$
$$\Delta H = -65.7 \text{ kJ mol}^{-1}$$

This indicates that when the reaction takes place

**A** a solid is formed and heat is evolved

**B** a solid is formed and heat is absorbed

**C** a solution is formed and heat is evolved

**D** a solution is formed and heat is absorbed

**E** the water originally present is turned into steam

**14** Some heats of combustion are as follows:

|  | $\Delta H$ (kJ mol$^{-1}$) |
|---|---|
| Ethane ($C_2H_6$) | $-1600$ |
| Propane ($C_3H_8$) | $-2250$ |
| Butane ($C_4H_{10}$) | $-2900$ |

The next member of this homologous series is pentane ($C_5H_{12}$). The most probable value of the heat of combustion (in kJ mol$^{-1}$) of pentane is

A $-3200$      D $-4500$

B $-3550$      E $-5150$

C $-3850$

**15** It was found that the heat required to evaporate 7.4 g of ethoxyethane [$(C_2H_5)_2O$, relative molecular mass 74] was 2602 J, whilst that required to evaporate 4.6 g of ethanol [$C_2H_6O$, relative molecular mass 46] was 3937 J. From this evidence, which of the following would you judge to be the most correct conclusion?

A The more carbon atoms there are in a molecule, the easier it is to evaporate the substance.

B The molecules of ethoxyethane are held together more strongly than those of ethanol.

C The higher the relative molecular mass of a liquid, the lower the heat of vaporization per mole.

D The molecules of ethanol are bound together more strongly than those of ethoxyethane.

E The atoms in ethanol molecules are more difficult to break apart than those in ethoxyethane molecules.

**16** One mole of magnesium atoms reacts with exactly one mole of copper(II) ions. In a series of experiments magnesium powder was added to copper(II) sulphate solution. In which of the following experiments would the temperature rise be the greatest? (Relative atomic mass: Mg = 24)

A 9.6 g of Mg and 1 dm$^3$ of 0.1 M aqueous $CuSO_4$

B 4.8 g of Mg and 1 dm$^3$ of 0.2 M aqueous $CuSO_4$

C 2.4 g of Mg and 1 dm$^3$ of 0.2 M aqueous $CuSO_4$

D 4.8 g of Mg and 1 dm$^3$ of 0.1 M aqueous $CuSO_4$

E 2.4 g of Mg and 1 dm$^3$ of 0.4 M aqueous $CuSO_4$

**17** Sodium oxide ($Na_2O$) is unaffected when hydrogen is passed over it at high temperatures because

A sodium oxide is a very reactive compound

B sodium and hydrogen do not react

C the bonding between oxygen and sodium is weak

D the sodium–oxygen bonding is stronger than the hydrogen–oxygen bonding

E the bonding between oxygen and hydrogen is strong

**18** For the reaction

$$H_2(g) + Cl_2(g) \rightarrow 2HCl(g)$$

some energy changes are shown below.

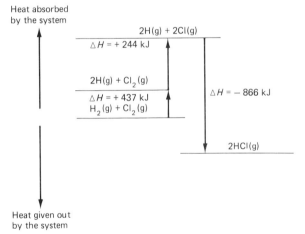

The heat of the reaction (in kJ) to form hydrogen chloride from hydrogen and chlorine molecules is:

A $437 + 244 + 866$

B $437 + 244 + 165$

C $681 - 437$

D $244 - 165$

E $-866 + 681$

| | First statement | Second statement | |
|---|---|---|---|
| | | | Directions summarized for questions 19 to 25 |
| **A** | True | True | Second statement is a correct explanation of the first |
| **B** | True | True | Second statement is NOT a correct explanation of the first |
| **C** | True | False | |
| **D** | False | True | |
| **E** | False | False | |

**19**

1 Heat is evolved when magnesium oxide is formed from its elements.

2 Magnesium oxide at room temperature contains more energy than magnesium and oxygen at the same temperature.

**20**

1 Heat is evolved when water is dropped onto anhydrous copper(II) sulphate.

2 The reaction between water and anhydrous copper(II) sulphate can be reversed.

**21**

1 The products of an endothermic reaction contain more energy than the reactants at the same temperature and pressure.

2 In an endothermic reaction where there is no overall change in temperature and pressure, energy is gained from the surroundings.

**22**

1 Heat energy is liberated when a fuel is burned in air or oxygen.

2 When a fuel is burned in air or oxygen, more energy is released in making bonds than is used in breaking bonds.

**23**

1 When heats of reaction in aqueous solution are measured, the conditions of the experiment should be arranged so that the temperature rises only a few degrees above room temperature.

2 If a reaction mixture becomes very hot, an appreciable part of the heat produced will be lost to the surroundings.

**24**

1 An exothermic reaction gives out heat energy to the surroundings.

2 The products of an exothermic reaction possess less total energy than the reactants at the same temperature.

**25**

1 Heat is absorbed by a liquid when it vaporizes.

2 The particles in a liquid exert an attractive force on each other that has to be overcome on evaporation.

**Questions 26–30** concern an investigation of the dissolving of sodium hydroxide in water.

4 g of sodium hydroxide were added to 50 cm³ of distilled water in an insulated beaker at room temperature. The mixture was stirred continuously and the temperature of the mixture was recorded every 10 seconds until the solution eventually reached room temperature again. The maximum temperature rise was 15°C.

**26** How many moles of sodium hydroxide and water were used in this experiment?
(Na = 23, O = 16, H = 1)

| | Sodium hydroxide | Water |
|---|---|---|
| **A** | 0.1 | 2.78 |
| **B** | 1 | 2.78 |
| **C** | 4 | 50 |
| **D** | 10 | 0.36 |
| **E** | 40 | 18 |

**27** The concentration of the sodium hydroxide solution formed in this experiment, in moles per litre, would be approximately

| | | | |
|---|---|---|---|
| **A** | 1 | **D** | 40 |
| **B** | 2 | **E** | 80 |
| **C** | 4 | | |

**28** Which of the following quantities could be calculated for sodium hydroxide from the results of this experiment?

**A** Heat of neutralization

**B** Heat of fusion

**C** Heat of solution

**D** Heat of precipitation

**E** Heat of vaporization

**29** Which graph would best indicate the result of plotting the temperature of the mixture against time?

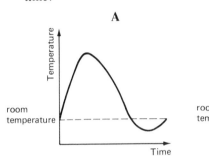
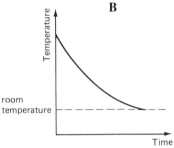
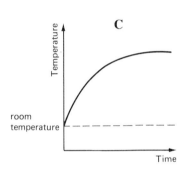

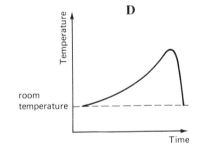
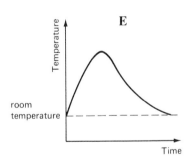

**30** If the experiment were repeated using the same quantity of sodium hydroxide but 100 cm$^3$ of distilled water instead of 50 cm$^3$, the maximum temperature rise would be close to

**A** 5°C

**B** 7.5°C

**C** 15°C

**D** 22.5°C

**E** 30°C

# Test 6 | Answers

1   The same number of moles of zinc are used on both occasions so the energy changes will be identical. The heat evolved is used to raise the temperature of equal volumes of solution, this making the temperature changes the same in both cases. The key must be **C**, **D** or **E**. The larger surface area of the zinc filings will make the second reaction take place more quickly and the key is **D**.

2   The numbers of moles of acid and base have been reduced to 2/5 of the original experiment. This will reduce the energy change in the second reaction to the same extent. The energy evolved is, however, used to raise the temperature of only 2/5 of the original volume of solution. The temperature change in the second reaction will therefore be equal to that in the first. The key will be **C**, **D** or **E**. The reactions involve the meeting of $H^+(aq)$ and $OH^-(aq)$ ions whose concentrations are identical in the two experiments. The reactions will occur at the same speed and the key is **C**.

3   100 cm³ of acid is a large excess which means that the acid can be regarded as remaining at 1.0 M throughout the experiment. The magnesium will not be able to distinguish between reacting with 100 cm³ and 200 cm³ of acid and the second reaction will occur at the same speed as the first. Be careful, however, because the key may not be **C**. 0.01 mole of magnesium atoms is used on both occasions so the energy changes will be the same. In the second experiment, this energy is being used to raise the temperature of a larger volume of solution. This will make the temperature change smaller and the only key where the word 'smaller' is used is **B**. **B** must be the correct key as **C** uses the word 'equal' which is incorrect.

4   The only difference between the two experiments is in the sodium and potassium ions which take no part in the reaction. They are 'spectator ions'. The energy changes will be the same in identical volumes of solution. The temperature changes will be equal and occur at the same speed, key **C**.

5   The energy change on neutralizing any strongly ionized acid with any strongly ionized base is always very close to the same value of $-56$ kJ per mole of water formed in the neutralization reaction. These reactions always involve the formation of water from $H^+(aq)$ and $OH^-(aq)$, this being the only constant reaction taking place. In **C**, for example, which is the correct key for this question, the $Na^+(aq)$ and $Cl^-(aq)$ ions are 'spectators' and take no part in the reaction. A similar situation arises when any other strongly ionised acid and base are used.

6   Combustion occurs in **B**, the hydrocarbon burning in air or oxygen to produce carbon dioxide and water.

7   The chemical equation gives the simplest ratio by moles in which the reactants interact and the products form. This equation indicates that one mole reacts with one mole to give one mole of each product. These quantities could be scaled up or down so long as the molar ratio is maintained. The scaling down could be to one atom or molecule of each but does not have to be – the limitation imposed by the first statement is not correct. The second statement is, however, correct. This information is contained in the equation.
   The equation contains no information about the rate of reaction and **3** is incorrect. Some equations contain information about the states of reactants and products. This equation does but the information is not tested in the question. An equation can also contain information about the energy change. This equation does and the negative sign of $\Delta H$ indicates that the reaction is exothermic. **4** is correct, as is **2** and the key is **C**.

8   In an endothermic change, the products contain more energy than was contained in the reactants at the same temperature. Energy is transferred from the surroundings to the chemicals during an endothermic change. This is not the case in **1** and **3** where the chemicals heat up as reaction takes place and in returning to room temperature energy is transferred from the chemicals to the surroundings.
   In **2** and **4**, however, energy is supplied by the surroundings as the changes take place. These reactions are endothermic and the key is **C**.

**9** The vertical scale shows the energy held by the chemicals. In this particular change, $H_2O(g)$ holds more energy than $H_2O(l)$. This energy has been supplied by the surroundings and the change is endothermic. **1** is not correct. The reaction is endothermic for the conversion of liquid to gas, while it will be exothermic for the reverse change. The magnitude of the energy change is the same in both directions so statement **2** is correct.

Steam and water at $100°C$ are at the same temperature, so it is not true to say that one is 'hotter' than the other. It is true to say that one contains more energy than the other and statement **4** says this correctly. **2** and **4** are correct statements and the key is **C**.

**10** The sign of $\Delta H$ is positive which indicates that energy is gained from the surroundings at constant temperature. **1** is not correct.

When the chemicals are mixed, the temperature does drop at first (**2** correct). Energy then flows in from the surroundings until the mixture has returned to room temperature. This energy change would be 15 kJ if one mole had been used and will be 1.5 kJ if only 0.1 mole is used. **4** is correct.

0.1 mole dissolved in $50 \text{ cm}^3$ of water is equivalent to 2.0 mole dissolved in $1000 \text{ cm}^3$ of water. This means that the solution will be about 2 M, not 0.5 M. Only **2** and **4** are correct and this is another **C**. Can you see why the solution will not be exactly molar?

**11** **1** is correct. These two substances are soluble in water and will supply the required aqueous ions. The chloride and potassium ions will just be spectators. Unless the initial temperatures are known, the experimenter will have to start again! Perhaps you have made a mistake like this. **2** is correct. **3** is also correct. You have probably made this assumption in experiments you have carried out. It is quite a good assumption because aqueous solutions tend to have a slightly lower heat capacity per gram than water but a slightly higher density than water. These two effects approximately cancel to give dilute solutions a heat capacity close to that of an equal volume of water.

**4** is also correct. Imagine the experiment was carried out first with a certain volume and then with double the volume of each. In the second case, the energy change would be doubled but this would be spread through twice the volume of solution. The temperature rise or fall would be the same. The key is **E**. Are the words 'equal' and '2 M' essential in **4**?

**12** To produce 112 kJ it is necessary for 2 moles of the acid and base to react. Only one mole of each can react in **1** and **3** because the substance present in the smaller number of moles will decide the extent of reaction. In **2** and **4**, two moles of each will react, leaving nothing over in **2**, while in **4**, one mole of HCl will remain unused. The key is **C**.

**13** This is the equation for the reaction met in question 4. Spectator ions have been removed. The state symbols show that two aqueous solutions produce a solid, so the key must be **A** or **B**. The additional information provided by this equation is that the products of the reaction contain less energy ($\Delta H$ negative) than the reactants at the same temperature. Heat is evolved in this reaction and the key is **A**. No information is provided in the equation about the statement made in **E**. To decide for certain on **E** one would need to know the concentrations of the original solutions, the ratio in which they were used, the heat capacity of the resulting mixture, and its boiling point. The relatively small value for $\Delta H$ suggests that **E** would never be true if aqueous solutions were mixed.

**14** The formula of the substance increases by one carbon atom and two hydrogen atoms between one compound and the next. If one mole of the substances is considered each time, there is an increase of one mole of carbon atoms and two moles of hydrogen atoms each time. This steady increase is reflected in the increase in the heat of combustion, which is 650 kJ per mole more exothermic each time. If this regular increase continues between butane and pentane, the $\Delta H$ value of $-2900$ kJ per mole for butane will become more exothermic by another 650 kJ per mole for pentane. This will give a value of $-3550$ kJ per mole for pentane, key **B**.

**15** The values given are for one-tenth of a mole, so the molar heats of vaporization are

ethoxyethane $(C_2H_5)_2O$    26.02 kJ per mole

ethanol $C_2H_6O$    39.37 kJ per mole

The compositions and relatively low molar heats of vaporization confirm that these compounds both have a molecular structure, the higher value for the heat of vaporization indicating that molecules of ethanol are more strongly bound to each other. The correct key is **D** rather than **B**. Vaporization does not involve breaking apart the atoms within a molecule (**E** incorrect). The statements made in **A** and **C** happen to be true for this particular pair of substances but are not generally true.

16 Luckily there is the same volume of solution to be heated up in all the experiments! It is only necessary to find the mixture in which the greatest number of moles of magnesium and copper(II) ions react. The numbers of moles taken are

|   | moles magnesium | moles copper(II) |
|---|---|---|
| **A** | 0.4 | 0.1 |
| **B** | 0.2 | 0.2 |
| **C** | 0.1 | 0.2 |
| **D** | 0.2 | 0.1 |
| **E** | 0.1 | 0.4 |

In **A**, **C**, **D** and **E**, only 0.1 mole of each can react because the extent of reaction is decided by the substance present in the smaller number of moles. The other substance is left in excess at the end of the reaction. In **B**, 0.2 mole of each can react and this mixture will produce the greatest heat output and the greatest temperature rise.

17 Sodium is a very reactive element but sodium oxide would not be described as a very reactive compound. Sodium has lost its activity when it combined with oxygen. This is similar to sodium losing its activity and chlorine losing its poisonous nature when the elements react to form sodium chloride. **A** is not correct.

The passing of hydrogen over heated sodium is an attempt to reduce the oxide to the metal. The hydrogen is meant to combine with the oxygen to make water but declines to do so. Statement **B** is not relevant. Can you discover whether **B** is correct?

The failure to react is either because there is an energy barrier preventing reaction (not mentioned in the possible answers) or because of the relative strengths of bonds. The reaction between sodium oxide and hydrogen would involve the breaking of the bonds between the sodium and oxide ions and in the hydrogen molecules. New bonds could then form in metallic sodium and between hydrogen and oxygen. The reaction is possible if it requires less energy to break the old bonds than is released on forming the new bonds. Statement **D** goes part of the way to saying that the old bonds are in fact stronger than the new bonds. This is the best explanation among those provided and would be regarded as the correct key.

18 In this reaction, the breaking of bonds requires 681 kJ per mole of $H_2/Cl_2$ while the formation of the new bonds releases 866 kJ per mole of $H_2/Cl_2$ used. Unlike the reaction in question 17, this one is possible. You probably know that it takes place readily once started. The reaction is exothermic, the energy change being 681 (put in) − 866 (given out). This is key **E**. $\Delta H$ for the reaction is − 185 kJ per mole of $H_2/Cl_2$.

19 You will know that the burning of magnesium in oxygen to produce magnesium oxide is a highly exothermic reaction. It is true to say that heat is evolved when magnesium oxide is formed from its elements. After the combustion, when the magnesium oxide has returned to room temperature, the surroundings have increased in energy. The surroundings contain more energy than before, this energy having been transferred from the chemicals which now contain less energy. While the first statement is true, the second is false and the key is **C**.

20 This is another reaction that you are almost certain to have seen or carried out yourself. It is used in the test for the presence of water which turns the white powder to a blue colour. At the same time, the powder warms up and it is true to say that heat is evolved. If the blue compound formed, which is hydrated copper(II) sulphate, is heated, the white, anhydrous form of the compound is produced. The reaction can be reversed but this is not the correct explanation for the reaction being exothermic. The key is **B** rather than **A**. A correct explanation for the reaction being exothermic could be given in terms of the energy needed to break the old bonds and the energy released on forming new bonds. This is discussed in the answers to questions 17 and 18.

21 In an endothermic reaction, the temperature first drops to below room temperature, then returns to room temperature as energy is transferred from the surroundings. The products do contain more energy than the reactants at the same temperature and pressure, this being due to the energy transfer from the surroundings. Both statements are true and the second is the explanation of the first. The key is **A**.

22 The first statement is true. It is similar to the first statements in questions 19 and 20. Unlike question 20, the second statement is a correct explanation for a reaction being exothermic. The key is **A**.

**23** If you have carried out any energy-change experiments involving solutions, you will know that the first statement is true. It is necessary to arrange for temperature changes to be relatively small, even if an insulated container is used. There is always an error in these experiments because of heat loss to the surroundings before the maximum temperature attainable has been reached by the reaction mixture. This error becomes more serious the greater the temperature rise in the experiment. Both statements are true, the second being a correct explanation of the first. The key is **A**.

**24** An exothermic reaction first increases in temperature but, given time, the temperature of the reaction mixture will eventually return to the original temperature. During this change, heat energy is given out to the surroundings (first statement true) and the products, once their temperature has dropped, contain less total energy than the reactants (second statement true and a correct explanation of the first). The key is **A**.

**25** The first statement is true. Energy has to be supplied to vaporize a liquid. An energy level diagram for such a change is given in question 9. The energy has to be supplied to overcome the attractive forces that exist between the molecules of liquid. The second statement is a correct explanation of the first and the key is **A**.

**26** Using the relative molar masses gives 40 g as the molar mass of NaOH.

$$4 \text{ g of NaOH contain } \frac{4 \text{ g}}{40 \text{ g/mole}} = 0.1 \text{ mole.}$$

The correct key must be **A** as this is the only one containing the expected number of moles of NaOH. As a further check, 50 g of water of molar mass 18 g must contain nearly 3 moles of water molecules, which agrees with the value given in **A**.

**27** The sodium hydroxide will contain 0.1 mole of NaOH in approximately 50 cm³ of solution. It is being assumed that 50 cm³ of water plus 4 g of NaOH will have the volume as the water used to dissolve the solid. This solution will have the same concentration as one of 2 moles of NaOH in 1000 cm³ (or 1 litre). The correct key is **B**.

**28** When sodium hydroxide dissolves in water, there is a temperature rise as described at the start of the set of questions. The quantity that can be calculated from the experimental results is the heat of solution, key **C**.

**29** The sodium hydroxide/water mixture is stirred continuously so a relatively rapid rise in temperature would be anticipated as the solid is being persuaded to dissolve as quickly as possible. **B** cannot be correct as that shows a temperature drop from the start.

The solution would reach its maximum temperature and then relatively slowly return to room temperature. In any container, the cooling would be expected to be slower than the initial increase in temperature. In the insulated container used in this experiment, the cooling would be expected to be even slower. Only **A**, **D** and **E** show cooling taking place. **A** cannot be correct as it shows cooling going to below room temperature, while **D** shows slow heating and rapid cooling which is the reverse of what is expected. **E** shows slower cooling than heating and also shows the cooling becoming even slower as room temperature is approached, this final effect being just what would be anticipated. The correct key is **E**.

**30** If it is assumed that the 4 g of sodium hydroxide dissolved completely in 50 cm³ of water, it will also dissolve completely in 100 cm³ of water. The interaction between sodium hydroxide and water would be expected to produce approximately the same quantity of heat however much water was used, provided the mass of solid is fixed. Both experiments should give about the same heat output but in the second experiment this heat will have to be used to raise the temperature of approximately twice as much solution. The heat capacities should be similar so the temperature rise in the second experiment should be about half the original value. This is 7.5°C which is key **B**.

# Rate of reaction and equilibrium

## Questions 1–5

The reaction between hydrochloric acid and EXCESS calcium carbonate is investigated to discover how the rate of reaction at the start, and the final amount of carbon dioxide formed, depend on various factors.

In a first experiment, 50 cm³ of 1.0 M hydrochloric acid is added to excess calcium carbonate at 20°C.

The factors are then varied one at a time with the following possible results:

| | Effect on rate at the start | Effect on the final amount of carbon dioxide collected at room temperature when the reaction has stopped |
|---|---|---|
| A | Increased | Unchanged |
| B | Increased | Increased |
| C | Decreased | Increased |
| D | Decreased | Unchanged |
| E | Decreased | Decreased |

Select, from A to E, the effect you would expect when

1  the temperature is lowered

2  an equal volume of 0.1 M acid is used instead of 1.0 M acid

3  an equal volume of 5.0 M acid is used instead of 1.0 M acid

4  50 cm³ of water is added to 50 cm³ of the 1.0 M acid

5  50 cm³ of 0.1 M acid is added to 50 cm³ of the 1.0 M acid

---

| Directions summarized for questions 6 to 14 | | | | |
|---|---|---|---|---|
| **A** 1, 2, 3 only correct | **B** 1, 3 only correct | **C** 2, 4 only correct | **D** 4 only correct | **E** Some other response or combination of responses is correct |

6  The rate of evolution of oxygen when a fixed volume of aqueous hydrogen peroxide is decomposed is increased by

1  adding a suitable catalyst

2  using a more concentrated solution of hydrogen peroxide

3  warming the solution

4  adding water to the solution

7  The reaction between zinc metal and a solution of copper(II) sulphate can be made to go more SLOWLY by

1  using a more finely divided sample of the zinc metal

2  cooling the reagents

3  stirring the mixture

4  diluting the copper(II) sulphate solution

8  The reaction

$$A(g) + B(g) \rightarrow C(g) + D(g)$$

is endothermic and slow at room temperature. When A(g) and B(g) are mixed at 100°C,

1  the reaction is now exothermic

2  collisions between molecules of A and B are more frequent than at room temperature

3  the energy possessed by A(g) and B(g) remains unaltered

4  the reaction is complete in less time than at room temperature

**9** Correct statements about a chemical reaction which has reached equilibrium include that

**1** all the reactants have been used up

**2** chemical reaction is no longer taking place

**3** the molecules in the system have stopped moving

**4** addition of some of the products to the equilibrium mixture would produce more of the reactants

**10** A state of dynamic equilibrium exists at constant temperature in

**1** an open pan of water boiling on a stove

**2** a stoppered bottle half full of a liquid

**3** a beaker in which zinc is dissolving in dilute sulphuric acid

**4** a stoppered bottle of saturated aqueous calcium hydroxide containing undissolved solid

**11** When hydrogen chloride is dissolved in water, the following reaction takes place:

$$HCl(g) + H_2O(l) \rightleftharpoons H_3O^+(aq) + Cl^-(aq)$$

Which of the following statements could suggest that this reaction is reversible?

**1** The solution conducts electricity.

**2** The solution gives off hydrogen when it reacts with magnesium.

**3** The solution neutralizes alkalis to form salts and water.

**4** A concentrated solution of hydrochloric acid gives off hydrogen chloride when exposed to the air.

**12** A solution of hydrogen sulphide gas in water reacts with aqueous zinc ions to produce a precipitate of zinc sulphide. The equation for this reaction, which is reversible, is

$$H_2S(aq) + Zn^{2+}(aq) \rightleftharpoons ZnS(s) + 2H^+(aq)$$

When the reaction has reached equilibrium, more zinc sulphide could be made to precipitate by adding

**1** extra hydrogen ions to increase the concentration of $H^+(aq)$

**2** extra zinc ions to increase the concentration of $Zn^{2+}(aq)$

**3** a suitable catalyst to increase the rate of reaction

**4** hydrogen sulphide to increase the concentration of $H_2S(aq)$

**13** When the system

$$Br_2(aq) + 3H_2O(l) \rightleftharpoons 2H_3O^+(aq) + BrO^-(aq) + Br^-(aq)$$

is in equilibrium

**1** the forward and backward reactions are both taking place at the same rate

**2** addition of water will displace the position of equilibrium to the left

**3** addition of acid will displace the position of equilibrium to the left

**4** addition to alkali will displace the position of equilibrium to the left

**14** The graphs below show the percentage of ammonia present in the gases at equilibrium when hydrogen and nitrogen react at various temperatures and pressures.

$$N_2(g) + 3H_2(g) \rightleftharpoons 2NH_3(g)$$

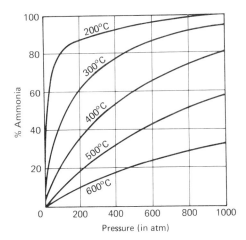

The gases would contain about 50% ammonia if they came to equilibrium at

**1** 300°C and 100 atm    **3** 500°C and 750 atm

**2** 400°C and 350 atm    **4** 600°C and 900 atm

Directions for questions 15 to 22. Each of the questions or incomplete statements in this section is followed by five suggested answers. Select the best answer in each case.

**15** The following quantities of sodium thiosulphate solution were added to five separate $10 \text{ cm}^3$ portions of 2 M hydrochloric acid. The mixture was made up to a total volume of $50 \text{ cm}^3$ in each case and the rate of reaction was determined. In which example does the reaction proceed the fastest?

A $10 \text{ cm}^3$ of 2 M sodium thiosulphate solution

B $20 \text{ cm}^3$ of 2 M sodium thiosulphate solution

C $10 \text{ cm}^3$ of 3 M sodium thiosulphate solution

D $20 \text{ cm}^3$ of 3 M sodium thiosulphate solution

E $10 \text{ cm}^3$ of 4 M sodium thiosulphate solution

**16** Which of the following systems would you expect to provide the fastest initial rate of reaction?

A $40 \text{ cm}^3$ of 1 M hydrochloric acid and 1.0 g of calcium carbonate lumps

B $40 \text{ cm}^3$ of 2 M hydrochloric acid and 1.0 g of calcium carbonate lumps

C $20 \text{ cm}^3$ of 1 M hydrochloric acid and 1.0 g of calcium carbonate lumps

D $20 \text{ cm}^3$ of 2 M hydrochloric acid and 1.0 g of calcium carbonate powder

E $40 \text{ cm}^3$ of 1 M hydrochloric acid and 1.0 g of calcium carbonate powder

**17** After an experiment to investigate the rate of reaction between calcium carbonate and hydrochloric acid, a graph was plotted showing the volume of gas evolved against time.

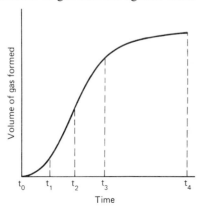

At which time was the rate of evolution of gas greatest?

A $t_0$        D $t_3$

B $t_1$        E $t_4$

C $t_2$

**18** When a few crystals of iodine are shaken with water, the water becomes pale yellow. This colour does not change in intensity after shaking for some time even though there is some solid iodine remaining in contact with the solution.

The explanation which best fits the observation of the constant yellow colour is that the

A iodine is going into solution so slowly that the colour does not appear to change

B iodine is precipitating from solution so slowly that the colour does not appear to change

C iodine is going into solution so slowly and is also precipitating so slowly that no change in colour is observed

D iodine is going into solution so rapidly and is also precipitating so rapidly that no change in colour is observed

E iodine is going into solution at the same rate as it is precipitating so that no change in colour is observed

**19** Which of the following is in a state of dynamic equilibrium?

A Water and water vapour in an open container of boiling water

B Alcohol and alcohol vapour in a closed flask maintained at a constant temperature

C Gas and air in a steadily burning bunsen flame

D Salt and water when the salt is dissolving in the water to form an unsaturated solution

E A piece of ice in a beaker of water at 20°C

**20** When an acidified solution of bismuth chloride is added to water, a precipitate of bismuth oxychloride (BiOCl) forms. The equation is

$$BiCl_3(aq) + H_2O(l) \rightleftharpoons BiOCl(s) + 2HCl(aq)$$

The reaction is reversible. To increase the yield of bismuth oxychloride from a given quantity of bismuth chloride, one should add

**A** a large volume of water

**B** a small volume of water

**C** dilute hydrochloric acid

**D** concentrated hydrochloric acid

**E** hydrogen chloride gas

**21** Ethanoic acid (acetic acid) and ethanol are mixed and allowed to reach a state of equilibrium.

$$CH_3COOH + C_2H_5OH \rightleftharpoons CH_3COOC_2H_5 + H_2O$$

ethanoic acid    ethanol    ethyl ethanoate    water

More ethyl ethanoate could be produced by

**A** removing ethanoic acid by neutralising with sodium hydroxide

**B** removing ethanol by distillation

**C** adding water

**D** adding dilute hydrochloric acid

**E** removing water by adding concentrated sulphuric acid

**22** When solutions containing $Fe^{3+}$ and $CNS^-$ ions are mixed a blood-red colour immediately appears, due to the formation of $FeCNS^{2+}$ ions. The following equilibrium is established:

$$Fe^{3+}(aq) + CNS^-(aq) \rightleftharpoons FeCNS^{2+}(aq)$$

Which operation would be most effective in decreasing the intensity of the blood-red colour?

**A** Filtering the mixture

**B** Evaporating some of the water

**C** Increasing the concentration of $Fe^{3+}(aq)$

**D** Increasing the concentration of $CNS^-(aq)$

**E** Adding ions which combine with $Fe^{3+}(aq)$ to precipitate them

| | Directions summarized for questions 23 to 27 | | |
|---|---|---|---|
| | First statement | Second statement | |
| **A** | True | True | Second statement is a correct explanation of the first |
| **B** | True | True | Second statement is NOT a correct explanation of the first |
| **C** | True | False | |
| **D** | False | True | |
| **E** | False | False | |

**23**

1 Chemical reactions take place more rapidly at higher temperatures.

2 Chemical bonds are broken more readily at high temperatures than at low temperatures.

**24**

1 The rate of combination of nitrogen and hydrogen to form ammonia is increased by increasing the pressure on the gases.

2 Increasing the pressure increases the concentration of gases and thereby increases the rate of collision of their molecules.

**25**

1 A catalyst is used in the manufacture of ammonia from nitrogen and hydrogen.

2 There are twice as many atoms in an ammonia molecule as in either a nitrogen molecule or a hydrogen molecule.

**26**

1 When ice is in equilibrium with water, there is no change in the amounts of ice and water present.

2 There is no transfer of molecules from ice to water or vice versa when the two are together at equilibrium.

**27**

1 In any equilibrium state the concentrations of all substances present are equal.

2 In a chemical reaction at equilibrium, the rates of the forward and backward reactions are equal.

**Questions 28–30**

When ethanoic acid (acetic acid, $CH_3COOH$) is dissolved in water, the two compounds react and come to equilibrium.

$$CH_3COOH(l) + H_2O(l) \rightleftharpoons$$
$$CH_3COO^-(aq) + H_3O^+(aq)$$

Solutions of ethanoic acid were made up with concentrations ranging from 1.0 M to 4.0 M. A strip of magnesium ribbon was cleaned and cut into lengths 2 cm long. For each experiment, a 2 cm strip of magnesium was added to $20 \text{ cm}^3$ of ethanoic acid solution in a boiling tube and the time taken for the magnesium to dissolve was measured. A graph was then plotted to show how the rate of the reaction varied with the concentration of the acid.

| Concentration | Time |
|---|---|
| 1.0 M | 490 s |
| 2.0 M | 190 s |
| 3.0 M | 100 s |
| 4.0 M | 95 s |

Concentration of the ethanoic acid

28 In order to plot the graph, a measure of the rate of the reaction had to be calculated for each concentration. Which number below is proportional to the rate of the reaction of magnesium with 2.0 M ethanoic acid?

A $\dfrac{190}{490}$

B $2 \times 190$

C $\dfrac{190}{2}$

D $\dfrac{1}{190}$

E $\dfrac{490 - 190}{2}$

29 In measuring the rate of reaction in a large excess of acid by observing the time taken for a 2 cm piece of magnesium to dissolve, it was assumed that

A one mole of magnesium reacted with one mole of $H_3O^+$ ions

B the rate of reaction was independent of the $H_3O^+$ ion concentration

C the rate of reaction was independent of the shape of the piece of magnesium

D magnesium reacted with $H_3O^+$ ions more rapidly than ethanoic acid reacted with water

E the concentration of the ethanoic acid did not change significantly during the reaction

30 The series of experiments was repeated, but the second time 2 g of sodium ethanoate ($CH_3COO^-Na^+$) was dissolved in the $20 \text{ cm}^3$ sample of ethanoic acid solution before the magnesium ribbon was added. The results obtained are given in the table.

| Ethanoic acid concentration | Time |
|---|---|
| 1.0 M | 650 s |
| 2.0 M | 290 s |
| 3.0 M | 175 s |
| 4.0 M | 160 s |

Which of the following explains the changes in the values of the reaction times compared to the first series of experiments?

A Ethanoate ions catalyze the reaction between magnesium and $H_3O^+$ ions.

B Adding ethanoate ions reduces the $H_3O^+$ ion concentration.

C Sodium ions form an insoluble precipitate with ethanoic acid.

D The temperature rises when sodium ethanoate is dissolved in water.

E Sodium ethanoate reacts with magnesium.

# Test 7 | Answers

1 Lowering the temperature will decrease the rate of reaction so **C**, **D** or **E** must be the correct key. The volume or molarity of the acid has not been changed which means the final amount of carbon dioxide will be unchanged although it will be formed more slowly. The key is **D**.

2 A lower concentration of acid is used which will decrease the rate of reaction. The smaller number of moles of acid used will decrease the final amount of carbon dioxide. The key is **E**.

3 The acid is more concentrated, which will increase the rate of reaction at the start. Five times as many moles of acid are used, which will produce a five-fold increase in the amount of carbon dioxide. The key is **B**.

4 The acid has been diluted. This will decrease the rate of reaction. The same volume of the original concentration acid was taken before dilution so the number of moles of acid is unchanged. This means that the final amount of carbon dioxide will be unchanged. The key is **D**.

5 The addition of the 0.1 M acid to the 1.0 M acid will produce an acid mixture of concentration below 1.0 M. This will decrease the rate of reaction at the start. The number of moles of acid has, however, been increased and this will increase the final amount of carbon dioxide. The key is **C**. Key **A** has not appeared in this set. Can you suggest a change that would produce the effects listed in **A**?

6 Addition of a suitable catalyst, increasing the concentration of the solution and raising the temperature will all increase the rate of reaction. Adding water to the hydrogen peroxide will have the opposite effect to **2**. **4** is the only change that will not increase the rate of evolution of oxygen. The key is **A**.

7 Using zinc that is more finely divided would increase the surface area that the copper(II) ions could attack. This would increase the rate of reaction. Stirring the mixture would prevent the copper(II) ions from dropping in concentration in an area close to the zinc as some copper ions were removed by reaction. The zinc would continually be meeting fresh copper(II) ions and this would certainly not make the reaction go more slowly.

Stirring would also remove reaction products, in this case zinc ions, from near the zinc surface. This would make it easier for fresh copper ions to approach the zinc surface. This second effect of stirring would also not make the reaction go more slowly. **1** and **3** are not suitable changes.

Diluting the copper(II) sulphate will reduce the number of collisions between copper(II) ions and the zinc surface. Cooling the reagents will make any collisions less effective. (See the answer to question 23 for a further discussion of this.) Both **2** and **4** will make the reaction go more slowly and the key is **C**.

8 A reaction that was endothermic at room temperature would be expected to remain endothermic at all temperatures, or certainly over a wide temperature range. The only situation in which this might not be true would be if the reaction was only slightly endothermic at room temperature when it might just possibly have become slightly exothermic by 100°C. This is, however, very unlikely and if you had to choose, as you do, between **1** being correct or incorrect, you would have to choose 'incorrect'.

One of the effects of a temperature rise is that molecules move more rapidly and collide more frequently. **2** is correct. Raising the temperature increases all the types of 'motion' of the molecules. They will move more rapidly and also rotate and vibrate more rapidly if they are molecules that rotate and vibrate. This means that the molecules will possess more energy at higher temperatures. **3** is incorrect.

Raising the temperature will increase the rate of reaction. This is connected with the increase in the energy held by the molecules at high temperature and will be discussed in the answer to question 23. As the rate of reaction is increased, the reaction will be complete in less time than at room temperature. **2** and **4** are correct and the key is **C**.

9 In a chemical reaction at equilibrium, both reactants and products are present (**1** incorrect). The situation is described as 'dynamic' because reaction is still taking place but at equal rates in both directions so that the concentrations no longer change (**2** incorrect). The molecules have not stopped moving and **3** is also incorrect.

When dynamic equilibrium has been reached, reactant molecules are forming product molecules and product molecules are forming reactant molecules at equal rates. Addition of extra product molecules disturbs this equilibrium state. There is a surplus of collisions between product molecules and this causes the equilibrium mixture to produce more reactants. **4** is correct and the key is **D**.

10 If a dynamic equilibrium is to have a chance of arising, there must be an opportunity for the particles concerned to move in both directions at equal rates. In an open pan of boiling water, the molecules move from liquid to vapour but once in the vapour they escape into the room and then from the room with very little chance of returning. Dynamic equilibrium would not be set up.

The only way to persuade **1** to reach dynamic equilibrium would be to use an enormous pan of water, 'stopper' the windows and insulate the room so that it could heat up to 100°C. This is, in effect, what is happening in **2**. In the stoppered bottle, molecules escape from the liquid into the vapour. They move round in the space above the liquid and have a chance of returning to the liquid state when they hit the surface. At first there are few molecules in the vapour to return to the liquid state but when the liquid has been in the bottle for some while, the pressure of the vapour molecules will have built up to a sufficiently high value for as many molecules to be returning to the liquid as are leaving it in the same time. This is a state of dynamic equilibrium.

The situation in **3** is similar to that in **1**. Hydrogen molecules are escaping into the room with little or no chance of returning to the liquid surface. They are not given the opportunity of showing whether they are capable of establishing a dynamic equilibrium.

The situation in **4** is similar to the one in **2**. Calcium and hydroxide ions escape from the solid into solution and return from solution to solid. Eventually, the ions move at equal rates in both directions and dynamic equilibrium is established. Only in **2** and **4** is dynamic equilibrium established and the key is **C**.

11 The first three statements are true but are not connected with the reaction being reversible. They are a result of the reaction moving from left to right and producing ions (statement **1**) and the $H_3O^+$ ion in particular (statements **2** and **3**). There is no indication in these statements that the reaction can move in both directions.

It is statement **4** which contains the hint that the reaction is reversible. To call the solution 'hydrochloric acid' implies that it has at least partly reacted to form $H_3O^+$ ions, the ions that make acids behave in the special ways that acids do. When given the chance, however, this 'hydrochloric acid' gives off HCl(g), showing that the reaction can move from right to left. Only **4** suggests that the reaction is reversible and the key is **D**.

12 When the reaction has reached equilibrium, the collisions between $H_2S$ and $Zn^{2+}$ are taking the reaction from left to right at the same rate as it is being taken from right to left by the collisions between ZnS and $H^+$. To make the reaction move over further to the right, various changes in conditions could be used. The collisions between $H_2S$ and $Zn^{2+}$ could be increased by increasing either or both of the concentrations of these substances. This is happening in **2** and **4** which are correct. Another way would be to reduce the collisions from right to left by reducing the concentrations of the substances on the right hand side of the equation. This suggestion is not made. In fact, the reverse is the case in **1** which would drive the reaction from right to left and produce less precipitate. The only other suggestion, use of a catalyst, will speed up the reaction equally in both directions and produce no change in the amount of precipitate. Only **2** and **4** are suitable and the key is **C**.

13 The reaction is a dynamic equilibrium so **1** is correct. The addition of water will produce more collisions with $Br_2$ and will therefore displace the equilibrium from left to right (**2** incorrect).

Acid contains $H_3O^+$(aq) ions and its addition will increase the collisions between particles on the right hand side of the equilibrium. This will displace the reaction from right to left and **3** is correct. Addition of alkali will, however, have the reverse effect as the $OH^-$ ions of the alkali will remove the $H_3O^+$ ions (**4** incorrect).

**1** and **3** are correct statements and the key is **B**.

14 Find the 50% level on the ammonia axis and run a horizontal line across until it cuts the curves corresponding to various temperatures.

The horizontal line cuts the 300°C curve at about 100 atm, the 400°C curve at about 350 atm, and the 500°C curve at about 750 atm. This makes **1**, **2** and **3** correct. The key is **A** because the line does not cut the 600°C curve until a temperature well in excess of 1000°C.

**15** This experiment is arranged so that the concentration of the hydrochloric acid remains the same in all cases. Can you see what this constant concentration is? It is only the concentration of the sodium thiosulphate that is varied, and it must be assumed that the reaction will be fastest in the mixture where the final concentration of this reagent is highest. As the total volume is always made up to the same value, it is only necessary to decide which of the original quantities of sodium thiosulphate solution contains the greatest number of moles. It is not necessary to calculate the actual number of moles. It is much quicker to convert each solution to an equivalent volume of, say, 1 M and then look to see which volume is the greatest. **A** is equivalent to 20 cm$^3$ of 1 M, **B** to 40 cm$^3$, **C** to 30 cm$^3$, **D** to 60 cm$^3$, and **E** to 40 cm$^3$. The correct key is **D** as this solution provides the greatest number of moles of sodium thiosulphate.

**16** What is wanted, if possible, is the acid of highest molarity reacting with powder rather than lumps. The volume of acid is not important because this problem is concerned with the rate of reaction at the start. The correct key is **D** because this provides the highest concentration of acid available and the solid is in the form of powder.

**17** Look for the time when the gradient of the curve is greatest. This is at t$_2$ and the correct key is **C**.

**18** The dissolving of a solid in a solvent involves dynamic equilibrium once the solution has become saturated. After this point, the solid is dissolving and precipitating at equal rates so that the concentration, indicated by colour intensity in this problem, does not change with time. **C** and **D** are better than **A** or **B** because they contain a suggestion of a dynamic state of affairs but they go too far in suggesting that the rate is slow or fast, and yet not far enough by failing to equate these rates. **E** is the correct explanation, containing as it does the indication that the rates are equal without comment on how fast the changes are taking place.

**19** This is a similar problem to that already met in question 10. **A** is incorrect because the water vapour is escaping without the opportunity to return and set up a dynamic equilibrium. The gas flame in **C** is steady because the flame is moving down at the same rate as the gas is moving up. There is a hint of 'equilibrium' here but it is not what a chemist would understand by the term. In **D**, the situation is on the way to dynamic

equilibrium but will only have arrived when the solution is saturated – then the salt will be dissolving and precipitating at equal rates. Ice in water at 20°C, key **E**, will soon all be liquid water. It would only be at 0°C that dynamic equilibrium would be possible in an open beaker.
   **B** is the correct key. It is similar to the 'stoppered bottle half full of liquid' in question 10. Alcohol liquid will be turning to alcohol vapour and vapour will be turning back to liquid at equal rates in the closed flask at constant temperature.

**20** If the yield of bismuth oxychloride (BiOCl) is being increased, the reaction is being driven from left to right. This could be achieved by increasing the collisions between particles on the left hand side of the reaction (increasing the concentration of BiCl$_3$ or H$_2$O). It could also be achieved by reducing the collisions between particles on the right hand side of the reaction equation. Of the suggestions made, only **A** is in line with the possibilities outlined, and this is the correct key.

**21** As with similar problems that have been met already, it is necessary either to increase the concentration of one or more reactants or to reduce the concentration of one or more products. This will alter the number of collisions in such a way that the reaction is forced to move over to the right. Removal of water is a correct suggestion and the key is **E**.

**22** Unlike earlier problems in this test, this one is concerned with driving an equilibrium from right to left. This would be achieved by increasing the concentration of FeCNS$^{2+}$(aq) or by reducing the concentration of either Fe$^{3+}$(aq) or CNS$^-$(aq). Filtering would alter nothing, evaporation would alter all the concentrations rather than the one or two which would produce the most favourable change, while **C** and **D** would produce an effect opposite to that required. Precipitation of Fe$^{3+}$(aq) would drive the equilibrium position from right to left, and **E** is the correct key.

**23** The first statement is true. Chemical reactions do take place more rapidly at higher temperatures. Raising the temperature of a substance increases the motion of its molecules in all possible ways. Usually the molecules will move faster, rotate faster and will also vibrate more fiercely. The faster movement of molecules means that there are more collisions between molecules at higher temperatures, but this is not the main reason for the increase in rate of reaction with rise in temperature. Chemical reactions involve the breaking of bonds and then the making of new bonds. Raising the temperature, by its influence on the vibration of molecules, increases the chance of bonds being broken, and so speeds up chemical reactions. The second statement is true and a correct explanation of the first. The key is **A**.

24 Increasing the pressure of the gases will increase the number of collisions between molecules and so increase the rate of combination. Both statements are true and the first is explained by the second, so the key is **A**.

25 Both statements are completely true but the second does not explain the first. The key is **B**.

26 Ice and water are in equilibrium at 0°C. The amounts of water and ice are not important so long as some of each is present. The ice can be turning to water or the water can be turning to ice. They are still described as 'being in equilibrium' and would continue to be described in this way until either all the ice had melted or all the water had frozen. Both statements in the question are in conflict with what has just been written and the key is **E**.

27 The second statement is true and summarizes what is understood by the term 'dynamic equilibrium'. The first statement is false and the key is **D**.

28 When measurements of this type are made, a large time means a small rate while a small time indicates a large rate. To obtain an indication of the rate it is necessary to calculate the reciprocal of the time taken for a certain amount of reaction to occur. If 2.0 M ethanoic acid is used, the time taken is 190 seconds. The rate of reaction is proportional to the reciprocal of this, which appears in key **D**.

29 **A** is incorrect and so would not have been assumed! **B** is unlikely to be correct but no assumption needs to be made about it in this experiment. **C** is also likely to be wrong but no assumption has to be made about it because the same shape was used each time. The experiment involves measuring a rate of reaction and finding how this rate changes with changing ethanoic acid concentration. Measurement of the rate comes from the times measured. There is no assumption necessary about the nature of the particles with which the magnesium reacts or about the relative speeds of reaction if more than one particle is involved. **D** is not assumed.

It is, however, necessary to assume that the acid remains at its original concentration throughout the experiment. **E** is the correct key. To justify this assumption the quantities would be chosen so that there was a considerable excess of ethanoic acid.

30 All the times are increased by the presence of sodium ethanoate. This means that the reactions are reduced in rate. **A**, **D** and **E** are explanations directed at explaining why the reaction becomes faster and so cannot be correct.

**C** is an explanation for a reduced rate of reaction because if the ethanoic acid is precipitated out, there will be less of the acid or its ions to react with the magnesium. The explanation is an improbable one because the precipitate would be most likely to be sodium ethanoate but, as the experiment uses a solution of that substance, it could not very well also be insoluble! Nor is the precipitate likely to be anything else because almost all sodium salts are highly soluble in water. This leaves **B** for consideration. Looking at the equation it will be seen that adding ethanoate ions will increase collisions with hydroxonium ions and so drive the equilibrium position to the left. This will reduce the rate of reaction and **B** is the correct key.

## Questions 1−4

The following equations represent chemical changes in which particles gain or lose electrons:

A    $X^{2+} + 2e^- \rightarrow X$

B    $X^+ + e^- \rightarrow X$

C    $X \rightarrow X^{2+} + 2e^-$

D    $2X^- \rightarrow X_2 + 2e^-$

E    $X^{2-} \rightarrow X + 2e^-$

Choose the equation which, with an appropriate symbol in place of X, best represents the change occurring

1    when bromide ions are discharged during the electrolysis of molten lead(II) bromide

2    when calcium is liberated by the electrolysis of molten calcium chloride

3    at a silver cathode (the negative electrode) during the electrolysis of aqueous silver nitrate

4    at the anode (the positive electrode) during the electrolysis of aqueous copper(II) sulphate using copper electrodes

| Directions summarized for questions 5 to 12 | | | | |
|---|---|---|---|---|
| **A** | **B** | **C** | **D** | **E** |
| 1, 2, 3 only correct | 1, 3 only correct | 2, 4 only correct | 4 only correct | Some other response or combination of responses is correct |

5    Liquids which conduct electricity include

1    aqueous sodium chloride

2    molten lead(II) bromide

3    mercury

4    hexane

6    Which of the following when dissolved in water increases its conductivity?

1    Hydrogen chloride

2    Oxygen

3    Copper(II) sulphate

4    Glucose

7    Electric current is carried through electrolytes by movement of

1    electrons          3    anions

2    cations           4    molecules

8    Elements which exist as positive ions in many compounds include

1    chlorine          3    carbon

2    sodium           4    magnesium

9    When a chlorine atom becomes a chloride ion, the

1    atomic number of chlorine increases by one

2    chlorine atom loses one electron

3    relative atomic mass of chlorine decreases by one

4    chlorine atom takes on one negative charge

10    Potassium iodide, which has a giant structure of ions, is converted into a system of free ions when it is

1    heated above its melting point

2    placed in contact with electrodes connected to a battery at room temperature

3    dissolved in water

4    finely powdered

11    Which of the following elements is/are obtained on a large scale by electrolysis?

1    Sulphur           3    Oxygen

2    Chlorine          4    Aluminium

12 Direct current was passed through concentrated aqueous lithium chloride (LiCl) using inert electrodes. You would expect

1 hydrogen to be evolved at the negative electrode

2 the aqueous lithium chloride to become alkaline

3 chlorine to be evolved at the positive electrode

4 lithium to be deposited at the negative electrode

| Metal | Solution | | | |
|---|---|---|---|---|
| | R sulphate | S sulphate | T sulphate | U sulphate |
| R | 0 | X | X | 0 |
| S | 0 | 0 | 0 | 0 |
| T | 0 | X | 0 | 0 |
| U | X | X | X | 0 |

Which of the following is the order of reactivity? (Place the most reactive first.)

A RSTU          D USTR

B SRTU          E URTS

C STRU

Directions for questions 13 to 21. Each of the questions or incomplete statements in this section is followed by five suggested answers. Select the best answer in each case.

13 Which of the following will occur when a solution of copper(II) sulphate is electrolyzed for some time using copper electrodes?

A The blue colour of the solution will become concentrated round the cathode.

B The sum of the masses of the two electrodes will remain unchanged.

C The separate masses of the two electrodes will remain unchanged.

D The total blue colour in the solution will become less and less.

E The pH of the solution will gradually decrease.

14 When a metal is being electroplated onto a cathode, the RATE at which the metal is being deposited by a steady current can be changed by using

A a clock          D a voltmeter

B a thermometer          E a rheostat

C an ammeter

15 The table shows what happened when pieces of four different metallic elements were put separately into aqueous solutions of their sulphates.

X    indicates that a coating or crystals could be seen on the metal

0    indicates that no change could be observed

16 The number of moles of electrons (faradays) needed to deposit 26 g of chromium [Cr = 52] from a solution containing $Cr^{3+}$(aq) is

A 0.5          D 3.0

B 1.0          E 6.0

C 1.5

17 During the electrolysis of an aqueous solution of a cerium salt, 70 g of cerium (relative atomic mass: Ce = 140) is deposited at the cathode by 2 moles of electrons (faradays). The formula of the cerium ion is probably

A $Ce^+$          D $Ce^{4+}$

B $Ce^-$          E $Ce^{4-}$

C $Ce^{2+}$

18 1 g of the element scandium (Sc = 45) is liberated in electrolysis by 1/15 mole of electrons (faraday).

1 g of oxygen (O = 16) is liberated by 1/8 mole of electrons (faraday).

What is the probable formula of scandium oxide?

A $ScO$          D $Sc_2O_3$

B $ScO_2$          E $Sc_3O_2$

C $Sc_2O$

**19** A silver ion carries one positive charge ($Ag^+$) and a sulphate ion two negative charges ($SO_4^{2-}$). The formula of silver sulphate is

A $AgSO_4$

B $Ag(SO_4)_2$

C $Ag_2SO_4$

D $Ag_2(SO_4)_2$

E $Ag_2(SO_4)_3$

**20** The products of the electrolysis of molten potassium bromide would be expected to be

| | At positive electrode | At negative electrode |
|---|---|---|
| A | hydrogen | oxygen |
| B | oxygen | hydrogen |
| C | hydrogen | oxygen and bromine |
| D | potassium | bromine |
| E | bromine | potassium |

**21** When a sodium atom becomes a sodium ion

A it loses one electron

B its atomic number changes by one unit

C it shares its electrons with other atoms

D it gains one electron

E its relative atomic mass changes by one unit

| Directions summarized for questions 22 to 25 | | |
|---|---|---|
| | First statement | Second statement | |
| A | True | True | Second statement is a correct explanation of the first |
| B | True | True | Second statement is NOT a correct explanation of the first |
| C | True | False | |
| D | False | True | |
| E | False | False | |

**22**

1 The ions of metals are positively charged.

2 Metal ions contain more protons than electrons.

**23**

1 Hydrogen is evolved when copper is added to dilute hydrochloric acid.

2 Hydrogen is below copper in the electro-chemical (activity) series.

**24**

1 One mole of electrons (faraday) of electricity will deposit 2 moles of atoms of calcium during the electrolysis of fused calcium chloride, $CaCl_2$.

2 Calcium ions each carry two positive charges.

**25**

1 Ionic substances conduct electricity when melted but not when solid.

2 Ions are free to move about in liquids but not in solids.

**Questions 26–30**

An experiment was carried out to measure the quantity of electricity required to deposit 1 mole of atoms of copper by electrolyzing an aqueous solution of copper(II) chloride between copper electrodes.

**26** Which of the following pieces of apparatus, in addition to those shown in the diagram, is needed to perform the experiment?

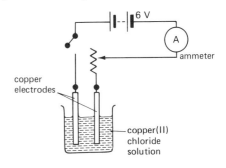

A Bulb

B Fixed resistance

C Voltmeter

D Bunsen burner

E Clock

**27** To obtain an accurate result the following precautions were taken. Which one was NOT necessary?

**A** The cathode was cleaned and dried before initial weighing.

**B** The cathode was completely immersed in the solution.

**C** The current was kept low and constant during the experiment.

**D** The cathode was washed by dipping it in distilled water after the electrolysis.

**E** The cathode was completely dried before final weighing.

**28** When a current of 0.2 ampere was passed through the solution for exactly one hour, 0.24 g of copper was deposited on the cathode. How much electricity is required to deposit 1 mole of atoms of copper? (Cu = 64)

**A** $\dfrac{64}{0.24} \times 0.2 \times 60 \times 60$ coulombs

**B** $\dfrac{64}{0.24} \times 0.2 \times 60$ coulombs

**C** $\dfrac{0.24}{64} \times 0.2 \times 60 \times 60$ coulombs

**D** $0.24 \times 64 \times 0.2 \times 60 \times 60$ coulombs

**E** $0.24 \times 60 \times \dfrac{0.2}{64}$ coulombs

**29** It was found that the mass of copper deposited on the cathode was slightly less than that calculated theoretically. Which of the following is the most likely explanation of the discrepancy?

**A** The electrolyte was not pure.

**B** The current was too small.

**C** The cathode was not completely dry before final weighing.

**D** Some of the copper did not adhere to the cathode.

**E** Copper was dissolved from the anode.

**30** If the anode had been made of carbon instead of copper, which of the following is the most likely composition of the mixture of gases evolved at the anode?

**A** Chlorine and oxygen

**B** Chlorine and hydrogen

**C** Chlorine and hydrogen chloride

**D** Chlorine and bromine

**E** Hydrogen and oxygen

---

# Test 8 | Answers

**1** Bromine is an element from Group 7 of the Periodic Table, a halogen. The ions formed by elements on the right-hand side of the Periodic Table, the non-metals, carry a negative charge. The magnitude of the charge for an element of Group 7 is 1. This is part of the pattern of valency, or bonding-power, which runs 1,2,3,4,3,2,1,0 on going from Group 1 to Group 8 (Group 0). You only need to be aware of this pattern, and to know the position of important elements in the Periodic Table, to be able to work out the likely charge carried by ions of non-transition elements. Thus, bromine, a member of Group 7, will have a charge on its ions of $-1$, while other non-metals such as oxygen (Group 6) or nitrogen (Group 5) will form negative ions $O^{2-}$ and $N^{3-}$.

In question 1 a bromide ion, $Br^-$, is having its charge neutralized at the positive electrode. This is described in the question as 'discharged'.

At the positive electrode electrons are either removed from incoming ions, such as $Br^-$, or from the electrode itself (see the answer to question 4). These electrons are moved round the external circuit, frequently by a battery, and are made available at the negative electrode (see the answers to questions 2 and 3).

The removal of electrons from the bromide ions can be represented by the equation

$$Br^- \rightarrow Br + e^-,$$

and is followed by combination of bromine atoms

to form molecules:

$$2Br \rightarrow Br_2.$$

These two steps can be gathered into one equation

$$2Br^- \rightarrow Br_2 + 2e^-.$$

This is the change shown in **D** which is the correct key.

2   Calcium is an element from Group 2. Being a metallic element, its ions carry a positive charge which has a magnitude of 2 because of the Group position. The ions are $Ca^{2+}$.

These positive ions will move to the negative electrode during electrolysis. If water had been present, you would have needed to decide whether calcium ions from the calcium chloride, or hydrogen ions from the water, were more likely to be discharged. As water is absent, however, it must be calcium ions which are discharged. This is indicated in the question where it says that 'calcium is liberated'. The positive charge on calcium is neutralized by the electrons that have been moved round the external circuit.

$$Ca^{2+} + 2e^- \rightarrow Ca.$$

This change is shown in key **A**.

3   Silver is a transition element so you cannot work out the charge on its ions from the position of the element in the Periodic Table. You either have to remember that silver ions are usually $Ag^+$, or work it out from a known formula of a silver compound, such as silver chloride. If you know that silver chloride has the formula AgCl, the charge of $-1$ on the chloride ion must be balanced by a charge of $+1$ on silver ions.

Aqueous silver nitrate contains the positive ions $H^+(aq)$ as well as $Ag^+(aq)$ ions. Both these ions will move to the negative electrode during electrolysis. It is usual for the ions of the element lower in the Electrochemical Series (ECS) to be discharged at the negative electrode (see the answers to questions 12 and 13). Silver is well below hydrogen in the ECS and its ions would be expected to be discharged by the electrons that have been made available at the negative electrode

$$Ag^+ + e^- \rightarrow Ag.$$

This change is shown in key **B**.

4   Any negative ions present move towards the anode during electrolysis. In question 4, both $OH^-(aq)$ and $SO_4^{2-}(aq)$ ions are present. The reaction at the anode involves removal of electrons and this could happen with hydroxide or sulphate ions, if the electrode were 'inert'.

$$OH^- \rightarrow OH + e^-; 4OH \rightarrow 2H_2O + O_2$$
$$SO_4^{2-} \rightarrow S_2O_8^{2-} + 2e^-$$

Copper is not, however, an inert electrode. You are likely to have met in your course the electrolysis described in question 4, and should know that the copper anode dissolves in the process. In this particular electrolysis electrons are removed, not from negative ions but from copper atoms which dissolve to produce positive copper ions. (This is tested further in question 13.)

$$Cu \rightarrow Cu^{2+} + 2e^-$$

This is shown correctly in key **C**.

5   Liquids that conduct electricity are aqueous solutions of ions (answer **1** correct), molten ionic compounds (answer **2** correct), and molten metals (answer **3** correct). Liquids with a simple molecular structure, such as hexane, do not conduct (answer **4** incorrect). **1, 2** and **3** are correct and the key is **A**.

6   Following on from question 5, another type of liquid that does not usually conduct electricity is a solution of one substance with a simple molecular structure dissolved in another substance with the same structure. Oxygen or glucose (simple molecular structures) dissolved in pure water (simple molecular structure) are examples of non-conducting solutions. **2** and **4** are incorrect.

If the solvent with the simple molecular structure is an ionizing solvent, such as water, it may react with the small molecule it is dissolving. This happens when hydrogen chloride is dissolved in water. The small molecule, HCl, reacts with the small molecule, water, to produce a solution of hydrochloric acid which contains oxonium ions (which may be called hydronium or hydroxonium ions on your course) and chloride ions. This solution conducts electricity and answer **1** is correct.

$$HCl(g) + H_2O(l) \rightarrow H_3O^+(aq) + Cl^-(aq)$$

Copper(II) sulphate is already composed of ions. Its giant structure is broken down on dissolving in water and the separated ions make the solution able to conduct electricity. Answer **3** is correct, as was answer **1**, and the key is **B**.

7   The electrical conductivity of an electrolyte is due to the movement of the negative and positive ions it contains. In this question, these ions are called cations and anions. Only the movement of the

particles in **2** and **3** is involved and the key is **E**.

The movement of electrons (answer **1**) is connected with the conductivity of metals, not the conductivity of electrolytes. During electrolysis, molecules (answer **4**) are carried along by moving cations and anions, and molecules may be involved in some reactions at electrodes, just as electrons always are. But neither electrons nor molecules are involved in the carrying of electric current through the electrolyte.

8   The elements that form ions with a positive charge are the metallic elements, as has been discussed in the answers to questions **1** to **4**. Sodium and magnesium are the only such elements in the list and the key is **C**.

9   The charged particles present in the atoms of any element are protons and electrons. A neutral atom contains an equal number of both particles, this number being the atomic number.

Chlorine, atomic number 17, contains 17 protons and 17 electrons. Being a non-metallic element, chlorine atoms form ions by gaining electrons. As it is an element of Group 7, chlorine atoms gain one electron and form $Cl^-$ ions. This is the only change that takes place.

When the chloride ion is formed from a chlorine atom, the number of protons remains at 17 (**1** is incorrect). One electron is gained (**4** correct) rather than lost (**2** incorrect). If **3** were to be correct, chlorine atoms would need to lose either one proton or one neutron on forming $Cl^-$ ions. This does not happen. Only **4** is correct and the key is **D**.

10  Giant structures of ions are only converted to freely moving ions by melting (**1** correct) or by dissolving in an ionizing solvent (**3** correct). The key is **B**.

11  Of the four substances in the list, only chlorine and aluminium are obtained on a large scale by electrolysis. The correct key is **B**. What processes are used to obtain sulphur and oxygen?

12  Aqueous lithium chloride contains the ions $Li^+(aq)$ and $Cl^-(aq)$ from ionization of the lithium chloride, and $H^+(aq)$ and $OH^-(aq)$ from slight ionization of water.

Both $Li^+(aq)$ and $H^+(aq)$ will move towards the negative electrode on electrolysis. The positive ion that is discharged will usually be that of the element lower in the ECS. This is hydrogen, and

the reaction taking place will be

$$2H^+ + 2e^- \rightarrow H_2$$

rather than

$$Li^+ + e^- \rightarrow Li$$

Statement **1** is correct while statement **4** is incorrect.

The discharge of hydrogen ions will disturb the ionization of water. This ionization can be written

$$H_2O \rightleftharpoons H^+ + OH^-.$$

If you have studied the topic 'equilibrium' you may realize that removal of $H^+$ ions would cause more water to ionize. This further ionization increases the concentration of $OH^-$ ions to a value higher than found in neutral water. The solution becomes alkaline around the positive electrode. Statement **2** is correct.

Statement **3** is concerned with the positive electrode. The negative ions $Cl^-$ and $OH^-$ move towards this electrode. The negative ion that is discharged will usually be the one that is higher in the ECS. The higher ion is $OH^-$ and one might expect the reactions

$$OH^- \rightarrow OH + e^-; \quad 4OH \rightarrow 2H_2O + O_2$$

to take place in preference to

$$2Cl^- \rightarrow Cl_2 + 2e^-.$$

This prediction would be correct if the concentrations of $OH^-(aq)$ ions and $Cl^-(aq)$ ions were the same. In water, however, the concentration of $OH^-(aq)$ ions is very low. In the concentrated lithium chloride of this question, the concentration of $Cl^-(aq)$ is high. Under these conditions, one would not be surprised to discover that chlorine was evolved at the positive electrode rather than oxygen. You should not be surprised because you should know that in a very similar situation with concentrated aqueous sodium chloride, chlorine is evolved. This point is tested again in question 30. Statement **3** is correct. **1** and **2** were also correct and the key is **A**.

13  The electrolysis of aqueous copper(II) sulphate with a copper anode results in dissolving of the anode as discussed in the answer to question 4.

$$Cu(s) \rightarrow Cu^{2+}(aq) + 2e^-$$

The positive ions present in the solution are $H^+(aq)$ and $Cu^{2+}(aq)$. Of these, copper is lower in the ECS, and the cathode reaction will be

$$Cu^{2+}(aq) + 2e^- \rightarrow Cu(s)$$

in preference to

$$2H^+(aq) + 2e^- \rightarrow H_2(g)$$

If two moles of electrons were passed in this electrolysis, one mole of copper would dissolve from the anode while one mole of copper would deposit on the cathode. As the anode decreases in mass, the cathode will increase in mass by exactly the same amount. This is covered by the statement in **B** which is the correct key.

From what has been said already, **A**, **C**, and **D** cannot be correct. The pH of the solution could only decrease if $OH^-(aq)$ ions were discharged at the anode. This does not happen with a copper anode and **E** is also incorrect.

**14** This question is concerned with changing the mass of metal deposited in a certain fixed time. This can only be achieved by changing the current, which alters the number of coulombs passed in the fixed time. The only piece of apparatus suitable for changing the current is the rheostat, which is a variable resistance. If the resistance in the circuit is increased, less current will flow, and a smaller number of coulombs will pass in the fixed time. This will decrease the rate at which the metal is deposited. A decrease in the resistance of the circuit would have the opposite effect. The correct key is **E**.

**15** Element $U$ displaces all the other metals from solution. $U$ must come first in order of reactivity. This means that the correct key is **D** or **E**.

$S$ displaces no metals from solution. $S$ must come last in order of reactivity, which makes the correct key **E** rather than **D**. Does key **E** place $R$ and $T$ the right way round?

**16** The equation for the discharge of $Cr^{3+}(aq)$ ions is

$$Cr^{3+}(aq) + 3e^- \rightarrow Cr(s).$$

This equation shows that 3 moles of electrons are needed to deposit 1 mole of chromium (52 g). 1.5 moles of electrons would be needed to deposit 26 g of chromium. The correct key is **C**.

**17** The information given is that 2 moles of electrons are needed to deposit 70 g of cerium. 4 moles of electrons would be needed to deposit 140 g of cerium, which is 1 mole of cerium. This suggests the equation for depositing cerium is

$$Ce^{4+}(aq) + 4e^- \rightarrow Ce(s).$$

The correct key is **D**.

**18** 1 g of scandium is liberated by 1/15 mole of electrons, so 15 g of scandium is liberated by 1

mole of electrons, and 45 g of scandium (1 mole) is liberated by 3 moles of electrons. The equation for discharging scandium must be

$$Sc^{3+} + 3e^- \rightarrow Sc.$$

1 g of oxygen is liberated by 1/8 mole of electrons, so 8 g of oxygen is liberated by 1 mole of electrons, and 16 g of oxygen (1 mole of atoms) is liberated by 2 moles of electrons. The equation for discharging oxide ions must be

$$O^{2-} \rightarrow O + 2e^-; 2O \rightarrow O_2.$$

This information shows that the charge on a scandium ion is $3+$ and on an oxide ion $2-$. To achieve a neutral (uncharged) compound, two scandium ions would be required for every 3 oxide ions. This could be represented by the formula $2Sc^{3+}, 3O^{2-}$ or by the formula $Sc_2O_3$. The second formula is given in key **D**.

**19** A neutral (uncharged) compound between silver ions and sulphate ions would need two silver ions for every sulphate ion. The formula could be written $2Ag^+, SO_4^{2-}$ or $Ag_2SO_4$. The second of these ways of writing the formula is shown in key **C**.

**20** Did you notice the word 'molten' in this question? When water is absent there is no problem of deciding whether $H^+(aq)$ and $OH^-(aq)$ ions from the water might be discharged. The only ions present in molten potassium bromide are $K^+$ and $Br^-$. The potassium ion will move to the negative electrode and potassium metal will be liberated.

$$K^+ + e^- \rightarrow K$$

The bromide ion will move to the positive electrode and bromine will be liberated.

$$2Br^- \rightarrow Br_2 + 2e^-$$

This is shown in key **D**.

**21** This question is concerned with the formation of a positive ion, while question 9 was concerned with the formation of the negative chloride ion.

A sodium atom contains equal numbers of protons and electrons, this number being 11, the atomic number of sodium. The formation of the sodium ion involves loss of one electron (key **A**). This leaves the sodium ion with only ten negative electrons whose combined charge does not completely balance the eleven positive charges of the protons. The excess positive charge is one unit and this is the reason for the sodium ion having a charge of $+1$.

**22** Both statements are true, as discussed for sodium ions in the answer to the previous question. The second statement is a correct explanation of the first and the key is **A**.

**23** Copper is below hydrogen in the ECS and this accounts for hydrogen not being evolved when copper is added to dilute hydrochloric acid. Both statements are false, key **E**.

**24** Calcium is a metallic element (ions positive) and is in Group 2 (magnitude of charge is 2). The second statement, is true.

Note the word 'fused' in the first statement. You are probably aware that this means the same as 'molten'. The equation for depositing calcium metal in electrolysis is

$$Ca^{2+} + 2e^- \rightarrow Ca.$$

The equation shows that the depositing of one mole of calcium atoms requires 2 moles of electrons. This is not what is said in the first statement. The first statement is false, the second true, and the key is **D**.

**25** The giant structure of ions found in the solid state of ionic substances does not allow free movement of ions. These substances are poor conductors when solid. They become good conductors when molten as the ions now have freedom of movement. Both statements in the question are true, although the 'not' in the first statement is rather too emphatic. The second statement is a correct explanation of the first and the key is **A**.

**26** Quantity of electricity is measured in coulombs which is obtained by multiplying current (in amperes) by time (in seconds). The circuit shown in this question is complete and includes an ammeter to measure the current. A clock to measure time is not, however, shown and it is this which is needed to carry out the experiment. The correct key is **E**.

Of the other four pieces of apparatus in the list, three could be included in the experiment but would not be needed. One piece of apparatus would not be included even if available. Can you spot which this is?

**27** Correct treatment of the cathode is described in **A**, **D** and **E**. The current must be kept low during the experiment to make certain that the deposit of copper sticks to the electrode as firmly as possible. It is also necessary to keep the current constant during the experiment unless a series of readings is going to be taken of the current at various times during the experiment. These two points are

covered by key **C**. What is not necessary is to completely immerse the electrode. Key **B** is the correct key here.

**28** 0.2 ampere for 1 hour, or $60 \times 60$ seconds, is $0.2 \times 60 \times 60$ ampere seconds (or coulombs). This number of coulombs deposits only 0.24 g of copper. To deposit 1 mole of copper, the quantity of electricity would need to be increased in the ratio 64/0.24. This gives an expected answer of $\frac{64}{0.24} \times 60 \times 60$ coulombs, which is key **A**.

**29** The five possible answers will be discussed in turn because it is necessary to decide between two very likely possibilities.

**A** It is probable that the electrolyte not being pure would have no influence on the result. The impurities would remain in solution and not affect the experiment. It is just possible that a particular impurity might be deposited, as well as copper, and produce the effect described. But you are asked to select the 'most likely explanation' and there is at least one that is far more likely than key **A**.

**B** There is some doubt over what is meant by 'too small'. If it means that a current of 0.2 amperes was too small and a current of, say, 0.4 amperes should have been used, then **B** is not correct. It does not matter what the current is, so long as it is low enough to make the deposit of copper adhere firmly to the electrode. Had **B** said 'too large', key **B** would certainly be in the running for being the correct key, because **B** would then be implying what is said in key **D**.

Perhaps 'too small' in key **B** means that the current of 0.2 ampere was not measured correctly and should really have been, say, 0.195 ampere. If this were the case, key **B** would be correct because a slightly lower than realized current would give a slightly lower than expected mass of copper. Don't forget you are looking for the most likely explanation. Key **B** would only be the best choice if one were to accept that it had been imperfectly worded. Had key **B** said 'the current was really slightly smaller than measured', then **B** would be a correct choice. Perhaps there is a more likely explanation among the remaining responses.

**C** There is no doubt that this is not the correct answer to the problem. A cathode that was not completely dry would make the mass of copper appear to be too large, not too small.

**D** A very frequent source of error in this experiment is failure of the copper to adhere completely to the electrode. You have probably carried out this experiment and may well have recognized immediately that key **D** is meant to be the 'most likely explanation'. It is far more likely than anything that has come before. The only remaining explanation, **E**, is a correct statement —copper does dissolve from the anode — but this is not relevant. The question is concerned with copper depositing on the cathode. The correct key must be **D**.

**30** If the electrode is copper, the reaction that produces electrons at the anode is (as discussed in the answer to question 4)

$$Cu \rightarrow Cu^{2+} + 2e^-.$$

If the electrode is inert, however, there are two negative ions present in solution which might be discharged, $Cl^-$ and $OH^-$. The reactions that could take place are

$$2Cl^- \rightarrow Cl_2 + 2e^-$$

or

$$OH^- \rightarrow OH + e^-; 4OH \rightarrow 2H_2O + O_2.$$

This question is concerned with mixtures of gases, and the only possible mixture that could be formed from the solution is chlorine and oxygen. The correct key is **A**.

You might like to carry out this experiment with a carbon electrode to see if chlorine and oxygen really are given off. You might find that the carbon electrode is not really 'inert'. What other gas should you be ready to test for if the carbon electrode is not completely inert?

# Test 9 | Acids and bases

**Questions 1–5** concern the following types of substance:

A  an acidic solid

B  an acidic liquid

C  an acidic gas

D  an alkaline liquid

E  an alkaline gas

From the above headings **A–E**, select the one which describes *most accurately* the product of the following reactions.

1  Burning sulphur in air

2  Burning phosphorus in oxygen

3  Reacting nitrogen and hydrogen in the presence of a catalyst

4  Reacting equal volumes of 1.0 M sodium hydroxide, NaOH, and 1.0 M sulphuric acid, $H_2SO_4$

5  Reacting aqueous solutions containing 1 mole of calcium hydroxide, $Ca(OH)_2$, and 1 mole of hydrogen chloride, HCl

| Directions summarized for questions 6 to 11 | | | | |
|---|---|---|---|---|
| A | B | C | D | E |
| 1, 2, 3 only correct | 1, 3 only correct | 2, 4 only correct | 4 only correct | Some other response or combination of responses is correct |

6  The pH of an aqueous solution of ethanoic acid (acetic acid) is found to be 2. The pH of this solution would be INCREASED by adding

1  sodium chloride crystals

2  aqueous sodium hydroxide

3  copper turnings

4  aqueous ammonia

7  Which of the following could be added to $10\ cm^3$ of 1.0 M HCl without changing the pH of the solution?

1  $5\ cm^3$ of 1.0 M HCl

2  $10\ cm^3$ of 0.1 M HCl

3  $10\ cm^3$ of 1.0 M HCl

4  $10\ cm^3$ of pure water

8  Correct statements about a 2 M solution of hydrogen chloride in water include:

1  the hydrogen chloride has donated protons to water molecules

2  the solution is a poor conductor of electricity

3  the pH of the solution is greater than 7

4  the solution will evolve hydrogen when treated with magnesium

9  Hydrochloric acid reacts with sodium hydroxide according to the equation

$$NaOH(aq) + HCl(aq) \rightarrow H_2O(l) + NaCl(aq).$$

After $20\ cm^3$ of 1.0 M hydrochloric acid has been added to $25\ cm^3$ of 1.0 M sodium hydroxide, the solution obtained will

1  have a pH greater than 7

2  yield only sodium chloride crystals on evaporation

3  conduct electricity

4  have a red colour

10 When $40 \text{ cm}^3$ of 0.2 M barium hydroxide are mixed with $80 \text{ cm}^3$ of 0.1 M sodium carbonate, a reaction takes place according to the equation

$$Ba(OH)_2(aq) + Na_2CO_3(aq) \rightarrow$$
$$BaCO_3(s) + 2NaOH(aq)$$

The resulting SOLUTION will

1  contain carbonate ions

2  be alkaline

3  contain barium hydroxide

4  conduct electricity

11 From the information given in the equations below

$$H_3O(l) + H_2O(l) \rightleftharpoons H_3O^+(aq) + OH^-(aq)$$
$$NH_3(aq) + H_2O(l) \rightleftharpoons NH_4^+(aq) + OH^-(aq)$$

it follows that

1  the water molecule can act as an acid

2  the $NH_4^+(aq)$ ion can act as an acid

3  the water molecule can act as a base

4  the $OH^-(aq)$ ion can act as an acid

Directions for questions 12 to 16. Each of the questions or incomplete statements in this section is followed by five suggested answers. Select the best answer in each case.

12 Which of the following solutions is most likely to react with solid barium carbonate?

A  Hydrogen chloride in dry toluene (methyl benzene)

B  Pure acetic acid (ethanoic acid) in dry toluene (methyl benzene)

C  Sodium hydroxide in water

D  Hydrogen chloride in water

E  Ammonia in water

13 A certain length of magnesium ribbon was reacted with an excess of dilute hydrochloric acid. The initial rate of production of hydrogen and the final volume of this gas were determined.

The experiment was repeated under identical conditions except that the hydrochloric acid was replaced by an equal volume of ethanoic acid (acetic acid) of the same molarity.

The initial rate of the second reaction and the final volume of hydrogen in the second reaction were compared with the values for the original experiment. Which result would be obtained?

|   | Initial rate of second reaction | Final volume of hydrogen in second reaction |
|---|---|---|
| A | Smaller | No change |
| B | Smaller | Smaller |
| C | No change | No change |
| D | No change | Smaller |
| E | Greater | Greater |

14 For which of the following pairs of acids and bases will complete neutralization occur between one mole of the acid and one mole of the base?

A  $HNO_3$ and NaOH

B  $H_2SO_4$ and NaOH

C  HCl and $Ca(OH)_2$

D  $H_2SO_4$ and $NH_3$

E  HCl and $Ba(OH)_2$

15 Barium hydroxide reacts with hydrochloric acid according to the equation

$$Ba(OH)_2 + 2HCl \rightarrow BaCl_2 + 2H_2O$$

$20 \text{ cm}^3$ of aqueous barium hydroxide were just neutralized by $30 \text{ cm}^3$ of 0.1 M hydrochloric acid. What was the molarity of the aqueous barium hydroxide?

A  0.300 M

B  0.150 M

C  0.100 M

D  0.075 M

E  0.033 M

16 Cold dilute hydrochloric acid will react readily with all of the following EXCEPT

A  magnesium

B  calcium carbonate

C  sodium hydroxide

D  magnesium oxide

E  copper

| Directions summarized for questions 17 to 21 | | |
|---|---|---|
| | First statement | Second statement | |
| **A** | True | True | Second statement is a correct explanation of the first |
| **B** | True | True | Second statement is NOT a correct explanation of the first |
| **C** | True | False | |
| **D** | False | True | |
| **E** | False | False | |

**17**

1  Effervescence usually occurs when a dilute acid is added to a metal oxide.

2  Hydrogen gas is usually formed when a dilute acid is added to a metal oxide.

**18**

1  A solution of hydrogen bromide in tetrachloromethane (carbon tetrachloride) will conduct electricity.

2  Hydrogen bromide can be a proton donor.

**19**

1  Aqueous solutions of weak acids have low electrical conductivities, compared with aqueous solutions of strong acids.

2  In aqueous solution, weak acids are ionized only to a small extent.

**20**

1  One mole of any acid will completely neutralize one mole of any base.

2  One mole of hydrogen ions will react with one mole of hydroxide ions.

**21**

1  A water molecule can act as an acid or as a base.

2  Water molecules can donate hydrogen ions as well as accept them.

**Questions 22–26** concern an experiment with barium hydroxide and sulphuric acid.

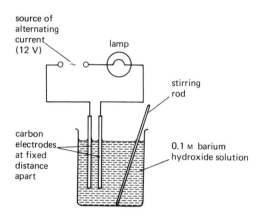

50 cm$^3$ of 0.1 M barium hydroxide, Ba(OH)$_2$(aq), were placed in the beaker. A few drops of phenolphthalein solution were added. The circuit was set up as shown in the diagram. 1.0 M sulphuric acid, H$_2$SO$_4$, was added slowly from a burette, with stirring, until an excess had been added. White barium sulphate, BaSO$_4$(s), formed in the beaker.

**22** The action which produces barium sulphate is

A  decomposition

B  ionization

C  precipitation

D  oxidation

E  electrolysis

**23** Which of the following best describes how the lamp would glow during the experiment?

| | Before acid was added | As the acid was gradually added | After excess acid had been added |
|---|---|---|---|
| **A** | Glows brightly | Glows brightly | Does not light up |
| **B** | Glows brightly | Gradually dims | Does not light up |
| **C** | Glows brightly | Gradually dims | Glows brightly |
| **D** | Does not light up | Gradually glows | Glows brightly |
| **E** | Does not light up | Glows brightly | Glows brightly |

**24** What volume of 1.0 M sulphuric acid will have been added when the amount of barium sulphate precipitate formed ceases to increase?

A  2.5 cm$^3$          D  50 cm$^3$

B  5 cm$^3$            E  500 cm$^3$

C  10 cm$^3$

**25** The final mixture in the beaker contains an excess of sulphuric acid. A sample of pure barium sulphate could best be obtained from this mixture by

**A** filtering, washing the precipitate with distilled water and drying it in an oven

**B** evaporating the mixture to dryness

**C** allowing the precipitate to settle, decanting the solution and drying the precipitate in an oven

**D** filtering, evaporating the filtrate to half its volume and leaving it to crystallize

**E** filtering and evaporating the filtrate to dryness

**26** Which of the following occurs during the course of the experiment?

**A** The pH of the solution changes from less than 7 to more than 7.

**B** Barium sulphate ionizes to form barium ions $Ba^{2+}$ and sulphate ions $SO_4^{2-}$.

**C** A white precipitate forms which then dissolves.

**D** Phenolphthalein changes from colourless to pink.

**E** $H_3O^+$ ions react with $OH^-$ ions to form water.

**Questions 27–30**

It is possible to titrate an acid with an alkali by measuring the temperature changes that occur when they are mixed in different proportions. In an experiment of this type, 5.0 cm³ of 2 M aqueous sodium hydroxide and 45.0 cm³ of hydrochloric acid of unknown concentration were measured out into separate beakers. The two solutions were then mixed and the temperature rose by 2.0°C. The experiment was repeated several times, using different volumes of acid and alkali but keeping the total volume 50.0 cm³ in each case.

The highest temperature rise was found to be 10.0°C, and occurred with 20.0 cm³ of sodium hydroxide and 30.0 cm³ of hydrochloric acid.

**27** Which pieces of apparatus would be most suitable for measuring out the solutions?

| | Sodium hydroxide | Hydrochloric acid |
|---|---|---|
| **A** | Burette | Burette |
| **B** | Burette | Measuring cylinder |
| **C** | Burette | Pipette |
| **D** | Measuring cylinder | Pipette |
| **E** | Pipette | Pipette |

**28** Which graph of the results (below) would be obtained?

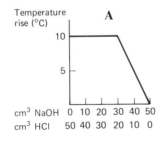

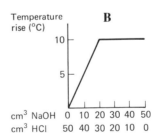

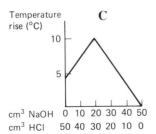

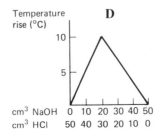

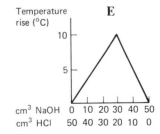

**29** The reaction may be represented by the equation

$$NaOH + HCl \rightarrow NaCl + H_2O.$$

What was the concentration of the hydrochloric acid?

**A** 0.75 M

**B** 1.00 M

**C** 1.33 M

**D** 2.00 M

**E** 3.00 M

**30** If the hydrochloric acid had been of exactly the same concentration (in moles per litre) as the sodium hydroxide, which mixture would have given the maximum temperature rise?

| | Volume of NaOH (cm³) | Volume of HCl (cm³) |
|---|---|---|
| **A** | 10.0 | 40.0 |
| **B** | 20.0 | 30.0 |
| **C** | 25.0 | 25.0 |
| **D** | 30.0 | 20.0 |
| **E** | 40.0 | 10.0 |

---

# Test 9 | Answers

**1** The main product of burning sulphur in air is sulphur dioxide, $SO_2$. This is a gas which dissolves in water to form a weakly ionized acid, sulphurous acid (which you may call sulphuric(IV) acid on your course). The reaction can be represented by the equation

$$SO_2(g) + H_2O(l) \rightarrow H_2SO_3(aq).$$

The burning of sulphur in air also produces small quantities of a white smoke which is composed of particles of an acidic solid, sulphur trioxide, $SO_3$.

You have to select the substance in the list which describes most accurately the product of burning sulphur in air. The choice has to be **C**, an acidic gas, rather than **A**, an acidic solid, because sulphur dioxide is the main product of the reaction.

**2** Phosphorus burns to form two oxides. One oxide has the formula $P_4O_6$ and is called either phosphorus trioxide or phosphorus(III) oxide. The other oxide has the formula $P_4O_{10}$ and is called phosphorus pentoxide or phosphorus(V) oxide.

The oxides of non-metals are usually acidic and this is the case with both $P_4O_6$ and $P_4O_{10}$, so the correct key must be **A**, **B** or **C**. $P_4O_{10}$ is a solid at room temperature (key **A**), while $P_4O_6$ is on the borderline between solid and liquid (key **A** or **B**). To avoid any doubt in this question, the phosphorus is burnt in oxygen rather than air. The use of pure oxygen makes $P_4O_{10}$ the likely product and the correct key is **A**.

**3** The reaction described would produce the gas ammonia, which dissolves in water to give an alkaline solution. The reactions could be represented by the following equations.

$$N_2(g) + 3H_2(g) \rightleftharpoons 2NH_3(g)$$
$$NH_3(g) + H_2O(l) \rightleftharpoons NH_4OH(aq) \rightleftharpoons$$
$$NH_4^+(aq) + OH^-(aq)$$

The correct key is **E** as the question is only concerned with the first stage of the reaction, the production of ammonia gas.

**4** There are two reactions possible between sodium hydroxide and sulphuric acid in aqueous solution, the first producing sodium hydrogen sulphate, the other producing sodium sulphate.

$$NaOH + H_2SO_4 \rightarrow NaHSO_4 + H_2O$$
$$2NaOH + H_2SO_4 \rightarrow Na_2SO_4 + 2H_2O$$

The equal volumes of 1.0 M solutions used in this question would result in reacting equal numbers of moles of sodium hydroxide and sulphuric acid. The reaction product is an aqueous solution of sodium hydrogen sulphate. The hydrogen sulphate ion is acidic, and this can be represented in the two ways below

$$HSO_4^- \rightarrow H^+ + SO_4^{2-}$$

or

$$HSO_4^- + H_2O \rightleftharpoons H_3O^+ + SO_4^{2-}.$$

The incomplete neutralization of sulphuric acid by sodium hydroxide produces an acidic solution which is described in key **B**, the correct key, as an acidic liquid.

5   The equation for the reaction described here is

$$2HCl(g) + Ca(OH)_2(aq) \rightarrow CaCl_2(aq) + 2H_2O(l)$$

This equation shows that the complete neutralization of 1 mole of calcium hydroxide requires 2 moles of hydrogen chloride. Only 1 mole of hydrogen chloride is available so neutralization will be incomplete. Calcium hydroxide will remain in aqueous solution, making the solution alkaline. This is described in key **D**.

6   An increase in pH is equivalent to a decrease in hydrogen ion concentration. Aqueous sodium hydroxide and aqueous ammonia contain hydroxide ions which will produce this effect by neutralizing hydrogen ions. Equations to represent this would be

$$H_3O^+(aq) + OH^-(aq) \rightarrow 2H_2O(l)$$

or

$$H^+(aq) + OH^-(aq) \rightarrow H_2O(l).$$

Only **2** and **4** would influence the pH (key **C**). Sodium chloride crystals would dissolve without further change, while copper, being below hydrogen in the electrochemical series (ECS), would not react at all.

7   The addition of either 0.1 M HCl or pure water will dilute the 1.0 M HCl and change the pH. Will the pH be raised or lowered?

In **1** and **3**, however, the addition of 1.0 M HCl to 1.0 M HCl will produce a larger volume of the same solution. There will be no change in pH and the key is **B**.

8   When hydrogen chloride dissolves in water, the change that takes place can be represented by the equation

$$HCl(g) + H_2O(l) \rightarrow H_3O^+(aq) + Cl^-(aq).$$

The 'dormant' or 'sleeping' acid, HCl, has been awakened by the water. A proton is donated to each water molecule (**1** correct) to give an acidic solution of oxonium ions, $H_3O^+$. This solution has a pH well below 7 (**3** incorrect). The solution contains ions which make it a good electrical conductor (**2** incorrect). Being acidic, the solution will react with metals such as magnesium to produce hydrogen (**4** correct). **1** and **4** are correct and the key is **E**.

9   The chemical equation given in the question shows that sodium hydroxide and hydrochloric acid react in the ratio 1 mole to 1 mole. The addition of only 20 cm³ of 1.0 M hydrochloric acid to 25 cm³ of 1.0 M sodium hydroxide does not give sufficient acid to neutralize all the sodium hydroxide. The solution will be alkaline, with a pH greater than 7 (**1** correct). On evaporation of this solution both sodium chloride and unneutralized sodium hydroxide will be left as solids (**2** incorrect). The solution contains two highly ionized substances, NaCl and NaOH, and will be a good conductor of electricity (**3** correct). All the substances involved are colourless (**4** incorrect). The correct key is **B**.

The solution in this problem would only have a colour if an indicator had been added. What indicators have you used on your course? Would any of these indicators have given a red solution in this particular experiment? Can you find in a reference book the name of any indicator which would be a red colour when sodium hydroxide has been incompletely neutralized by hydrochloric acid?

10  40 cm³ of a 0.2 M solution contain the same number of moles as 80 cm³ of a 0.1 M solution. The barium hydroxide and sodium carbonate are being mixed in the ratio 1 mole to 1 mole which is just the ratio required by the chemical equation.

The resulting solution will not contain carbonate ions as there is no sodium carbonate remaining (which would have provided these ions) while the barium carbonate formed is insoluble — shown by the state symbol (s), so answer **1** is incorrect. Answer **3** is also incorrect because the barium hydroxide has been completely consumed. The sodium hydroxide formed will make the solution alkaline and conducting by its ionization. **2** and **4** are correct and the key is **C**.

11  The higher equation shows one water molecule (acting as an acid) giving a proton to another water molecule, which, in accepting the proton, acts as a base. **1** and **3** are correct.

When the reaction in the lower equation goes from right to left, the $NH_4^+(aq)$ ion gives a proton (acting as an acid) to the $OH^-$ ion. The $NH_4^+(aq)$ ion is acting as an acid. **2** is correct.

Only **1**, **2** and **3** are correct (key **A**) as in both reactions, on going from right to left, the $OH^-(aq)$ ion accepts a proton and acts as a base.

**12** You are given a choice of acids and bases to react with barium carbonate. The correct choice must be an acid as these react with carbonates to form salts, water, and carbon dioxide. In **A** and **B**, however, the dormant acid will not be awakened by the solvent. The best answer is **D** because here is a dormant acid, hydrogen chloride, which has been awakened by use of water as the solvent, as discussed in the answer to question 8.

**13** Ethanoic acid is a weakly ionized acid while hydrochloric acid is fully ionized. The initial rate (which means the rate at the start) will be slower if the weakly ionized acid is used. The correct key must be **A** or **B**.

The final volume of hydrogen will be decided by the quantity of magnesium used. This is because the acid is in excess. As the same quantity of magnesium was used in both experiments, there would be no change in the final volume of hydrogen. The correct key is **A**.

**14** Equal numbers of moles of acid and base are being used. It is necessary to find a pair where the acid provides the same number of hydrogen ions as the base provides hydroxide ions. This is only the case in **A**, which is the correct key.

Another pair which could be made up from the substances in the question, and which would give complete neutralization, is $H_2SO_4$ and $Ca(OH)_2$. How many further pairs can you find which would be expected to give complete neutralization? (There are three or four more — is it three or is it four?)

**15** 30 cm$^3$ of 0.1 M HCl contain $\frac{30}{1000} \times 0.1$ mole of HCl. The chemical equation shows that 1 mole of $Ba(OH)_2$ reacts with 2 moles of HCl.

20 cm$^3$ of $Ba(OH)_2$ must, therefore, contain $\frac{1}{2} \times \frac{30}{1000} \times 0.1$ mole, so 1000 cm$^3$ of $Ba(OH)_2$ must contain $\frac{1000}{20} \times \frac{1}{2} \times \frac{30}{1000} \times 0.1$ mole. This is 0.075 mole.

A solution that contains 0.075 mole in 1000 cm$^3$ is 0.075 M. This is key **D**.

**16** Dilute acids, for example dilute hydrochloric acid, react readily with bases, such as sodium hydroxide and magnesium oxide, to form a salt and water. These acids also react readily with carbonates to give a salt, water, and carbon dioxide. The third general reaction of acids is to form hydrogen and a salt with reactive metals such as magnesium. Copper, however, is too low

in the ECS to react in this way, and the correct key is **E**.

You might like to try an experiment where you leave some copper (lumps or sheet) in dilute hydrochloric acid for several days or longer. If you do try the experiment, set up a second one using dilute sulphuric acid and make certain that some of the copper is exposed to the air above the acid. Can you explain what has happened by the end of the experiment? Is it true to say that copper does not react with dilute acids such as hydrochloric or sulphuric acid?

**17** Hydrogen is formed when a dilute acid is added to a suitable metal, such as magnesium. When a metal oxide is used in place of the metal, the presence of the oxygen results in the formation of hydrogen oxide, or water, rather than hydrogen. There is no effervescence. Both statements are false and the key is **E**.

$$Mg + H_2SO_4 \rightarrow H_2 + MgSO_4$$
$$MgO + H_2SO_4 \rightarrow H_2O + MgSO_4$$

You might like to try this reaction using magnesium oxide or calcium oxide from an old bottle. It is likely that you will see bubbles appearing, especially if the bottle has spent a lot of its life with the lid off. Can you collect enough of this gas to carry out a lime water test on it? What is the gas and why does it appear?

**18** Hydrogen bromide can act as a proton donor. It does this when it forms white clouds of ammonium bromide on reacting with ammonia or when it dissolves in water to form hydrobromic acid

$$HBr(g) + NH_3(g) \rightarrow NH_4Br(s)$$
$$HBr(g) + H_2O(l) \rightarrow H_3O^+(aq) + Br^-(aq).$$

Hydrogen bromide is composed of small molecules and in the solvent tetrachloromethane, which has a similar structure, the hydrogen bromide dissolves without reaction. No ions are formed and the solution is a poor conductor. The first statement is false, the second true, and the key is **D**.

**19** The term 'weak acid' means that the acid, for example ethanoic acid, is only weakly ionized in water. This is unlike a strong acid, such as sulphuric acid, which is strongly ionized in water. Incomplete ionization

$$CH_3CO_2H + H_2O \rightleftharpoons H_3O^+ + CH_3CO_2^-.$$

Complete ionization

$$H_2SO_4 + H_2O \rightarrow H_3O^+ + HSO_4^-.$$

The small extent of the ionization of a weak acid means that there are few ions available to carry current through the aqueous solution. The electrical conductivity is low compared with an aqueous solution of a strong acid. Both statements are true, the second being a correct explanation of the first. The key is **A**.

20 One mole of hydrogen ions reacts with one mole of hydroxide ions (second statement true).

$$H^+(aq) + OH^-(aq) \rightarrow H_2O(l)$$

or

$$H_3O^+(aq) + OH^-(aq) \rightarrow 2H_2O(l).$$

This does not, however, mean that one mole of any acid will completely neutralize one mole of any base (first statement false). Refer to question 14 and its answer if you are not certain on this point. The correct key is **D**.

21 Both statements are true, the second being a correct explanation of the first. The key is **A**. The situation has already been discussed in the first paragraph of the answer to question 11.

22 The reaction that produces sulphate involves neutralization.

$$Ba(OH)_2(aq) + H_2SO_4(aq) \rightarrow$$
$$BaSO_4(s) \text{ and } 2H_2O(l)$$

or

$$2OH^-(aq) + 2H^+(aq) \rightarrow 2H_2O(l).$$

The word 'neutralization' is not included in the possibilities in the question so you must look further. The state symbol (s), which is given in the question, indicates that solid, insoluble, barium sulphate is formed during the reaction. This is precipitation and the key is **C**.

23 Before any sulphuric acid has been added, the ionization of the aqueous barium hydroxide will make the solution a good conductor. The lamp should glow and the key must be **A**, **B** or **C**.

As the reaction progresses, unionized water and insoluble barium sulphate are formed. Neither of these will assist conduction and the lamp will dim as the barium hydroxide is consumed. The key must be **B** or **C**, not **A**.

After excess sulphuric acid has been added, the ions produced by the acid will make the solution a good conductor and the lamp should glow again. Only key **C** fits all the three situations.

24 The reaction began with 50 cm$^3$ of 0.1 M barium hydroxide. This reacts in the ratio 1 mole to 1 mole with sulphuric acid. 50 cm$^3$ of 0.1 M sulphuric acid would be just sufficient to react with all the barium hydroxide and form the maximum amount of barium sulphate precipitate. The sulphuric acid is, in fact, 1.0 M. 5 cm$^3$ of 0.1 M sulphuric contains the same number of moles of sulphuric acid as are present in 50 cm$^3$ of 0.1 M solution. This is the value given in key **B**.

25 The required barium sulphate is an insoluble solid in aqueous sulphuric acid. To purify the solid, this mixture first requires filtering to isolate the solid. The dilute sulphuric acid on the surface of the solid needs to be washed away with distilled water. The only impurity is then water, which could be removed by careful drying. This is outlined in **A** which is the correct key. **B**, **C**, and **E** would produce barium sulphate contaminated with sulphuric acid. Would **D** produce any barium sulphate?

26 Acid is being added to alkali so the pH change will be the reverse of that described in **A** and the colour change of phenolphthalein will be the reverse of that described in **D**. The reaction involves the coming together of Ba$^{2+}$ and SO$_4^{2-}$ ions, so **B** is incorrect. A white precipitate of BaSO$_4$ does form, but this does not then dissolve (**D** incorrect). The reaction is a precipitation and a neutralization as discussed in the answer to question 22. Key **E** correctly describes the reaction as a neutralization.

27 The investigation requires both the sodium hydroxide and the hydrochloric acid to be measured out with 0.1 cm$^3$ accuracy. It must also be possible to measure out a variety of volumes of both solutions, always with the same accuracy. A burette would be the most convenient piece of apparatus to use for both solutions and the key is **A**.

A measuring cylinder would be quick to use but would not give the required accuracy. It would be possible to use pipettes graduated in 0.1 cm$^3$ divisions but these are normally only available with a maximum volume of 10 cm$^3$. Several fillings would be needed to obtain most of the volumes used. What is required is a graduated pipette which holds 50 cm$^3$ but the equivalent is already available and is called a 'burette'!

28 There will be no reaction and no temperature rise if the experiment uses no sodium hydroxide (**A** and **C** incorrect), or no hydrochloric acid (**B**

incorrect). Only **D** and **E** show the expected rise to a temperature peak, when equal numbers of moles of acid and base are taken, followed by a temperature drop as an excess of sodium hydroxide is added.

The information at the start of this set of questions indicates that the maximum temperature was obtained with 20 cm$^3$ of sodium hydroxide and 30 cm$^3$ of hydrochloric acid. This is shown correctly in **D**.

**29** The experiment shows that 20 cm$^3$ of 2 M NaOH reacts with 30 cm$^3$ of hydrochloric acid.

The quick way to find the molarity of the hydrochloric acid is to first realize that more of it is needed so it must be weaker than the sodium hydroxide. The acid must be weaker in the ratio 20/30, so the molarity of the acid is 2 M × 20/30 which is 1.33 M. The correct key is **C**.

You may not trust this method! If this is the case, calculate the number of moles of NaOH used.

1000 cm$^3$ of 2 M NaOH contain 2 mole,

20 cm$^3$ of 2 M NaOH contain 2 mole × $\dfrac{20 \text{ cm}^3}{1000 \text{ cm}^3}$ which is $\dfrac{4}{100}$ mole.

The equation given shows that NaOH reacts with HCl in the ratio 1 mole to 1 mole.

30 cm$^3$ of HCl must also contain $\dfrac{4}{100}$ mole

so 1000 cm$^3$ of HCl contain $\dfrac{4}{100}$ mole × $\dfrac{1000 \text{ cm}^3}{30 \text{ cm}^3}$

which is $\dfrac{40}{30}$ mole or 1.33 mole.

This means that the concentration of the acid is 1.33 M.

**30** If the acid and base had been of equal concentration, the maximum amount of reaction would have been achieved when using equal volumes, and hence equal numbers of moles, of each. This is the 25 cm$^3$ to 25 cm$^3$ mixture of key **C**, which would have given the maximum temperature rise.

# Test 10 | Carbon chemistry

## Questions 1–4

Which of the lettered headings most accurately describes each of the numbered substances?

A Polymer      D Hydrocarbon

B Carbohydrate      E Protein

C Vegetable oil

1 Nylon      3 Octane

2 Glucose      4 Ethene

| Directions summarized for questions 5 to 11 | | | | |
|---|---|---|---|---|
| **A** | **B** | **C** | **D** | **E** |
| 1, 2, 3 only correct | 1, 3 only correct | 2, 4 only correct | 4 only correct | Some other response or combination of responses is correct |

5 Ethene, which is formed by cracking hydrocarbons,

1 is a gas

2 can be polymerized

3 reacts with bromine water

4 is very soluble in water

6 Methane is a member of a homologous series of hydrocarbons called the alkanes. It is characteristic of any homologous series that all the members are IDENTICAL in

1 relative molecular mass

2 physical properties

3 structural formula

4 general formula

7 A hydrocarbon, containing 8 hydrogen atoms per molecule, has a relative molecular mass of 56. The substance decolorizes bromine water. ($C = 12$, $H = 1$)

From this information it may be deduced that the hydrocarbon

1 has the formula $C_4H_8$

2 is an unsaturated organic compound

3 may be represented as $\begin{array}{c} H_2C{-}CH_2 \\ | \quad\quad | \\ H_2C{-}CH_2 \end{array}$

4 will probably form a solid product by polymerization

8 A polymer could be directly obtained from

1 $\begin{array}{c} CH_2 \\ | \\ H{-}C{=}CH_2 \end{array}$

2 a reaction between compounds of formulae

$CH_3{-}\boxed{\phantom{xxx}}{-}COOH$

and

$CH_3{-}\boxed{\phantom{xxx}}{-}OH$

3 a reaction between compounds of formulae

$HOOC{-}\boxed{\phantom{xxx}}{-}COOH$

and

$HO{-}\boxed{\phantom{xxx}}{-}OH$

4 $CH_3CH_2CH_3$

9 In which of these changes are larger molecules broken down to smaller molecules?

1 Starch is hydrolyzed with dilute hydrochloric acid.

2 Castor oil is hydrolyzed with sodium hydroxide solution.

3 Paraffin vapour is cracked by passing over red-hot chips of porcelain.

4 Ethene (ethylene) is polymerized by heating under pressure in the presence of a catalyst.

**10** Which of the following are typical of plastics such as polyethene (polythene) or nylon?

**1** They are composed of long chain molecules.

**2** They conduct electricity.

**3** They are made from monomers.

**4** They dissolve in dilute acids.

**11** Which of the following fibres occur naturally?

**1** Wool

**2** Nylon

**3** Cotton

**4** Terylene

---

Directions for questions 12 to 17. Each of the questions or incomplete statements in this section is followed by five suggested answers. Select the best answer in each case.

---

**12** The alkane $C_8H_{18}$ could be obtained from the higher member of the family, $C_{15}H_{32}$, by the process of

**A** cracking

**B** dehydrogenation

**C** depolymerization

**D** fractional distillation

**E** fermentation

**13** There are two compounds with the molecular formula $C_4H_{10}$. The structure of one of these compounds may be represented:

The other compound could be represented.

**14** A gaseous organic compound readily decolorizes bromine water and aqueous potassium permanganate. The compound is likely to be

A an unsaturated hydrocarbon

B a protein

C a sugar

D an alcohol

E starch

**15** The polymer PVC has the structure

$-CHCl-CH_2-CHCl-CH_2-CHCl-$

The formula of the single substance, which is polymerized to form PVC, would be

A $CH_3-CH_3$  D $CHCl=CHCl$

B $CH_2=CH_2$  E $CH_2=CHCl$

C $CH_3-CH_2Cl$

**16**

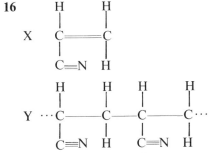

X and Y represent the starting material and product, respectively, for an organic reaction. This reaction illustrates

A hydrogenation  D depolymerization

B dehydrogenation  E polymerization

C cracking

**17** Which statement is true of a soapless detergent but is untrue of soap?

A It consists of molecules containing many carbon atoms.

B It reduces the surface tension of water.

C It forms a lather when shaken with distilled water.

D It does not form a precipitate with hard water.

E It is made by reacting a fat with an alkali.

Directions summarized for questions 18 to 21

| | First statement | Second statement | |
|---|---|---|---|
| A | True | True | Second statement is a correct explanation of the first |
| B | True | True | Second statement is NOT a correct explanation of the first |
| C | True | False | |
| D | False | True | |
| E | False | False | |

**18**

1 Ethane, $C_2H_6$, and ethene $C_2H_4$, are both gases at room temperature and atmospheric pressure.

2 Ethane and ethene are members of the same homologous series.

**19**

1 Starch is a carbohydrate.

2 Carbon dioxide and water are produced when starch is heated with copper(II) oxide.

**20**

1 Glucose undergoes fermentation to ethanol and carbon dioxide.

2 Glucose is a compound of carbon, oxygen and hydrogen.

**21**

1 A vegetable oil forms a soap when it is boiled with aqueous sodium hydroxide.

2 The alkaline hydrolysis of vegetable oils produces sodium salts of acids having long chains of carbon atoms.

| Name | Formula | Melting point in °C | Boiling point in °C |
|---|---|---|---|
| Pentene | CH$_3$—CH$_2$—CH$_2$—CH=CH$_2$ | −165 | 30 |
| 2-methylbutane | H$_3$C\\<br>  $\,$CH—CH$_2$—CH$_3$<br>H$_3$C⁄ | −129 | 38 |
| Pentane | CH$_3$—CH$_2$—CH$_2$—CH$_2$—CH$_3$ | −129 | 38 |
| Hexane | CH$_3$—CH$_2$—CH$_2$—CH$_2$—CH$_2$—CH$_3$ | −94 | 69 |
| Heptane | CH$_3$—CH$_2$—CH$_2$—CH$_2$—CH$_2$—CH$_2$—CH$_3$ | −91 | 98 |
| Octane | CH$_3$—CH$_2$—CH$_2$—CH$_2$—CH$_2$—CH$_2$—CH$_2$—CH$_3$ | −57 | 125 |

**Questions 22–23** are concerned with an investigation of a mixture of hydrocarbons. Some properties of the components of this mixture are given in the table.

**22** This mixture of hydrocarbons is fractionally distilled. The first liquid collected from the top of the fractionating column is likely to consist mainly of

**A** pentene  **D** heptane

**B** 2-methylbutane  **E** octane

**C** pentane

**23** At the end of the distillation six fractions have been collected. Portions of each of these are shaken with bromine water. One of these fractions decolorizes the bromine water rapidly. The boiling point of this fraction, in °C, is most likely to be

**A** 125  **D** 30

**B** 98  **E** 28

**C** 38

**Questions 24–26** concern an experiment in which a mixture of bromine (Br$_2$) and excess hexane (C$_6$H$_{14}$) was exposed to a bright light. Hexane is a colourless liquid and is a member of the same homologous series as methane. When the light was switched on, the mixture in the flask began to bubble, giving off a colourless, acidic gas which produced steamy fumes on contact with the air.

The main product of the reaction remaining in the flask was the colourless liquid monobromohexane (C$_6$H$_{13}$Br).

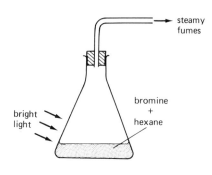

**24** The most likely method of separating the bromohexane (C$_6$H$_{13}$Br) from the excess hexane (C$_6$H$_{14}$) would be by

**A** chromatography  **D** electrolysis

**B** filtration  **E** crystallization

**C** distillation

**25** The colourless gas given off is most likely to be

**A** hydrogen

**B** hydrogen bromide

**C** water vapour

**D** bromine vapour

**E** bromohexane vapour

**26** What type of reaction has occurred in the flask?

**A** Substitution

**B** Addition

**C** Esterification

**D** Polymerization

**E** Cracking

**Questions 27–30**

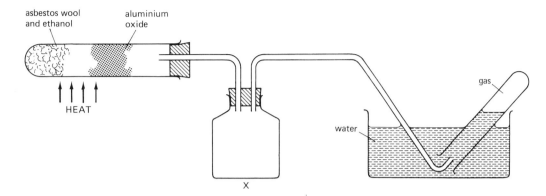

The above apparatus was used in an investigation of the effect of heat on ethanol. The aluminium oxide catalyses the decomposition of the ethanol to ethene and water.

$$C_2H_5OH \rightarrow C_2H_4 + H_2O$$

**27** The reaction provides an example of

**A** depolymerization

**B** esterification

**C** dehydrogenation

**D** oxidation

**E** dehydration

**28** The most probable reason for including the bottle X was to

**A** give the reaction time to take place

**B** provide a supply of air to oxidize the ethanol completely

**C** cool the gas produced during the reaction

**D** act as a safety device to prevent sucking back into the heated tube

**E** prevent too rapid an evolution of gas

**29** The ethene was added to bromine vapour. You would observe

**A** the brown colour of the bromine becoming darker

**B** steamy acidic fumes of hydrogen bromide

**C** a fine black deposit of carbon

**D** the ethene burning with a smoky flame

**E** the brown colour of the bromine vapour disappearing

**30** The equation for the reaction which takes place between ethene and bromine is

**A** $C_2H_4 + Br_2 \rightarrow C_2H_3Br + HBr$

**B** $C_2H_4 + Br_2 \rightarrow C_2H_4Br_2$

**C** $C_2H_4 + 2Br_2 \rightarrow C_2H_2Br_2 + 2HBr$

**D** $C_2H_4 + 2Br_2 \rightarrow 2C + 4HBr$

**E** $C_2H_4 + 2Br_2 \rightarrow C_2H_4Br_4$

# Test 10 | Answers

**1** Nylon is a polymer and the correct key is **A**. Is nylon an addition polymer or a condensation polymer? In nylon the group

links its monomer units together.

This group is present in proteins where it is called the 'peptide link'. In proteins, the peptide links are separated by only one carbon atom each time. This is shown below

Find the structure of nylon in a reference book. Would it be correct to call nylon a protein?

**2** Glucose, $C_6H_{12}O_6$, is a compound of carbon, hydrogen, and oxygen in which the hydrogen and oxygen atoms are present in the ratio 2 to 1. As this ratio is identical to that found in water, $H_2O$, glucose is called a carbohydrate. The correct key is **B**.

**3** Octane is an alkane with eight carbon atoms, $C_8H_{18}$. It is a hydrocarbon, key **D**.

**4** Ethene, $C_2H_4$, is also a hydrocarbon, key **D**.
    This set of questions has not tested whether you can recognize a vegetable oil or a protein. Can you name a substance which fits each of these descriptions? (Questions 9 and 11 each contain one such substance.)

**5** Ethene, $C_2H_4$, is a small molecule and is a gas (**1** correct). It is unsaturated and hence reacts with itself by addition, this reaction being called 'polymerization' (**2** correct). Ethene also reacts with other substances by addition, including bromine (**3** correct). Like all hydrocarbons, and the majority of small molecules, ethene is not very soluble in water (**4** incorrect). **1**, **2** and **3** are correct and the key is **A**.

**6** The only way in which members of a homologous series are identical is in their general formula. Only **4** is correct and the key is **D**. What is the general formula of the members of the alkane series?

**7** The presence of 8 hydrogen atoms per molecule will make a contribution of 8 units to the relative molecular mass of 56. The remaining contribution of 48 units must come from 4 carbon atoms, so the formula is $C_4H_8$ (**1** correct).
    The hydrocarbon contains two hydrogen atoms less than are found in the saturated chain-alkane with four carbon atoms, $C_4H_{10}$. The absence of two hydrogen atoms could be accounted for by the presence of a ring of three or four carbon atoms, as shown for cyclobutane in **3**. The absence of two hydrogen atoms could also be accounted for by the presence of the smallest possible ring of carbon atoms, a ring containing only two carbon atoms. This is not normally regarded as a ring but as a pair of carbon atoms joined by a 'double bond'. One test for the double bond is that it decolorizes bromine by addition, as the substance in this question does. Such compounds are said to be 'unsaturated' and **2** is correct.
    As the substance contains a double bond, it cannot have the structure given in **3** which is incorrect.
    Addition reactions are the principal reactions of unsaturated hydrocarbons. A special type of addition is 'self addition'. This is usually called 'polymerization'. The unsaturated hydrocarbon in this question would be expected to polymerize, so **4** is correct, as are **1** and **2**. The key is **E**.

**8** Molecules must have two active ends if they are going to be capable of polymerization. If molecules have only one active end, as in **2**, they are as capable of polymerizing as a group of one-armed chemists is capable of singing Auld Lang Syne properly! In addition polymerization, these active ends are the carbon atoms on either side of the double bond, so the molecule in **1** would polymerize (to polypropene).
    **2** represents 'one-armed chemists' and **4** represents 'no-armed chemists', neither of which is capable of polymerizing. The molecules in **3**, however, each possess two active ends and would

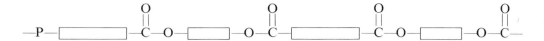

be capable of polymerizing by loss of water between the groups COOH and OH. This type of polymerization where a small molecules, such as water, is lost, is called 'condensation polymerization'. The polymer would have the structure shown above. It is a polyester.

The molecules in **1** and **3** are capable of direct polymerization and the key is **D**. Is it an oxygen atom or a carbon atom at point P in the polymer shown above?

**9** **1**, **2**, and **3** all involve the formation of smaller molecules from larger molecules, and the key is **A**. Do you know what type of smaller molecule is produced in each case? The reaction described in **4** forms larger molecules from smaller ones.

**10** Plastics such as polythene and nylon are composed of long-chain molecules and are made from monomers. **1** and **3** are correct and the key is **B**. Two very useful properties of most plastics are that they conduct electricity poorly and resist attack by dilute acids (**2** and **4** incorrect).

**11** Wool (a protein) and cotton (a carbohydrate) are naturally occurring fibres. Nylon and terylene can be made in the form of fibres but they do not occur naturally. The correct key is **B**.

**12** The change described here could be achieved by the process called cracking, key **A**.

**13** **A**, **B**, **C**, and **D** are all different ways of representing the original molecule. Note that they all contain a chain of 4 carbon atoms. **E** is different and is the correct key. **E** contains a chain of only 3 carbon atoms. Can you name the two compounds?

**14** Ready decolorization of bromine and acidified aqueous potassium permanganate (or turning alkaline aqueous potassium permanganate green) are reactions of an unsaturated compound, key **A**.

**15** The repeating unit in the structure given is $-CH_2-CHCl-$. This shows that the original monomer was $CH_2=CHCl$ (key **E**) which formed PVC by addition polymerization. The polymer is named after the monomer, vinyl chloride (chloroethane). What does PVC stand for?

**16** The small molecule, X, has produced a large molecule, Y, in which the repeating unit is X without its double bond. This is polymerization, key **E**. Is the polymerization of the addition or of the condensation type?

**17** A soap, frequently called a soapy detergent, forms a lather with distilled water but produces a precipitate, called scum, with hard water. The advantage of a soapless detergent is that it forms a lather with hard water without first forming scum. This is correctly described in key **D**.

**A**, **B**, and **C** are true of both substances. **E** is true of soap but untrue for a soapless detergent, which is the opposite way round to the requirement of the question.

**18** The first statement is true. Ethane and ethene are both gases at room temperature and pressure. They are not, however, members of the same homologous series. The key is **C**. What names are given to the homologous series of which these two substances are members?

**19** Starch is a polymer with the formula $(C_6H_{10}O_5)_n$. It is a compound of carbon, hydrogen, and oxygen in which the ratio of hydrogen atoms to oxygen atoms is 2 to 1. This makes it a carbohydrate and the first statement is true.

The second statement is also true. Heating starch with copper(II) oxide would produce carbon dioxide and water. This shows that starch contains carbon and hydrogen. It does not prove that starch contains oxygen, as all the oxygen might have come from the copper(II) oxide. It certainly does not prove that starch contains hydrogen atoms and oxygen atoms in the ratio 2 to 1. Both statements are true but the second is not a correct explanation of the first. The key is **B** rather than **A**.

**20** The first statement is true. Glucose does undergo fermentation and does produce ethanol and carbon dioxide. The second statement is also true but is not the reason why glucose undergoes fermentation. There are many compounds composed of carbon, oxygen, and hydrogen, such as ethanol, ethanoic acid, or ethyl ethanoate, but only very special compounds containing these elements are capable of fermentation to ethanol and carbon dioxide. The key, once again, is **B** rather than **A**.

21 You have probably seen a soap made by the method described in the first statement and recognize that this is true. The second statement is a correct explanation of what happens during the reaction and the key is **A**.

22 In fractional distillation, the substance with the lowest boiling point leaves the top of the column first. The table shows that 2-methylbutane has the lowest boiling point. This is key **B** in the question.

23 The fraction that decolorizes bromine water must contain an unsaturated compound. The only unsaturated compound in the table is pentene whose boiling point is 30°C. The correct key is **D**.

24 Bromohexane and hexane are both composed of small molecules and hence are likely to be fairly volatile liquids. They would be separated by distillation, key **C**. Which of the two substances is likely to distill over first?

25 The reaction taking place involves a halogen attacking a saturated compound, hexane. This is not addition but substitution. Substitution takes place when a hydrogen atom is replaced by another group, in this case bromine.

$$C_6H_{14} + Br_2 \rightarrow C_6H_{13}Br + HBr$$

One half of the bromine molecule takes the place of the displaced hydrogen atom. The other half of the bromine molecule combines with the hydrogen to form hydrogen bromide gas. This is key **B**.

26 As has been discussed above, the reaction involves substitution. The correct key is **A**.

27 The reaction might be called 'cracking' but this term does not appear in the question. The equation given shows that a water molecule is lost from the ethanol molecule. This could be called 'dehydration', key **E**.

28 If you have ever carried out this reaction without the bottle X, or a safety valve, you will know how hazardous the experiment can be! Without a safety device, water frequently sucks back into the heated tube, often with dramatic results. **D** is the correct key.

29 Bromine vapour will take part in an addition reaction with the unsaturated ethene:

$$CH_2{=}CH_2 + Br_2 \rightarrow CH_2Br{-}CH_2Br$$

The brown colour of the bromine will disappear, key **E**.

30 The equation given in **B**, the correct key, is another way of writing the equation given above in the answer to question 29.

# Test 11　Inorganic chemistry 1

**Questions 1–3** are concerned with the following chlorides:

  **A** Calcium chloride

  **B** Silver chloride

  **C** Copper(II) chloride

  **D** Ammonium chloride

  **E** Phosphorus pentachloride

Choose from these the chloride which

1 is frequently used as a drying agent in the laboratory

2 gives off an alkaline gas when warmed with aqueous sodium hydroxide

3 is insoluble in water but dissolves in aqueous ammonia

**Questions 4–7** concern the following chemical tests:

  **A** adding dilute nitric acid, followed by aqueous silver nitrate

  **B** adding dilute acid, followed by passage of the evolved gas through lime water

  **C** heating with aqueous alkali and testing the evolved gas with litmus paper

  **D** adding acidified potassium permanganate solution

  **E** carrying out a flame test using a suitable wire

Select the test which would clearly distinguish between

4 sodium carbonate and sodium sulphate

5 potassium chloride and potassium sulphate

6 sodium chloride and potassium chloride

7 ammonium chloride and magnesium chloride

**Questions 8–11** concern the following ionic equations. In these equations, the letters G to Z are NOT the usual symbols of the particles concerned.

  **A**   $G^+ + J^- \rightarrow GJ$

  **B**   $L + M^{2+} \rightarrow L^{2+} + M$

  **C**   $Q + 2R^+ \rightarrow Q^{2+} + R_2$

  **D**  $T^+ + X^{2+} \rightarrow T + X^{3+}$

  **E**  $Y^{2+} + Z^{2-} \rightarrow YZ$

Choose, from **A** to **E**, the equation which most closely represents each of the following reactions.

8 The test for chloride ions in aqueous solution, using aqueous silver nitrate

9 The test for sulphate ions in aqueous solution, using aqueous barium chloride

10 The reaction between zinc and aqueous copper(II) ions

11 The oxidation of iron(II) ions in aqueous solution, using aqueous silver ions

| Directions summarized for questions 12 to 17 | | | | |
|---|---|---|---|---|
| **A** 1, 2, 3 only correct | **B** 1, 3 only correct | **C** 2, 4 only correct | **D** 4 only correct | **E** Some other response or combination of responses is correct |

12 Chlorine gas can be obtained by

  1 passing bromine vapour through sodium chloride solution

  2 electrolyzing aqueous sodium chloride solution

  3 heating concentrated hydrochloric acid with manganese(IV) oxide

  4 adding concentrated sulphuric acid to sodium chloride

13 Copper atoms can be readily converted into $Cu^{2+}$ ions by the action of

 1 dilute sulphuric acid

 2 concentrated aqueous sodium hydroxide

 3 dilute hydrochloric acid

 4 concentrated nitric acid

14 Hydrogen chloride is a gas which

 1 is more dense than air

 2 is very soluble in water

 3 turns damp universal indicator paper red

 4 burns in air

15 Powdered samples of copper(II) oxide and carbon could be distinguished by

 1 heating them separately in a stream of dry hydrogen

 2 heating each in a plentiful supply of air

 3 mixing each with lead(II) oxide and heating

 4 their appearance

16 Ammonia is a gas which

 1 forms salts with acids

 2 is a compound of hydrogen and nitrogen

 3 reacts with hot copper(II) oxide to give copper, nitrogen, and water

 4 forms a solution in water with a pH less than 7

17 Which pairs of aqueous solutions would produce a white precipitate when mixed?

 1 Silver nitrate and lithium chloride

 2 Iron(II) sulphate and sodium hydroxide

 3 Barium chloride and potassium sulphate

 4 Silver nitrate and sodium iodide

Directions for questions 18 to 23. Each of the questions or incomplete statements in this section is followed by five suggested answers. Select the best answer in each case.

18 All of the following equations correspond to a reaction which takes place at room temperature EXCEPT

 A $Mg + 2HCl \rightarrow MgCl_2 + H_2$

 B $2Na + 2H_2O \rightarrow 2NaOH + H_2$

 C $Zn + H_2SO_4 \rightarrow ZnSO_4 + H_2$

 D $Fe + H_2SO_4 \rightarrow FeSO_4 + H_2$

 E $Cu + 2HCl \rightarrow CuCl_2 + H_2$

19 Phosphine $(PH_3)$ is a gas which behaves like ammonia $(NH_3)$.
 You would expect phosphine to react with hydrogen iodide (HI) to form a compound of formula

 A $PH_2I$

 B $PH_4I$

 C $PH_5I_2$

 D $PI_3$

 E $PI_5$

20 Which of the following compounds is LEAST soluble in water at room temperature?

 A Calcium hydroxide

 B Calcium hydrogencarbonate

 C Calcium carbonate

 D Sodium carbonate

 E Sodium hydrogencarbonate

21 Compound P is a white solid. On adding aqueous barium chloride to an aqueous solution of P, a white precipitate is formed which does not dissolve when dilute hydrochloric acid is added. When P is warmed with aqueous sodium hydroxide, a gas is evolved which turns moist red litmus paper blue. From these observations, it can be concluded that P is

 A ammonium chloride

 B sodium sulphite

 C potassium sulphate

 D ammonium sulphate

 E lithium chloride

22 Which of the following substances gives off only one gaseous product on heating?

   A Lead(II) nitrate

   B Zinc nitrate

   C Copper(II) nitrate

   D Potassium nitrate

   E Silver nitrate

23 Which of the following powdered substances would gain in mass when heated in air?

   A Calcium hydroxide

   B Copper(II) sulphate

   C Iodine

   D Iron

   E Sodium nitrate

| Directions summarized for questions 24 to 26 | | |
|---|---|---|
| First statement | Second statement | |
| A True | True | Second statement is a correct explanation of the first |
| B True | True | Second statement is NOT a correct explanation of the first |
| C True | False | |
| D False | True | |
| E False | False | |

24

   1 When chlorine is bubbled into aqueous potassium iodide, iodine is liberated.

   2 Chlorine forms chloride ions more readily than iodine forms iodide ions.

25

   1 Sulphur forms two oxides of formulae $SO_2$ and $SO_3$.

   2 Sulphur exists in two crystalline forms.

26

   1 Carbon dioxide is as soluble as hydrogen chloride in water at room temperature.

   2 Carbon dioxide and hydrogen chloride both dissolve in water to produce acidic solutions.

**Questions 27–30** concern the following observations made during experiments on a white powder which is an ore of a metal.

   I    The white powder changed to a yellow powder on heating.

   II   The yellow powder, on heating in a stream of hydrogen, produced molten, shiny globules. After cooling to room temperature, the globules solidified and then gave a thick grey streak when rubbed across paper.

   III  The white powder was insoluble in hot or cold water. When added to dilute hydrochloric acid, it gave off a colourless gas for a short time but the reaction soon stopped and most of the ore remained undissolved.

   IV   On warming the mixture of hydrochloric acid with the white powder, a gas was given off again and the solid dissolved. The gas given off gave a white precipitate with lime water.

27 The gas evolved from the ore and dilute hydrochloric acid is most likely to be

   A chlorine

   B hydrogen chloride

   C hydrogen

   D carbon dioxide

   E oxygen

28 The ore is most likely to be a compound of

   A calcium

   B copper

   C lead

   D potassium

   E sodium

29 When hydrogen was passed over the hot yellow powder, the powder was

   A catalyzed

   B decomposed

   C neutralized

   D oxidized

   E reduced

**30** Which one of the following is the most probable explanation of the fact that only a small proportion of the ore dissolved in the hydrochloric acid at room temperature?

**A** There was not enough acid to dissolve all the ore.

**B** There was only a small amount of pure ore present to dissolve.

**C** The reaction produced an insoluble layer around the ore particles which dissolved on heating.

**D** The reaction was endothermic and so the reaction soon stopped.

**E** The acid was too dilute to dissolve all the ore.

---

# Test 11 | Answers

**1** Calcium chloride (key **A**) in the anhydrous form is frequently used as a drying agent in the laboratory. It acts as a drying agent by turning to the hydrated compound as it reacts with water vapour or water present in organic liquids. Anhydrous copper(II) chloride would act in a similar way but is not frequently used as a drying agent.

$$CaCl_2(s) + 6H_2O(g) \rightarrow CaCl_2.6H_2O(s)$$

One of the other substances in the list reacts very readily with water vapour but forms another gas in the process. What is this other substance and what is the gas it forms with water?

**2** The alkaline gas that you are likely to have met on your course is ammonia, $NH_3$. This is formed when ammonium compounds (key **D**) are warmed with alkali.

$$NH_4Cl + NaOH \rightarrow NH_3 + H_2O + NaCl$$

or

$$NH_4^+ + OH^- \rightarrow NH_3 + H_2O$$

**3** Silver chloride (key **B**) is the only chloride in the list that is insoluble in water. There are only two other chlorides that are regarded as insoluble in water, those of lead(II) and mercury(I). You are likely to have come across the insolubility of silver chloride in the test for chlorides, bromides, and iodides, using aqueous silver nitrate. How does aqueous ammonia feature in this test?

**Questions 4–7**

Test **A** is a test used to distinguish a chloride, bromide, or iodide from other negative ions. It could be used to distinguish the pair in question 5.

Chlorides, bromides, and iodides produce white, pale yellow, and deeper yellow precipitates of the silver halides. These three ions can be further distinguished by the action of aqueous ammonia on the precipitate of silver halide (see the answer to question 3). Silver chloride dissolves in ammonia, silver bromide is partially soluble in ammonia, while silver iodide shows no sign of dissolving in ammonia.

Test **B** is a test for carbonates and hydrogencarbonates where the evolved carbon dioxide turns lime water milky. It could be used to distinguish the pair in question 4.

Test **C** is a test for ammonium compounds in which the evolved ammonia turns indicators to their alkaline colour (see the answer to question 2). This test could be used to distinguish the pair in question 7.

Test **D** is a test for reducing agents or for unsaturation in carbon compounds. The purple permanganate ion is reduced to almost colourless $Mn^{2+}(aq)$ ions. This test would not be used to distinguish between any of the substances in this set of questions. Permanganate may be called manganate(VII) on your course.

Test **E** is frequently used to recognize individual elements of Group 1 and Group 2 (from calcium to higher atomic numbers). This test could be used to distinguish between the substances in question 6. Do you know the flame colours given by sodium compounds and potassium compounds?

**8** This test for aqueous chloride ions involves the precipitation of white, insoluble, silver chloride.

$$Ag^+(aq) + Cl^-(aq) \rightarrow AgCl(s).$$

This is represented by the equation in key **A**.

**9** The test for aqueous sulphate ions produces a white precipitate of barium sulphate

$$Ba^{2+}(aq) + SO_4^{2-}(aq) \rightarrow BaSO_4(s).$$

This is represented by the equation in key **E**.

**10** Zinc is higher than copper in the electrochemical series (ECS). Zinc is oxidized to zinc ions and, at the same time, copper(II) ions are reduced to copper metal.

$$Zn(s) + Cu^{2+}(aq) \rightarrow Zn^{2+}(aq) + Cu(s)$$

Key **B** shows this change.

**11** If iron(II) ions are being oxidized, they must be changing to iron(III) ions. As iron(II) is being oxidized, silver ions must be reduced. The only possible product of reducing $Ag^+$ is silver metal. The equation must be

$$Ag^+(aq) + Fe^{2+}(aq) \rightarrow Ag(s) + Fe^{3+}(aq).$$

Key **D** shows this change.

**12** Bromine does not displace chlorine from aqueous chloride ions (**1** incorrect). The reaction that does take place is the reverse of this — chlorine displaces bromine from bromide ions.

$$Cl_2(g) + 2Br^-(aq) \rightarrow 2Cl^-(aq) + Br_2(aq)$$

The electrolysis of aqueous chloride ions can give chlorine as the main anode product if the concentration of the solution is high (**2** correct).

$$2Cl^-(aq) \rightarrow Cl_2(g) + 2e^-$$

The electrolysis of chloride ions involves electron loss, which is oxidation. Chloride ions can also be chemically oxidized to chlorine by:
i) reacting cold concentrated hydrochloric acid with potassium permanganate,
ii) heating a mixture of sodium chloride and concentrated sulphuric acid with manganese(IV) oxide,
iii) heating concentrated hydrochloric acid with manganese(IV) oxide (**3** correct).

The action of concentrated sulphuric acid on solid chlorides forms hydrogen chloride gas, not chlorine (**4** incorrect).
**2** and **3** are correct and the key is **E**.

**13** Copper is low in the ECS and this means that oxidation of copper metal to copper(II) ions is difficult.

Dilute sulphuric acid and dilute hydrochloric acid do not oxidize copper unless air is also present, and then the reaction is very slow (**1** and **3** incorrect). Have you carried out the experiments

suggested in the answer to question 16 of test 9? Sodium hydroxide at any concentration does not convert copper to copper(II) ions (**2** incorrect).

Concentrated nitric acid does readily oxidize copper to copper(II) ions. Here it is the nitrate ion which is assisting in the oxidation. The nitrate ion is reduced to $NO_2$. Only 4 is correct and the key is **D**.

$$Cu + 4HNO_3 \rightarrow Cu(NO_3)_2 + 2NO_2 + 2H_2O$$

Can you find out whether dilute nitric acid oxidizes copper? Does sulphuric acid, when it is concentrated, have any action on copper? See the answer to question 28 of the next test if you need help here.

**14** **1**, **2**, and **3** are all correct properties of hydrogen chloride gas. Hydrogen chloride does not burn in air (**4** incorrect). The key here is **A**.

Although hydrogen chloride does not burn in air, it is oxidized by air if the two gases are passed over a hot catalyst. This reaction can be used in the large scale production of chlorine.

$$4HCl(g) + O_2(g) \rightarrow 2Cl_2(g) + 2H_2O(g)$$

**15** Copper(II) oxide and carbon are both black, so they cannot readily be distinguished by their appearance (**4** incorrect). The first three methods would, however, all be suitable, as only one of the two substances reacts in each case. The key is **A**.

Which substance is it that reacts in **1**, **2**, and **3** and what is seen in each case?

**16** The first three properties are correct for ammonia and the key is **A**. **4** is incorrect because the aqueous solution of ammonia is alkaline, which means that its pH is greater than 7.

**17** All four pairs produce precipitates when mixed, but only in **1** and **3** are the precipitates white. The correct key is **B**.

What are the precipitates in **1** and **3**? What are the precipitates in **2** and **4** and what are their colours?

**18** The elements in **A**, **B**, **C**, and **D** are all above hydrogen in the ECS. These elements will displace hydrogen from water (if they are sufficiently high in the series) or from dilute acids. Copper, however, is below hydrogen in the ECS and the reaction given in **E**, the correct key, does not take place. This reaction is discussed further in the answer to question 13.

**19** The reaction of ammonia with hydrogen halides gives ammonium halides.

$$NH_3(g) + HCl(g) \rightarrow NH_4Cl(s)$$

$$NH_3(g) + HI(g) \rightarrow NH_4I(s)$$

Ammonium iodide has the formula $NH_4I$. The expected formula for the equivalent compound formed from phosphine would be $PH_4I$, key **B**.

**20** You will have met calcium carbonate (key **C**) as chalk and as the substance precipitated in the lime water test for carbon dioxide. This is the only really insoluble compound in the list.

**21** The test with barium chloride, followed by acid, shows that P is a sulphate. Warming P with alkali gives an alkaline gas which is probably ammonia released from an ammonium salt. P is likely to be ammonium sulphate, key **D**.

**22** Nitrates of metals very high in the ECS decompose to oxygen and the nitrite. Potassium nitrate (which may be called potassium nitrate(V) on your course) decomposes to oxygen and potassium nitrite (which may be called potassium nitrate(III)).

$$2KNO_3(s) \rightarrow O_2(g) + 2KNO_2(s)$$

Nitrates of metals lower in the ECS decompose on heating to the oxide and two gases, oxygen and nitrogen dioxide. Nitrates of metals very low in the ECS may decompose to the metal rather than the oxide.

The substance in the question which gives only one gaseous product is potassium nitrate, key **D**.

**23** Calcium hydroxide is likely to lose water on heating and become the oxide. Copper(II) sulphate will lose water of crystallization, if it is hydrated, and will then begin to lose sulphur trioxide on further heating. Iodine will not change chemically but will vaporize. Sodium nitrate will decompose to oxygen and sodium nitrite, with some further decomposition to the oxide, oxygen, and nitrogen dioxide.

All the substances discussed up to this point will lose mass on heating. Iron (key **D**) will certainly not lose mass on heating, and if heated strongly enough will gain in mass as the surface becomes converted to oxides of iron.

**24** When chlorine is bubbled into aqueous iodide ions, the reaction that takes place can be represented by the equation:

$$Cl_2(g) + 2I^-(aq) \rightarrow 2Cl^-(aq) + I_2(aq)$$

Both statements are true, the second being a correct explanation of the first, and the key is **A**.

**25** Sulphur does form two oxides, $SO_2$ and $SO_3$, and it does exist in two crystalline forms, rhombic and monoclinic. It is just a coincidence that the number of oxides and the number of crystalline forms are the same. The second statement is not a correct explanation of the first and the key is **B**.

**26** Carbon dioxide and hydrogen chloride do both dissolve in water to produce acidic solutions (second statement true). Hydrogen chloride is, however, far more soluble than carbon dioxide. The first statement is false and the key is **D**.

**27** The gas is described in IV as giving a white precipitate with lime water. This is a positive test for carbon dioxide, key **D**.

**28** All the elements in this question are metals. The information in II indicates that a metal is found as 'shiny globules' on heating with hydrogen. This appears to be a reduction reaction. Calcium, potassium, and sodium are too high in the ECS to be formed from their compounds by reduction with hydrogen. Copper and lead are low enough in the ECS to be formed in this way. The metal is likely to be lead (rather than copper) as it appears to melt fairly easily, form shiny globules, and also mark paper in the way lead does. The correct key is **C**.

**29** The yellow powder is probably lead(II) oxide, formed by the decomposition of its ore. The conversion of a compound, such as the oxide, to the metal by passing hydrogen over it, is an example of reduction, key **E**.

**30** **A**, **B**, and **E** cannot be correct. In IV, on warming the mixture which only slightly reacted in III, carbon dioxide is evolved and the ore converted to a new compound, soluble in hot hydrochloric acid. There must have been sufficient ore and hydrochloric acid available all the time. The heat was necessary to overcome a barrier to further reaction.

**D** is also incorrect. Endothermic reactions are far less frequent than exothermic reactions, but such reactions do not stop after a short while just because they are endothermic.

**C** seems to be the most likely possibility, particularly as by this time you will probably suspect that the ore is lead(II) carbonate, which in II is converted to yellow lead(II) oxide by loss of carbon dioxide. One reaction common to all carbonates is the reaction with acid to form a salt, water, and carbon dioxide. The salt here would be lead(II) chloride which, as mentioned in the answer to question 3, is one of the few insoluble chlorides. Lead(II) chloride, however, is far more soluble when hot than when cold. An insoluble layer of lead(II) chloride is formed on the ore in the cold. This prevents further attack by acid until the lead(II) chloride is dissolved away on raising the temperature. **C** is the correct key.

# Test 12  Inorganic chemistry 2

**Questions 1–3**

When dilute sulphuric acid is added to a substance, the following results may be obtained:

**A** no reaction occurs

**B** a salt and hydrogen are formed

**C** a salt and water only are formed

**D** a salt, water, and carbon dioxide are formed

**E** a salt, water, and sulphur dioxide are formed

Choose, from **A** to **E**, the result you would expect when warm dilute sulphuric acid is added to

1  magnesium

2  magnesium oxide

3  zinc carbonate

**Questions 4–6** are concerned with aqueous solutions of the following compounds:

**A** iron(III) chloride

**B** copper(II) chloride

**C** magnesium nitrate

**D** silver nitrate

**E** sodium carbonate

Select the salt which in aqueous solution most closely fits the description given in each question below.

4  Gives a white precipitate with dilute hydrochloric acid.

5  Gives a white precipitate with sodium hydroxide solution but no precipitate with sodium chloride solution.

6  Gives a white precipitate with barium chloride solution but no precipitate with sodium chloride solution.

**Questions 7–10** concern the following ways in which substances may be classified:

**A** all compounds

**B** all elements

**C** each substance reacts with water to give an alkaline solution

**D** all good conductors of electricity when liquid and solid

**E** all good conductors of electricity when liquid but NOT when solid

Classify the group of substances in each question below into one of the categories **A** to **E**.

7  Sodium, calcium oxide, lithium, potassium

8  Glucose, ice, lead(II) bromide, silver nitrate

9  Argon, lead, rhombic sulphur, tin

10  Aluminium, brass, silver, steel

| Directions summarized for questions 11 to 16 | | | | |
|---|---|---|---|---|
| **A** 1, 2, 3 only correct | **B** 1, 3 only correct | **C** 2, 4 only correct | **D** 4 only correct | **E** Some other response or combination of responses is correct |

11  A solid gave off a brown gas when it was gently heated. The solid could have been

1  potassium nitrate

2  sodium bromide

3  ammonium nitrate

4  lead(II) nitrate

12 A precipitate would be formed on mixing aqueous solutions of

1 silver nitrate and barium chloride

2 silver nitrate and sodium chloride

3 sodium sulphate and barium chloride

4 sodium chloride and barium nitrate

13 The properties of the transition metals include that they

1 have a higher density than the alkali metals

2 form salts which, when dissolved in water, give coloured solutions

3 conduct electricity well

4 react rapidly with water to produce hydrogen

14 The gas evolved when sodium chloride reacts with concentrated sulphuric acid

1 can also be produced by burning a stream of hydrogen in chlorine

2 produces hydrogen when passed over heated iron

3 is very soluble in water

4 bleaches moist universal indicator paper

15 Chemical changes which give chlorine gas as one of the products include the

1 electrolysis of molten potassium chloride

2 action of hydrogen chloride on heated iron

3 electrolysis of sea water

4 addition of bromine to potassium chloride solution

16 In which of the following mixtures would the metal oxide be reduced when the mixture was heated?

1 Aluminium oxide and copper

2 Zinc oxide and aluminium

3 Chromium(III) oxide and potassium

4 Iron(III) oxide and copper

Directions for questions 17 to 22. Each of the questions or incomplete statements in this section is followed by five suggested answers. Select the best answers in each case.

17 Which of the following elements does NOT react with water or steam?

A Sodium    D Iron

B Zinc      E Copper

C Calcium

18 It could be shown that chlorine is a more powerful oxidizing agent than bromine by

A determining the relative molecular masses of hydrogen chloride and hydrogen bromide

B bubbling chlorine into a solution containing bromide ions

C bubbling chlorine into liquid bromine

D determining the volume of hydrogen halide produced on reacting each element with hydrogen

E adding bromine to a solution containing chloride ions

19 A solid element conducts electricity. When a sample of the element was heated with copper(II) oxide in the absence of air, copper was produced and a gas was given off. The element could have been

A calcium     D magnesium

B sulphur     E carbon

C iron

20 All of the following would be decomposed by strong heating in a bunsen flame EXCEPT

A calcium carbonate

B hydrated copper(II) sulphate

C lead(II) nitrate

D magnesium oxide

E sugar

**21** When excess 1 M hydrochloric acid is added to calcium carbonate, a rapid evolution of carbon dioxide occurs. If 1 M sulphuric acid is used in place of hydrochloric acid, a small quantity of the gas is given off for a few seconds but the action then stops.

This difference in behaviour is because

**A** 1 M sulphuric acid contains twice as many hydrogen ions as 1 M hydrochloric acid

**B** calcium sulphate is only slightly soluble in water

**C** calcium chloride is used as a dehydrating agent

**D** dilute hydrochloric acid is made by dissolving hydrogen chloride gas in water

**E** calcium is high in the electrochemical series of the elements

**22** Which of the following pieces of evidence indicates that a metal $R$ is more reactive than a metal $Q$?

**A** The salts of $R$ are coloured but those of $Q$ are white.

**B** The metal $R$ is a better conductor of electricity than the metal $Q$.

**C** The oxide of $Q$ can be reduced by carbon.

**D** A sample of $Q$ can be obtained by heating the oxide of $Q$ with $R$.

**E** The oxide of $R$ can be reduced by carbon.

| Directions summarized for questions 23 to 25 | | |
|---|---|---|
| First statement | Second statement | |
| **A** True | True | Second statement is a correct explanation of the first |
| **B** True | True | Second statement is NOT a correct explanation of the first |
| **C** True | False | |
| **D** False | True | |
| **E** False | False | |

**23**
**1** When sulphur is burned in oxygen, the oxide formed dissolves in water to give a solution of pH less than 7.

**2** Oxides of sulphur dissolves in water to produce alkaline solutions.

**24**
**1** A precipitate is formed when aqueous sodium hydroxide is added to aqueous iron(II) sulphate.

**2** Iron(II) hydroxide is insoluble in water.

**25**
**1** When concentrated sulphuric acid is added to a mixture of sodium chloride and manganese(IV) oxide, the only gaseous product is hydrogen chloride.

**2** Manganese(IV) oxide is an oxidizing agent.

**Questions 26–30**

The apparatus shown below was used to study the products formed when ammonia is passed over heated copper(II) oxide in the tube *R*.

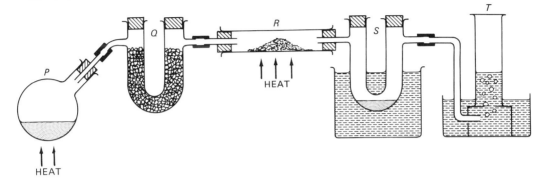

**26** Ammonia is generated in the flask *P* by heating ammonium chloride with a second substance. The second substance could be

**A** pure water

**B** concentrated sulphuric acid

**C** calcium chloride

**D** calcium hydroxide

**E** ammonium sulphate

**27** The U-tube *Q* is filled with silica gel. The most likely purpose of the silica gel is to

**A** ensure an even flow of gas

**B** dry the ammonia

**C** prevent the ammonia sucking back

**D** crack the ammonia

**E** catalyze the reaction

**28** At the end of the experiment the black copper(II) oxide in *R* is converted to a reddish brown solid. This solid is found to conduct electricity, which suggests that it is metallic copper. Evidence for the fact that the solid is copper could be obtained by its reaction with

**A** hot water

**B** litmus solution

**C** zinc sulphate solution

**D** sodium hydroxide solution

**E** dilute nitric acid

**29** A liquid is collected in the U-tube *S* which proves to be alkaline to litmus, dissolves anhydrous copper(II) sulphate giving a blue solution, and boils at about 100°C. It may be concluded that this liquid is

**A** pure ammonia

**B** water containing dissolved ammonia

**C** pure water

**D** water containing dissolved hydrogen chloride

**E** water containing dissolved ammonium chloride

**30** The gas collected over water in the gas jar *T* is colourless (even when mixed with air) and extinguishes a lighted taper without burning itself. The gas is most likely to be

**A** carbon dioxide

**B** ammonia

**C** nitrogen

**D** hydrogen

**E** nitrogen oxide (NO)

1  Dilute sulphuric acid reacts with metals reasonably high in the electrochemical series (ECS), such as magnesium, to form a salt and hydrogen. The key is **B**.

$$H_2SO_4 + Mg \rightarrow MgSO_4 + H_2$$

2  If oxygen is combined with the metal to form a base, such as magnesium oxide, then a salt and the oxide of hydrogen (water) are formed on reaction with dilute sulphuric acid. The key is **C**.

$$H_2SO_4 + MgO \rightarrow MgSO_4 + H_2O$$

3  A carbonate can be regarded as a compound of a metal oxide (basic oxide) with carbon dioxide (acidic oxide). When sulphuric acid reacts with such a compound, a salt and the oxide of hydrogen (water) are formed and carbon dioxide is released. The key is **D**. The equation below is given for magnesium carbonate to complete the sequence of reactions discussed in the answers to 1 and 2 above.

$$H_2SO_4 + MgCO_3 \rightarrow MgSO_4 + H_2O + CO_2$$

Can you write the equation for zinc carbonate reacting with dilute sulphuric acid?

4  Aqueous silver nitrate, key **D**, gives a white precipitate of silver chloride with dilute hydrochloric acid

$$AgNO_3(aq) + HCl(aq) \rightarrow$$
$$AgCl(s) + HNO_3(aq)$$

or

$$Ag^+(aq) + Cl^-(aq) \rightarrow AgCl(s)$$

Which is the one other substance in the list that will react with dilute hydrochloric acid?

5  The hydroxide ions in aqueous sodium hydroxide precipitate the insoluble hydroxides of metals. This would not happen with salts of Group 1 where the hydroxides are very soluble. **A** will produce a yellow precipitate of iron(III) hydroxide, **B** a blue precipitate containing copper(II) hydroxide, and **C** a white precipitate of magnesium hydroxide. **D** might, for a moment, give a white precipitate of silver hydroxide but this rapidly darkens as the hydroxide decomposes to silver oxide by loss of water. **E** will give no precipitate.

The correct key must be **C**. The information about the lack of precipitate with aqueous sodium chloride is included to eliminate any possibility of the substance being silver nitrate.

$$Mg^{2+}(aq) + 2OH^-(aq) \rightarrow Mg(OH)_2(s).$$

6  A, B, and C will not react with aqueous barium chloride. Silver nitrate, key **D**, will give a white precipitate of silver chloride. Sodium carbonate, key **E**, will give a white precipitate of barium carbonate. The salt cannot be silver nitrate as it gives no precipitate (of silver chloride) with aqueous sodium chloride. The salt must be sodium carbonate, key **E**.

7  This group contains three reactive metallic elements and the oxide of a reactive metallic element. All four of these will react with water to give a solution of the metal hydroxide, which will make the water alkaline. The correct key is **C**.

8  The substances in this group have little in common beyond all being compounds, key **A**.

9  This group is composed only of elements, key **B**.

10  There are two metallic elements in this list and two metallic alloys. Being metallic, they will all conduct electricity in both the solid and the molten state. The correct key is **D**.

11  Potassium nitrate is the nitrate of a metal very high in the ECS. As discussed in the answer to question 22 of test 11, such nitrates decompose to oxygen and the nitrite. No brown gas would be evolved.

Sodium bromide would not change at all on gentle heating. It would melt at a high temperature without any other change taking place.

Ammonium nitrate decomposes on heating, often violently, in a unique way for a nitrate, to give two colourless gases.

$$NH_4NO_3 \rightarrow N_2O + 2H_2O.$$

Lead(II) nitrate, like other nitrates of metals lower in the ECS, decomposes to the oxide, oxygen, and brown fumes of $NO_2$.

$$2Pb(NO_3)_2 \rightarrow 2PbO + O_2 + 4NO_2$$

Only **4** is correct and the key is **D**.

**12** **1** and **2** will produce white precipitates of silver chloride. **3** will produce a white precipitate of barium sulphate. **4** will give no reaction. The key is **A**.

**13** Properties **1** and **3** are common to all transition metals while property **2** is true for nearly all transition metals. Can you find any element in the transition series scandium to zinc for which property **2** is not true?

In the series scandium to zinc, the only element that is below hydrogen in the ECS is copper. Some of these elements are high in the ECS but are protected from water attack by a water-tight layer of oxide. This is similar to the situation with aluminium, which should react with water but is protected by a layer of $Al_2O_3$. The highest of these elements in the ECS is scandium, which comes between magnesium and aluminium. Can you find in a reference book whether scandium shows any sign of reacting with water?

**1**, **2**, and **3** would be regarded as general properties of transition elements, while **4** would not. The key is **A**.

**14** The gas formed when sodium chloride reacts with concentrated sulphuric acid is hydrogen chloride. **1**, **2**, and **3** are all properties of this gas, while **4** is a property of chlorine, not of hydrogen chloride. The key is **A**.

**15** Chlorine is among the products of **1** and **3**. Hydrogen is the product of reaction **2**, together with iron(II) chloride. The reaction suggested in **4** does not take place because bromine is a less reactive halogen than chlorine. The correct key is **B**.

**16** If the metal oxide is to be reduced, the metal acting as the reducing agent must be higher in the ECS than the metal combined with oxygen. This is only the case in **2** (where aluminium is above zinc) and in **3** (where potassium is above chromium). The key is **E**.

**17** Sodium and calcium react with liquid water, while zinc and iron will react if they are strongly heated in steam. As discussed in the answer to question 13, copper is the only transition metal, in the series scandium to zinc, that is below hydrogen in the ECS. Copper has no reaction with liquid water or steam and the key is **E**.

**18** The displacement of bromine from a bromide using chlorine (key **B**) shows that chlorine is a more powerful oxidizing agent than bromine

$$Cl_2(g) + 2Br^-(aq) \rightarrow 2Cl^-(aq) + Br_2(aq)$$

Chlorine molecules accept electrons and chlorine acts as an oxidizing agent. Bromide ions supply these electrons and the bromide acts as a reducing agent.

$$2Br^- \rightarrow Br_2 + 2e^-$$
$$2e^- + Cl_2 \rightarrow 2Cl^-$$

**19** You may have found that this is a tricky one! 'Solid element' gives nothing away. All five are solid elements. 'Conducts electricity' shows that the element cannot be sulphur. The element could be one of the three metals or the conducting form of carbon, graphite.

The three metals are all well above copper in the ECS and so would be expected to convert copper(II) oxide to copper and the metal oxide on heating. But metal oxides are all solids. Carbon, however, would also reduce copper(II) oxide to copper and in the process a gaseous oxide of carbon would be formed. The correct key is **E**.

Can you write two possible equations for the reaction of copper(II) oxide with carbon? If you have the opportunity, make a small quantity of very well mixed copper(II) oxide and carbon. Heat this in an ignition tube and test to see which of your two equations represents the reaction that really takes place.

**20** Magnesium oxide is the product of the very exothermic reaction between magnesium and oxygen. The temperature produced by this reaction is so high that a burn from it does not only affect the skin, it also cooks the flesh underneath. Once the oxide has been formed it is impossible to decompose it by strong heating in a bunsen flame. The key is **D**.

Do you know what is formed in the decompositions of **A**, **B**, **C**, and **E**?

**21** If it were fully ionized, 1 M sulphuric acid would contain twice as many hydrogen ions as 1 M hydrochloric acid, in a given volume of solution. This is not the reason for the difference in behaviour because it should make the sulphuric acid more reactive, not less reactive. **A** is not the correct key.

The formation of an invisible layer protecting a carbonate from further attack has been discussed in the answer to question 30 of the previous test. There it was insoluble lead chloride that was protecting the carbonate, here it is a layer of almost insoluble calcium sulphate.

$$CaCO_3(s) + 2HCl(aq) \rightarrow$$
$$CaCl_2(aq) + H_2O(l) + CO_2(g)$$
$$CaCO_3(s) + H_2SO_4(aq) \rightarrow$$
$$CaSO_4(s) + H_2O(l) + CO_2(g)$$

**B** is the correct key. **C, D**, and **E** are all correct statements but are not explanations for the observed difference in behaviour.

**22** $R$ would be more reactive than $Q$ if it could remove oxygen from the oxide of $Q$.

$$R + QO \rightarrow RO + Q$$

This is what is described in **D**, which is the correct key. In a competition between $R$ and $Q$ to hold on to oxygen, $R$ wins.

**23** The first statement is true. Sulphur is a non-metal and, as expected, its oxides dissolve in water to form an acidic solution. The pH of an acidic solution is less than 7. The second statement is false and the key is **C**.

**24** Hydroxide ions react with iron(II) ions to give a precipitate of iron(II) hydroxide. Both statements are true and the second is a correct explanation of the first. The key is **A**.

$$Fe^{2+}(aq) + 2OH^-(aq) \rightarrow Fe(OH)_2(s)$$

**25** When concentrated sulphuric acid is added to sodium chloride, hydrogen chloride is formed. If manganese(IV) oxide is also present it acts as an oxidizing agent (second statement true). The manganese(IV) oxide oxidizes some of the hydrogen chloride to a second gaseous product, chlorine. The first statement is false. The key is **D**.

**26** The second substance must provide hydroxide ions for the reaction

$$NH_4^+(s) + OH^-(s) \rightarrow NH_3(g) + H_2O(g)$$

The only substance that can provide these ions is calcium hydroxide, key **D**.

Would you expect any gas to be formed if ammonium chloride were heated with calcium oxide rather than with calcium hydroxide? Try this reaction on a small scale and see if you can detect any gas formed. Can you write an equation for the formation of any gas you do detect?

**27** Water vapour is formed with the ammonia gas as shown in the equation given in the answer to question 26. Silica gel is a very useful drying agent as it is both efficient and almost completely inert. It can be used to dry any gas except fluorine or hydrogen fluoride. Its function is to dry the ammonia, key **B**.

**28** Copper has no reaction with the substances **A, B, C**, or **D**. Because of its position in the ECS, copper has no reaction with the hydrogen ions present in all acids. The negative ions in certain acids are, however, capable of oxidizing copper to copper(II) ions.

The nitrate ion in dilute nitric acid is able to oxidize copper in this way, the nitrate ion being reduced to nitrogen monoxide gas, NO, in the process. The dissolving of copper to form a blue solution and the production of a colourless gas (NO) which turns to brown $NO_2$ on contact with air would be evidence that a solid is copper. The correct key is **E**.

$$3Cu + 8HNO_3 \rightarrow 6Cu(NO_3)_2 + 2NO + 4H_2O$$

Concentrated nitric acid acts in a similar way (see the answer to question 13 of the previous test). Nitric acid, when dilute or concentrated, reacts with copper in the cold. Concentrated sulphuric acid is able to oxidize copper to copper(II) but only when hot. Sulphur is reduced to sulphur dioxide in the process. A simplified equation for the reaction is

$$Cu + 2H_2SO_4 \rightarrow CuSO_4 + 2H_2\dot{O} + SO_2.$$

**29** The expected reaction of ammonia with hot copper(II) oxide is

$$2NH_3(g) + 3CuO(s) \rightarrow$$
$$3Cu(s) + 3H_2O(l) + N_2(g)$$

The detection of the copper has already been discussed in question 28. The liquid collected would be expected to be water, slightly contaminated with unchanged, alkaline ammonia. This is confirmed by the properties given and the key is **B**.

**30** The gas collected would be expected to be nitrogen and this is confirmed by its lack of colour, lack of reaction with air (which shows it is not NO) and failure to support combustion. The correct key is **C**.

**Questions 1–4** concern the following substances which are manufactured industrially:

  A  ammonia

  B  ethene

  C  nitric acid

  D  sodium hydroxide

  E  sulphuric acid

Choose the substance which is

1  capable of being converted to a solid polymer

2  manufactured by the reaction between two gases at about 200 atm and 500°C in the presence of a catalyst

3  used in the manufacture of soap

4  manufactured by an electrolytic process

| Directions summarized for questions 5 to 11 | | | | |
|---|---|---|---|---|
| **A** 1, 2, 3 only correct | **B** 1, 3 only correct | **C** 2, 4 only correct | **D** 4 only correct | **E** Some other response or combination of responses is correct |

5  The element silicon is a constituent of

  1  limestone          3  clay

  2  sand               4  rock salt

6  The manufacture of iron from iron ore involves the use of

  1  coke as a reducing agent

  2  a fractionating column

  3  limestone to remove impurities from the ore

  4  large quantities of liquid air

7  Carbon is often used as a reducing agent to obtain metals from their oxides because

  1  it can take away oxygen from all metal oxides

  2  it is available in large quantities in the form of coke

  3  it has a low melting point

  4  the oxides of carbon are gases and do not remain mixed with the metal product

8  Which metals are normally extracted from their ores by the stated process?

  1  Aluminium by electrolysis of a molten compound.

  2  Sodium by reduction of the oxide with carbon.

  3  Iron by reduction with coke at high temperature.

  4  Copper by electrolysis of a molten salt.

9  Hydrogen gas is used as a starting material in the large scale production of

  1  sulphuric acid

  2  polythene

  3  chlorine

  4  ammonia

10  Using only air, water, and carbon as raw materials, and any catalysts, apparatus, and equipment that are necessary, it would be possible to manufacture on a large scale

  1  ammonia

  2  hydrochloric acid

  3  nitric acid

  4  sulphuric acid

11  Important synthetic polymers include

  1  starch             3  protein

  2  nylon              4  perspex

Directions for questions 12 to 20. Each of the questions or incomplete statements in this section is followed by five suggested answers. Select the best answer in each case.

12 The major component of natural gas (North Sea Gas) is

A methane      D nitrogen

B carbon monoxide      E oxygen

C hydrogen

13 The following materials are all natural sources of chemicals. Which one of them is essentially a single chemical compound?

A Sea water      D Coal

B The atmosphere      E Petroleum oil

C Limestone

14 The Haber process is concerned with the manufacture of all of the following EXCEPT

A ammonium sulphate

B ammonia

C nitric acid

D ammonium nitrate

E nitrogen

15 Aluminium is obtained industrially by

A reduction of aluminium oxide with carbon

B reduction of aluminium oxide with carbon monoxide

C electrolysis from aqueous aluminium chloride

D electrolysis from molten aluminium sulphate

E electrolysis from aluminium oxide

16 Which one of the chemicals listed below can be manufactured industrially from one of the others in the list?

A Ammonia

B Nitric acid

C Sulphuric acid

D Sodium hydroxide

E Chlorine

17 The most reactive metals are extracted by

A electrolysis of solid compounds

B reduction of ores with carbon

C oxidation of ores in a furnace

D electrolysis of molten compounds

E heating unstable compounds in air

18 Which of the following metals is readily obtained by heating its oxide with hydrogen or carbon?

A Aluminium      D Magnesium

B Copper      E Sodium

C Calcium

19 Ammonium sulphate is used as

A an antiseptic      D a fungicide

B a fertilizer      E an insecticide

C a cleaning agent

20 The apparatus shown in the diagram can be used to test for the presence of nitrogen in foodstuffs.

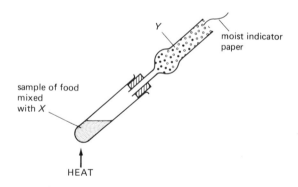

Which pair of reagents is suitable for use as $X$ and $Y$ in this apparatus?

| | X | Y |
|---|---|---|
| A | Soda lime | Charcoal |
| B | Copper(II) oxide | Soda lime |
| C | Calcium oxide | Copper(II) oxide |
| D | Charcoal | Calcium chloride |
| E | Calcium chloride | Calcium oxide |

| Directions summarized for questions 21 to 24 | | |
|---|---|---|
| First statement | Second statement | |
| **A** True | True | Second statement is a correct explanation of the first |
| **B** True | True | Second statement is NOT a correct explanation of the first |
| **C** True | False | |
| **D** False | True | |
| **E** False | False | |

**21**

1 Sulphuric acid can be prepared from sulphur dioxide, oxygen and water.

2 Sulphur dioxide burns readily in air or oxygen to form sulphur trioxide.

**22**

1 Limestone is used in the blast furnace for the production of iron.

2 Limestone removes some impurities as a slag in the blast furnace for the production of iron.

**23**

1 Chlorine is a valuable by-product of the manufacture of sodium hydroxide by the electrolysis of brine.

2 Chlorine is used to make brine for the manufacture of sodium hydroxide.

**24**

1 Synthetic detergents readily form a scum with hard water.

2 Synthetic detergents precipitate calcium ions from hard water.

**Questions 25–30** relate to the following description of the extraction of zinc in a blast furnace. Read through the passage before answering the questions which follow.

One of the major zinc ores is *blende* (zinc sulphide) which is often found mixed with *galena* (lead sulphide). The ore is first roasted in air, and the sulphur dioxide produced is used to manufacture sulphuric acid which is a valuable by-product. The roasted ore is then mixed with coke, sand and limestone and is loaded into the blast furnace. When the hot air blast is in operation, the gases emerging are at a temperature of 1000°C and contain nitrogen, argon, carbon monoxide, carbon dioxide and zinc vapour. In order to prevent the re-oxidation of the zinc, the hot gases are cooled rapidly in a shower of molten lead droplets at 560°C which is thrown up by rotating paddles. This part of the furnace is called the 'lead-splash condenser'. The zinc vapour dissolves in the molten lead which is then run off into a cooling tank and cooled to 440°C. At this temperature liquid zinc separates out and is run off into casting trays where it solidifies on further cooling into bars with a purity of 98.5%. The molten lead is piped back to the lead-splash condenser to be used again.

**25** The ore is roasted before being loaded into the blast furnace in order to

A dry it

B separate the blende from the galena

C convert the sulphides into oxides

D manufacture sulphur

E preheat it

**26** Which is the reducing agent in this blast furnace?

A Air

B Limestone

C Sand

D Carbon monoxide

E Carbon dioxide

**27** Which is most liable to bring about the reoxidation of the zinc as the hot vapours cool?

A Nitrogen

B Argon

C Carbon monoxide

D Molten lead

E Carbon dioxide

**28** The operation of this process depends on a crucial factor concerning the solubility of zinc in lead. Which is this crucial factor?

**A** Zinc is soluble in lead at all temperatures.

**B** Zinc is soluble in lead at both 440°C and 560°C.

**C** Zinc is more soluble in lead at 440°C than at 560°C.

**D** Zinc is more soluble in lead at 560°C than at 440°C.

**E** Zinc is insoluble in lead at all temperatures.

**29** From the information given the melting point of zinc could be

**A** 420°C

**B** 440°C

**C** 560°C

**D** 640°C

**E** 1000°C

**30** Since there is galena mixed with the blende, lead is produced as well as zinc in this process. From what part of the plant is this lead recovered? (The boiling-point of lead is 1550°C.)

**A** From the lead-splash condenser

**B** From the residue of the blast furnace

**C** From the cooling tanks

**D** From the casting trays

**E** From the gases emerging from the condenser

# Test 13 || Answers

**1** Ethene (key **B**) is the substance that is capable of being converted into a solid polymer. What is the name of the polymer?

**2** Nitric acid is manufactured by the reaction between two gases, ammonia and oxygen, at a high temperature in the presence of a catalyst. However, the temperature, used is far higher than 500°C while the pressure is only atmospheric or a little above.

Ethene is manufactured by cracking a hydrocarbon vapour at high temperature over a catalyst. There is only one gas involved, unless air or steam are added, and if a pressure above atmosphere is used it never approaches the 200 atmosphere value in this question.

All the conditions outlined in the questions are those used in the manufacture of ammonia, key **A**.

**3** Sodium hydroxide in aqueous solution is heated with animal or vegetable fats in the production of soap. The correct key is **D**.

**4** The only substance in the list that is manufactured by an electrolytic process is sodium hydroxide, key **D**. What electrolyte is used in the manufacture of sodium hydroxide?

**5** Limestone is calcium carbonate, $CaCO_3$. Rock salt is sodium chloride, NaCl. Neither of these contain the second most abundant element in the Earth's crust, silicon.

Silicon is a constituent of sand, silicon dioxide, and of clay. Clay is made up of very small particles formed by the weathering of certain rocks. It is a complex substance containing as its principal elements the three which are most abundant in the Earth's crust. These are, with the most abundant first, oxygen, silicon, and aluminium. **2** and **3** are correct and the key is **E**.

**6** The manufacture of iron from iron ore uses **1** and **3**. The key is **B**. Air is also used but not as liquid air.

**7** If the temperature is high enough, carbon will remove the oxygen from all metal oxides. For the metals at the top of the electrochemical series (ECS) the temperature involved is frequently well above any viable operating value. Although **1** is a correct statement if the temperature is high enough, it is not a reason for often using carbon as a reducing agent to obtain metals from their oxides. **1** is incorrect in the context of the question.

**3** is completely incorrect. **2** and **4** are good reasons for the frequent use of carbon as a reducing agent. The key is **C**. Two other reasons for using carbon are that it is relatively inexpensive and that it is often a very effective reducing agent.

8 Aluminium is normally extracted by electrolysis of aluminium oxide dissolved in molten sodium aluminium fluoride. **1** is correct. This process consumes a vast quantity of electricity. It may eventually be replaced by one of two processes using aluminium chloride. This compound can either be reduced by an electrolytic process that is less wasteful or it can be reduced in a non-electrolytic process using manganese.

Sodium is normally extracted by electrolysis of the molten chloride. **2** is not correct in this question. Sodium can be extracted by reduction of the oxide with carbon but the temperature involved is high and there is a considerable danger of forming an explosive compound of sodium.

Iron is normally extracted by reduction with coke at a high temperature (**3** correct).

Copper is frequently purified by electrolysis of an aqueous solution. It is never extracted by electrolysis of a molten salt. **4** is incorrect.

**1** and **3** are normal methods of extraction and the key is **B**.

9 Hydrogen is only used as a starting material in the large-scale production of ammonia. The key is **D**.

10 The manufacture of ammonia requires nitrogen and hydrogen. Nitrogen is obtained from the air by liquefaction followed by fractional distillation. Steam passed over white-hot carbon forms hydrogen. With these two gases available, ammonia can be manufactured by the Haber process. Once ammonia is available, it can be oxidized by mixing with air and passing over a catalyst at high temperature. A series of changes beyond this point using air and water give nitric acid as the end-product.

The manufacture of hydrochloric acid requires a source of chlorine, which is not available. The manufacture of sulphuric acid requires a source of sulphur, which is not available. Only ammonia and nitric acid can be manufactured from the raw materials given and the key is **B**.

11 Starch and protein are natural polymers. Nylon and perspex are synthetic polymers. The key is **C**.

12 The major component of natural gas is methane, $CH_4$. The key is **A**.

13 The only material in the question that is a single

chemical compound is limestone, calcium carbonate. The key is **C**. All the other substances are mixtures, with coal being an extremely complex mixture.

14 The Haber process forms ammonia, which can then be used to manufacture ammonium salts by neutralization and nitric acid by oxidation. Nitrogen is a raw material in the Haber process, not a product of it. The correct key is **E**.

15 The industrial process for obtaining aluminium has been discussed in the answer to question 8. The correct key is **E**.

16 Nitric acid (key **B**) is manufactured industrially by oxidation of ammonia, which is also listed.

17 The usual method for extracting the most reactive metals is electrolysis of molten compounds, key **D**.

18 Metals that are in the middle of the ECS, or lower, can usually be obtained by reducing an oxide with hydrogen or carbon. Calcium, sodium, magnesium, and aluminium are too high to be readily obtained by this method. Copper is low in the ECS and its oxide is readily reduced by heating with hydrogen or carbon. The correct key is **B**.

19 Ammonium sulphate is used as a fertilizer, key **B**. What is the most important element that this fertilizer supplies to plants?

20 The nitrogen-containing substance in food is protein. This is broken down into ammonia by heating with alkali. *X* must be an alkali, so the key is either **A** or **C**. If you have ever carried out this test without the tube containing *Y*, you will be aware of the very unpleasant smell produced and will probably have found that the indicator paper turns brown in the smoke and vapour that pours out of the end of the test-tube. The tube containing *X* has been added in this experiment to trap the particles of smoke and the absorb the large molecules in the vapour that turn the indicator paper brown and smell so unpleasant. A good absorber for this purpose would be charcoal, so the correct key is **A**.

21 The first statement is true but the second is false — sulphur dioxide does not burn readily in air or oxygen. The gases must be passed over a hot catalyst to bring about reaction. The key is **C**.

22 Limestone, a form of calcium carbonate, is used in the blast furnace for the production of iron. Some iron ores already contain calcium carbonate, so

limestone does not always have to be added. This does not make the first statement false.

The calcium carbonate removes some impurities by forming a slag of calcium silicate. The second statement is true and a correct explanation of why limestone is used. The key is **A**.

23 The first statement is true. Chlorine is a valuable by-product of this electrolytic process. The second statement is false and the key is **C**. Brine, aqueous sodium chloride, is one of the most readily available of raw materials. Chlorine does not have to be used to make it.

24 Both statements are false for synthetic detergents, often called 'soapless detergents'. The key is **E**. Both statements would be true for soapy detergents, often called 'soaps'.

25 The ore is a mixture of zinc sulphide and lead sulphide. The information given in the second sentence shows that the roasting in air produces sulphur dioxide. This cannot leave the free metals or the rest of the process would be unnecessary! It is likely that the sulphides are being converted into oxides which then have to be reduced by carbon. The key is **C**.

$$2ZnS(s) + 3O_2(g) \rightarrow 2ZnO(s) + 2SO_2(g)$$

26 The reducing agent in this blast furnace is likely to be either carbon or carbon monoxide. Possible reactions involving these reducing agents are

$$2ZnO + C \rightarrow 2Zn + CO_2$$

$$ZnO + C \rightarrow Zn + CO$$

$$ZnO + CO \rightarrow Zn + CO_2$$

Carbon does not appear as a possible answer so the key must be **C**.

27 Of the alternatives listed, only carbon monoxide and carbon dioxide contain oxygen. You have already been forced to decide that the reducing agent in the process must be carbon monoxide

$$ZnO + CO \rightarrow Zn + CO_2$$

The oxidizing agent for the zinc must be carbon dioxide, which partially reverses the original reaction. The key is **E**.

$$ZnO + CO \leftarrow Zn + CO_2$$

28 The important information taken from the description is 'the hot gases are cooled rapidly in a shower of molten lead droplets at 560°C... The zinc vapour dissolves in the molten lead... Cooled to 440°C. At this temperature liquid zinc separates out...' Zinc is dissolving in molten lead at 560°C and is coming out of solution on cooling to 440°C. This is summarized in key **D**.

29 The description contains the information 'cooled to 440°C. At this temperature liquid zinc separates out and is run off... where it solidifies on further cooling...' This shows that zinc has to be cooled to below 440°C before it solidifies. The only temperature in the list that is below 440°C is 420°C, key **A**.

30 The gases emerge from the blast furnace at 1000°C. This is below the boiling point of lead, so the lead must either remain inside the blast furnace as solid or run to the bottom as liquid. You do not have to decide between these two because the only description that could be correct is 'From the residue of the blast furnace', key **B**.

If the gases emerging from the blast furnace are at 1000°C, what can you say about the temperature inside the furnace? How does this temperature compare with the melting point of lead? Do you think that the lead will remain inside the furnace as solid, or will it run to the bottom as liquid?

# Test 14 | Chemistry in society 2

**Questions 1–6** concern the following terms:

**A** naturally occurring starting material

**B** synthetic starting material

**C** catalyst

**D** main product

**E** by-product

Select, from **A** to **E**, the term which best describes the first named substance in each of the following industrial processes.

**1** Basic slag (calcium silicate) in the blast furnace.

**2** Phenylethene (styrene) in a polymerization process.

**3** Sodium chloride in the electrolytic production of sodium hydroxide.

**4** Iron in the Haber process for the manufacture of ammonia.

**5** Sulphur in the contact process for the manufacture of sulphuric acid.

**6** Sodium hydroxide in the manufacture of soap.

| Directions summarized for questions 7 to 15 | | | | |
|---|---|---|---|---|
| **A** 1, 2, 3 only correct | **B** 1, 3 only correct | **C** 2, 4 only correct | **D** 4 only correct | **E** Some other response or combination of responses is correct |

**7** Oxides which may be formed by burning fuels include

**1** $CO_2$      **3** CO

**2** $H_2O$      **4** $SiO_2$

**8** Limestone is used as a raw material in the large scale production of

**1** chlorine

**2** aluminium from bauxite

**3** copper

**4** iron in a blast furnace

**9** The product(s) removed from the blast furnace is (are)

**1** pure iron (wrought iron)

**2** pig iron (cast iron)

**3** steel

**4** slag

**10** Methods of isolating metals from their compounds include

**1** displacement from salt solutions by another metal higher in the activity series

**2** reduction of oxides with coke

**3** electrolysis of molten salts

**4** using a powerful magnet

**11** Oxide ores are the main source for the manufacture of

**1** sodium

**2** sulphur

**3** chlorine

**4** aluminium

**12** When custard powder is mixed thoroughly with dry black copper(II) oxide and heated in a test tube, water vapour is formed and a gas which turns lime water milky. Some of the copper(II) oxide changes to metallic copper. This shows that custard powder MUST contain

**1** carbon      **3** hydrogen

**2** oxygen      **4** water

13 Which of the following elements may be found inside an electric light bulb?

1 An inert (noble) gas.

2 Magnesium.

3 Tungsten.

4 Hydrogen gas.

14 Soaps and detergents help to clean fabrics by

1 forming bubbles which dissolve the dirt

2 raising the surface tension of the water

3 destroying grease molecules by chemical attack

4 attaching themselves to grease particles

15 A balanced fertilizer usually contains compounds of

1 nitrogen

2 barium

3 phosphorus

4 iodine

Directions for questions 16 to 26. Each of the questions or incomplete statements in this section is followed by five suggested answers. Select the best answer in each case.

16 Which of the following is NOT a naturally occurring raw material?

A Sulphur

B Water

C Sodium chloride

D Calcium carbonate

E Sodium hydroxide

17 Oxygen is produced commercially mainly by

A electrolysis of brine

B boiling liquid air

C electrolysis of water

D heating a metal oxide

E photosynthesis

18 The manufacture of aluminium from purified bauxite involves reduction using

A carbon

B electrolysis

C carbon monoxide

D magnesium

E hydrogen

19 Calcium is a metal high in the activity series; it reacts vigorously with water, and burns fiercely when heated in air or chlorine. By which of the following methods do you think it could be most conveniently extracted?

A Reduction of its oxide by carbon.

B Reduction of its oxide by hydrogen.

C Electrolysis of an aqueous solution of calcium chloride.

D Electrolysis of molten calcium chloride.

E Displacement from an aqueous solution of calcium chloride by iron.

20 The substances used to make iron in a blast furnace are iron ore and

A air and coke only

B air, coke, and sand

C air, sand, and limestone

D air, coke, and limestone

D coke, limestone, and sand

21 The reaction between nitrogen oxide (NO) and oxygen is an important stage in the industrial production of

A ammonia

B sulphuric acid

C ammonium sulphate

D nitric acid

E nitrogen

22 Chlorine is used for all of the following EXCEPT

A as a bleaching agent

B as a disinfectant

C as a raw material for producing hydrochloric acid

D in the production of bromine from potassium bromide

E in the manufacture of sodium chloride

23 Sewers and rivers are sometimes polluted by 'non-biodegradable' detergents which cause large amounts of froth to build up. What is a 'non-biodegradable' detergent?

A One that contains only inorganic compounds.

B One that does not contain any added enzymes.

C One that is not broken down by naturally occurring bacteria.

D One that decomposes quickly after use.

E An advertiser's way of describing soap for washing machines.

24 From which component of our diet does the body obtain essential nitrogen?

A Fats            D Mineral salts

B Carbohydrates  E Drinking water

C Proteins

25 The production of ammonia is important because

A ammonia can be used to make artificial food directly

B proteins can be made by polymerizing ammonia

C ammonia is used to make fertilizers containing nitrogen

D ammonia is used in the production of soap

E sulphuric acid is made by the catalytic oxidation of ammonia

26 The term 'cracking' in the petroleum industry is used to describe the

A distillation of crude oil to obtain fractions of different boiling point ranges

B breaking down of large hydrocarbon molecules into smaller molecules

C conversion of a polymer into a monomer

D manufacture of polythene

E addition of hydrogen to unsaturated hydrocarbons

| Directions summarized for questions 27 to 30 | | |
|---|---|---|
| | First statement | Second statement |
| A | True | True | Second statement is a correct explanation of the first |
| B | True | True | Second statement is NOT a correct explanation of the first |
| C | True | False | |
| D | False | True | |
| E | False | False | |

27

1 The Haber process was developed to make ammonia from nitrogen in the air.

2 Nitrogen occurs only in air.

28

1 The electrolysis of the molten chlorides of metals high in the electrochemical series is used as a method of extraction of these metals.

2 The chlorides of metals high in the electrochemical series have high melting points.

29

1 Superphosphate is a quicker acting fertilizer than calcium phosphate.

2 Superphosphate is much more soluble in water than calcium phosphate.

30

1 The presence of excessive calcium ions in natural waters makes it difficult to get a lather with soap.

2 Calcium ions react with soap to form a precipitate.

# Test 14 | Answers

1 Basic slag is a by-product (key **E**) of the production of iron in the blast furnace.

2 Styrene (phenylethene) is the starting material in the production of polystyrene. The styrene does not occur naturally, it has to be produced from petroleum. The key is **B**.

3 Sodium chloride is the starting material in this electrolysis. It occurs naturally and the key is **A**.

4 Iron is the main catalyst (key **C**) in the manufacture of ammonia.

5 Sulphur is a naturally occurring starting material (key **A**) in the manufacture of sulphuric acid by the contact process. Is sulphur the only naturally occurring starting material in the process? Is it necessary for the sulphur to start in the form of the element?

6 Sodium hydroxide is a synthetic starting material (key **B**) in the manufacture of soap. With what is the sodium hydroxide reacted in the process and is this second substance also a synthetic starting material?

7 Carbon-containing fuels burn to carbon dioxide, with carbon monoxide as an additional product if the combustion is incomplete. Hydrogen-containing fuels burn to water vapour. Silicon dioxide is not formed by burning fuels — oxidized silicon is present as an impurity in some fuels and remains in the ash after combustion. The key is **A**.

8 Limestone is used in only one of the production processes, that of iron in a blast furnace so the key is **D**.

9 The products removed from the blast furnace are pig iron and slag. Pure iron and steel have to be made from pig iron. The key is **C**.

10 Metals can be isolated from their compounds by displacement, this method being used if the higher metal is cheap and the lower metal is expensive (**1** correct). Reduction of oxides with coke and electrolysis of molten salts are frequently used to isolate metals from their compounds (**2** and **3** also correct). Use of a powerful magnet only separates magnetic materials from non-magnetic ones, it does not isolate a metal from its compounds (**4** incorrect). The key is **A**.

11 An oxide ore is the main source in the manufacture of aluminium, key **D**, only. What are the main sources for the manufacture of the other elements?

12 The hydrogen in the water vapour must have come from the custard powder but the oxygen in the water vapour might all have come from the copper(II) oxide. Similarly, the carbon in the carbon dioxide must have come from the custard powder, while the oxygen might all have come from the copper(II) oxide. The experiment provides no evidence of whether custard powder contains water. Only carbon and hydrogen are certain components and the key is **B**.

13 An electric light bulb contains a noble gas such as argon and a tungsten filament (key **B**). The presence of the noble gas makes it more difficult for tungsten atoms to vaporize and weaken the filament. This lengthens the life of the bulb. It is said that a trace of oxygen is also added to shorten the life of the bulb. Perhaps this is just a malicious rumour spread by candle manufacturers!

14 The only correct statement about the action of soaps and detergents is the one made in **4**. The key is **D**.

15 A balanced fertilizer contains compounds of nitrogen and phosphorus, key **B**. What is the other element that a balanced fertilizer must contain?

16 Sodium hydroxide, key **E**, is not a naturally occurring raw material. Was your answer to question 6 correct?

17 Oxygen is produced commercially by liquefying air and then boiling off the gases (key **B**). Does oxygen boil off first? (See Introduction, question 7).

18 The manufacture of aluminium from aluminium oxide (purified bauxite) involves reduction by electrolysis, key **B**.

$$Al^{3+} + 3e^- \rightarrow Al$$

At the high operating temperature, some of the oxygen released at the anode oxidizes the carbon electrode to oxides of carbon. Because the carbon is oxidized, the oxygen must be reduced by the carbon. The key is **B** rather than **A**, however, because the reduction of oxygen by carbon is an unwanted reaction which consumes the electrode. It is an unfortunate occurrence rather than the basis of the manufacture of aluminium.

19 The high reactivity of calcium makes it necessary to obtain the metal by electrolysis of a molten salt, key **D**. **A** might give calcium, but only at temperatures too high to be economically competitive. **B** takes place in the opposite direction, even at high temperatures — the calcium reacts with steam to give calcium oxide and hydrogen. What, if anything, would happen on carrying out the reactions given in **C** and **E**?

20 The other substances used to make iron in a blast furnace are air, coke, and limestone, key **D**. Sand is not deliberately added — it is an impurity, as is aluminium oxide, which has to be removed by the limestone. At which places in the blast furnace are the air, coke, and limestone added?

21 The reaction between NO and oxygen, to make $NO_2$, is an essential stage in the oxidation of ammonia to produce nitric acid, key **D**. How is the $NO_2$ converted to nitric acid?

   NO used to be important in a now obsolete process for making one of the other substances in the list. Look up the 'Lead Chamber process' in an old reference book. Which substance in the list used to be made by a method where NO played a part? Was NO oxidized by oxygen in this process?

22 Chlorine has all of the uses given except the last. The key is **E**. Chlorine is manufactured from sodium chloride, not sodium chloride from chlorine.

23 A biodegradable detergent is one that is broken down by bacteria present in the water system. A non-biodegradable detergent is one that is not broken down in this way, as stated in key **C**.

24 The nitrogen is supplied by proteins in the diet, key **C**. What protein-containing foods have you eaten today? Have you eaten any vegetable protein?

25 One of the important uses of ammonia is given in key **C**. Can you name any fertilizers that are made from ammonia? On a commercial scale, it is not necessary to convert the ammonia into anything else. Ammonia can be applied direct to the soil and it is becoming fairly usual to see ammonia tankers in farming areas. Can you see any problems that might have to be overcome when applying ammonia direct to the soil?

26 The meaning of the term 'cracking' is correctly given in key **B**.

27 The first statement is true, the second false (key **C**). Can you name at least one further way in which nitrogen occurs, other than in the air?

28 Both statements are true, but the reason for using electrolysis of the molten chlorides is not that the chlorides have high melting points. The key is **B**. The reason for using electrolysis (electrical reduction) is that chemical reduction is difficult or impossible. The reason for using the molten compounds rather than aqueous salts is to avoid the almost certain evolution of hydrogen instead of the metal. If you wonder why 'almost' was included in the previous sentence, look up 'Castner-Kellner cell' in an older reference book.

29 Both statements are true, the higher solubility of superphosphate being the reason for its quicker action. The key is **A**.

30 Both statements are true, and the second is the correct explanation of why excessive calcium ions make it difficult to form a lather. The key is **A**. What term is used to describe water that contains calcium ions? What is the other common ion, found in water, that produces the same effect?

# Test 15 | All topics 1

Questions 1–5 concern the possible combinations of properties of substances shown in headings A–E below.

| Electrical conductance of the pure substance at room temperature | Solubility in water | Properties of the solution in water |
|---|---|---|
| A Non-conductor | Soluble | A neutral solution which conducts electricity |
| B Non-conductor | Soluble | A neutral solution which does NOT conduct electricity |
| C Non-conductor | Insoluble | — |
| D Conductor | Insoluble | — |
| E Non-conductor | Soluble | An alkaline solution which conducts electricity |

Select, from A to E, the appropriate set of properties for each of the following substances.

1   Silicon dioxide          4   Lead

2   Sodium hydroxide         5   Sugar

3   Potassium nitrate

Questions 6–9 concern the following substances:

   A Ethanol                 D A protein

   B Ethene (ethylene)       E Starch

   C Glucose

For each question below select, from A to E, the substance which will give the stated result for the test described.

6   A dark blue colour forms when iodine solution is added to a solution of the substance.

7   A red precipitate forms when a solution of the substance is warmed with Fehling's solution or Benedict's reagent.

8   Bromine water becomes colourless when shaken with the substance.

9   An alkaline gas is given off when the substance is mixed with soda lime and heated.

Questions 10–13 concern the following practical methods:

   A Chromatography          D Electrolysis

   B Crystallization         E Filtration

   C Distillation

Choose from A to E, the method which would be used in each of the following separations.

10   The extraction of aluminium from its purified ore.

11   The isolation of oxygen from liquid air.

12   The separation of coloured dyes in a sample of coloured ink.

13   The separation of paraffin oil from natural crude oil.

| Directions summarized for questions 14 to 17 | | | | |
|---|---|---|---|---|
| A 1, 2, 3 only correct | B 1, 3 only correct | C 2, 4 only correct | D 4 only correct | E Some other response or combination of responses is correct |

14   When ammonia is added to an acid, the following reaction occurs

$$H_3O^+(aq) + NH_3(aq) \rightleftharpoons NH_4^+(aq) + H_2O(l)$$

In the above equilibrium, which substance(s) is/are acting as bases?

1   $NH_3(aq)$                3   $H_2O(l)$

2   $H_3O^+(aq)$              4   $NH_4^+(aq)$

110

15 The following equilibrium is established when sulphur dioxide dissolves in water:

$$SO_2(aq) + 2H_2O(l) \rightleftharpoons H_3O^+(aq) + HSO_3^-(aq)$$

Which reagent(s) would cause the equilibrium position to shift towards the left-hand side when added to the solution?

1 $NaHSO_3(aq)$      3 $HCl(aq)$

2 $NH_3(aq)$      4 $NaOH(aq)$

16 Which of the following gaseous hydrocarbons would burn completely in oxygen to produce a total volume of carbon dioxide and steam which is three times the volume of the original gaseous hydrocarbon? (Assume that all the volumes were measured at the same temperature and pressure).

1 $C_2H_2$      3 $CH_4$

2 $C_2H_4$      4 $C_2H_6$

17 The graph shows how the proportion of ammonia formed in an equilibrium mixture of nitrogen, hydrogen, and ammonia varies with pressure at two different temperatures, in the presence of a catalyst.

$$N_2 + 3H_2 \rightleftharpoons 2NH_3$$

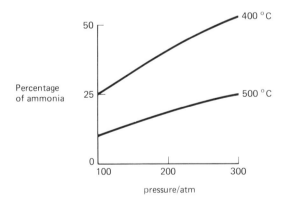

The graph shows that the percentage of ammonia

1 depends on the type of catalyst used

2 decreases as the temperature increases at constant pressure

3 is a maximum at 500°C and 300 atmospheres

4 increases when the pressure increases at constant temperature

Directions for questions **18** to **21**. Each of the questions or incomplete statements in this section is followed by five suggested answers. Select the best answers in each case.

18 How many protons, neutrons and electrons are there in an atom represented by the symbol $^{35}_{17}Cl$?

|   | Protons | Neutrons | Electrons |
|---|---------|----------|-----------|
| A | 17 | 18 | 17 |
| B | 35 | 17 | 35 |
| C | 17 | 35 | 35 |
| D | 35 | 17 | 17 |
| E | 18 | 17 | 18 |

19 Cracking a hydrocarbon mixture results in the

A formation of small molecules from larger ones

B combustion of the hydrocarbons

C formation of ethanol

D formation of polymers

E separation of the hydrocarbons into groups of similar boiling-point

20 What mass of water is formed when 0.72 g of pentane, $C_5H_{12}$, is burned in excess oxygen? (Relative atomic masses: H = 1, C = 12, O = 16)

A 0.12 g      D 1.08 g

B 0.18 g      E 2.16 g

C 0.54 g

21 100 cm³ of nitrogen oxide gas (NO) combine with 50 cm³ of oxygen to form 100 cm³ of a single gaseous compound, all volumes being measured at the same temperature and pressure. Which of the following equations fits these facts?

A    $NO(g) + O_2(g) \rightarrow NO_3(g)$

B    $NO(g) + 2O_2(g) \rightarrow MO_5(g)$

C    $2NO(g) + O_2(g) \rightarrow 2NO_2(g)$

D    $2NO(g) + O_2(g) \rightarrow N_2O_4(g)$

E   $2NO(g) + 2O_2(g) \rightarrow N_2O_6(g)$

| | Directions summarized for questions **22** to **26** | | |
|---|---|---|---|
| | First statement | Second statement | |
| **A** | True | True | Second statement is a correct explanation of the first |
| **B** | True | True | Second statement is NOT a correct explanation of the first |
| **C** | True | False | |
| **D** | False | True | |
| **E** | False | False | |

**22 1** Sodium is placed in Group I of the Periodic Table.

   **2** Sodium forms an ionic compound with chlorine.

**23 1** Phosphorus is classed as a non-metallic element.

   **2** Phosphorus burns in air to form two oxides.

**24 1** Hydrogen chloride gas reacts with heated iron to form hydrogen.

   **2** Hydrogen chloride gas contains hydrogen ions.

**25 1** When steam is passed over heated magnesium, hydrogen is produced together with a white residue.

   **2** In the reaction between magnesium and steam, magnesium acts as an oxidizing agent.

**26 1** Nitric acid produces hydrogen when it reacts with copper.

   **2** Nitric acid can be used to convert $Cu^{2+}$ ions to copper.

**Questions 27–30** concern the large scale production of ammonia from nitrogen and hydrogen by the Haber process.

$$N_2 + 3H_2 \rightleftharpoons 2NH_3$$

**27** A dynamic equilibrium is reached in the reaction between nitrogen and hydrogen.

$$N_2 + 3H_2 \rightleftharpoons 2NH_3$$

This equilibrium will be reached more quickly at high pressure because

**A** the reacting molecules collide with greater energy

**B** the reacting molecules collide more often

**C** there is a decrease in volume during the reaction

**D** an ammonia molecule is smaller than a hydrogen molecule

**E** the temperature of a gas under pressure is higher than at normal pressure

**28** The large quantities of nitrogen for this process are obtained by

**A** electrolysis of sea water

**B** electrolysis of nitric acid

**C** fractional distillation of liquid air

**D** bacterial action on air at room temperature

**E** heating potassium nitrate

**29** Which of the following is the most important large scale use of ammonia?

**A** As a convenient source of nitrogen gas

**B** For neutralizing waste acids from chemical processes

**C** As a component of cleaning fluids

**D** As a reagent for detecting aqueous copper(II) ions

**E** For making fertilizers containing nitrogen

**30** Which of the following is most likely to be used industrially for the production of hydrogen for the Haber process?

**A** Treating iron with dilute hydrochloric acid

**B** Adding sodium to water

**C** Passing steam over hot magnesium

**D** Cracking alkanes obtained from crude oil

**E** Electrolyzing water containing dilute sulphuric acid

# Test 16 | All topics 2

**Questions 1–6** concern the following elements and compounds:

   **A** Sodium chloride    **D** Copper

   **B** Ethanol           **E** Diamond

   **C** Rhombic sulphur

Which substance

1  is composed of a giant structure of positive and negative ions?

2  would give an X-ray diffraction pattern indicating an irregular arrangement of particles at room temperature?

3  has the lowest melting point?

4  is expected to have a structure similar to that of silicon?

5  conducts electricity when molten but NOT when solid?

6  when melted and allowed to cool, gives crystals of one structure which slowly change into a different structure?

| Directions summarized for questions **7** to **12** | | | | |
|---|---|---|---|---|
| **A** <br> 1, 2, 3 only correct | **B** <br> 1, 3 only correct | **C** <br> 2, 4 only correct | **D** <br> 4 only correct | **E** <br> Some other response or combination of responses is correct |

7  Sand may be obtained from a mixture of powdered calcium carbonate and sand by

   1  washing the mixture with water

   2  heating the mixture

   3  adding aqueous sodium hydroxide and filtering

   4  heating with dilute hydrochloric acid and filtering

8  Iodine, like chlorine, forms a gaseous compound with hydrogen. Hydrogen iodide would be expected to

   1  be soluble in water

   2  cause damp indicator paper to register an acidic pH value

   3  react with heated iron to form iron(II) iodide and hydrogen

   4  support the combustion of a lighted splint

9  Titanium is a transition metal. Titanium would be expected to

   1  form ions carrying different numbers of positive charges

   2  form a hydroxide that is readily soluble in water

   3  form compounds which give coloured solutions in water

   4  react with cold water to form hydrogen

10 The reaction between methane and oxygen may be represented by the equation

$$CH_4(g) + 2O_2(g) \rightarrow CO_2(g) + 2H_2O$$
$$\Delta H = -x \text{ kJ mol}^{-1} \text{ of methane.}$$

(Relative atomic masses: $C = 12$, $H = 1$, $O = 16$)

From this information it can be deduced that

1 16 moles of methane would require 64 moles of oxygen molecules for complete combustion

2 the reaction is endothermic

3 1 mole of methane would give one mole of gaseous products at 20°C

4 the complete conversion of one mole of methane to carbon dioxide and water vapour results in $x$ kJ of heat being produced

11 Sodium hydroxide, NaOH, reacts with nitric acid, $HNO_3$, as in the equation

$$HNO_3(aq) + NaOH(aq) \rightarrow NaNO_3(aq) + H_2O(l)$$

When $20 \text{ cm}^3$ of $1.0 \text{ M } HNO_3$ is mixed with $30 \text{ cm}^3$ of $0.5 \text{ M NaOH}$, which of the following statements is/are true?

1 The pH of the mixture is greater than 7.

2 If the mixture is evaporated to dryness, both sodium hydroxide and sodium nitrate crystallize.

3 The reaction can be classed as neutralization.

4 Nitric acid accepts protons from sodium hydroxide.

12 The reaction between silver nitrate solution and iron(II) sulphate solution can be represented by the equation

$$Ag^+(aq) + Fe^{2+}(aq) \rightleftharpoons Ag(s) + Fe^{3+}(aq)$$

If the system is in equilibrium

1 addition of iron(III) sulphate will increase the number of iron(II) ions ($Fe^{2+}$)

2 addition of iron(II) sulphate will reduce the number of silver ions

3 addition of silver nitrate will increase the number of iron(III) ions ($Fe^{3+}$)

4 addition of sodium nitrate will increase the amount of silver precipitate

*Directions for questions* **13** *to* **18**. *Each of the questions or incomplete statements in this section is followed by five suggested answers. Select the best answer in each case.*

13 A sample of pure dry ammonia was passed over red hot iron wire and was found to break down to nitrogen and hydrogen according to the equation

$$2NH_3(g) \rightarrow 3H_2(g) + N_2(g)$$

The volume of mixed gases formed from $60 \text{ cm}^3$ of ammonia after cooling to room temperature is likely to be

A $120 \text{ cm}^3$      D $30 \text{ cm}^3$

B $60 \text{ cm}^3$      E $20 \text{ cm}^3$

C $40 \text{ cm}^3$

14 Soap can be made by boiling a fat with aqueous sodium hydroxide. At a later stage in the process the resulting mixture is boiled with concentrated aqueous sodium chloride. The reason for using sodium chloride is

A to bring about complete conversion of the fat to soap

B to act as a catalyst in the reaction

C to increase the rate of conversion of fat to soap

D to make the soap less soluble in the mixture

E to make the water produce a better lather

15 Which substance reacts most rapidly with bromine?

A $CH_3-\overset{\displaystyle\|}{\underset{\displaystyle O}{C}}-OH$      D $H_2C=CH_2$

                                       E $CH_4$

B $CH_3CH_2OH$

C $CH_3-\overset{\displaystyle\|}{\underset{\displaystyle O}{C}}-OC_2H_5$

**16** Silver chloride is precipitated when solutions of sodium chloride and silver sulphate are mixed:

$$2NaCl(aq) + Ag_2SO_4(aq)$$
$$\rightarrow 2AgCl(s) + Na_2SO_4(aq)$$

How many $cm^3$ of 1.0 M sodium chloride would just react with 1 $dm^3$ of 0.01 M silver sulphate?

**A** 1             **D** 20

**B** 2             **E** 100

**C** 10

**17** The following table shows the solubilities of a number of substances

| Substance | Solubility in water |
|---|---|
| Barium sulphate | Insoluble |
| Calcium sulphate | Slightly soluble |
| Lead(II) nitrate | Soluble |
| Lead(II) oxide | Insoluble |
| Lead(II) sulphate | Insoluble |
| Sodium sulphate | Soluble |

Which of the following pairs of substances would be most suitable for preparing lead sulphate?

**A** Lead(II) oxide and sodium sulphate

**B** Lead(II) nitrate and sodium sulphate

**C** Lead(II) oxide and barium sulphate

**D** Lead(II) nitrate and barium sulphate

**E** Lead(II) nitrate and calcium sulphate

**18** A metallic element X reacts with oxygen according to the following equation:

$$2X + \tfrac{1}{2}O_2 \rightarrow X_2O$$

The number of the Group in the Periodic Table where this element is most likely to be found is

**A** I             **D** IV

**B** II            **E** V

**C** III

| Directions summarized for questions **19** to **22** | | |
|---|---|---|
| | First statement | Second statement |
| **A** | True | True | Second statement is a correct explanation of the first |
| **B** | True | True | Second statement is NOT a correct explanation of the first |
| **C** | True | False | |
| **D** | False | True | |
| **E** | False | False | |

**19**

**1** Copper is deposited on the cathode when an aqueous solution of copper(II) sulphate is electrolyzed.

**2** Aqueous copper(II) sulphate solution contains $Cu^{2+}(aq)$ ions.

**20**

**1** A solution of hydrogen chloride in toluene does NOT conduct electricity.

**2** Hydrogen chloride does NOT ionize in toluene.

**21**

**1** 20 $cm^3$ of hydrogen reacts with exactly 10 $cm^3$ of oxygen, both volumes being measured at the same temperature and pressure.

**2** Hydrogen molecules contain twice as many atoms as do oxygen molecules.

**22**

**1** The progress of the neutralization of sulphuric acid by barium hydroxide can be followed by measuring the electrical conductivity of the mixture.

**2** The reaction between sulphuric acid and barium hydroxide causes a change in the number of ions in the solution.

**Questions 23–26**

In a series of experiments to determine the formula of magnesium oxide, some students each took a different length of clean magnesium ribbon and heated it strongly in a crucible fitted with a lid.

The following masses (in grams) were found:
Mass of empty crucible, $m_1$

Mass of empty crucible + magnesium, $m_2$

Mass of empty crucible + magnesium oxide after heating, $m_3$

**23** The mass of oxygen combined with the magnesium is given by

**A** $m_2 - m_1$        **D** $m_3 - (m_2 - m_1)$

**B** $m_3 - m_1$        **E** $(m_2 - m_1) - (m_3 - m_1)$

**C** $m_3 - m_2$

**24** If the masses (in grams) from each experiment were plotted on the graph below, which result would be obtained?

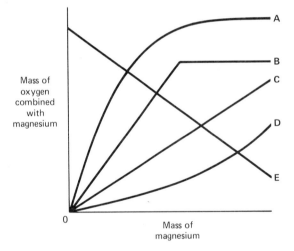

**25** In one experiment 0.12 g of oxygen appeared to combine with 0.24 g of magnesium. From these results, how many moles of oxygen atoms combine with one mole of magnesium atoms? (Relative atomic masses: $O = 16$, $Mg = 24$)

**A** 0.34

**B** 0.50

**C** 0.67

**D** 0.75

**E** 0.95

**26** The number of moles of oxygen atoms calculated in question **25** is less than expected. (Relative atomic masses: $N = 14$, $O = 16$, $Mg = 24$). This low result could be due to any of the following EXCEPT

**A** The balance was read incorrectly so that $m_3$ was larger than it should have been

**B** the magnesium reacted to some extent with nitrogen rather than with oxygen

**C** the magnesium did not react completely with oxygen

**D** the crucible was damp so that water evaporated during the heating

**E** some of the oxide escaped when the crucible lid was lifted during the heating

The answers for tests **15** and **16** appear at the end of the introduction on page xii

# The Story of
## the
# CHAMPIONS
## of the
# Round Table

**ir Launcelot of the Lake.**

The Story of
the
CHAMPIONS
of the
Round Table

Written and Illustrated
by
HOWARD PYLE.

Dover Publications, Inc.

NEW YORK

Published in Canada by General Publishing
Company, Ltd., 30 Lesmill Road, Don Mills, Toronto,
Ontario.
Published in the United Kingdom by Constable
and Company, Ltd., 10 Orange Street, London
W.C. 2.

This Dover edition, first published in 1968, is an
unabridged and unaltered republication of the work
originally published by Charles Scribner's Sons in
1905.

*Library of Congress Catalog Card Number:* 68—16470

Manufactured in the United States of America
Dover Publications, Inc.
180 Varick Street
New York, N. Y. 10014

# Foreword.

✠                                                 ✠

*IN a book which was written by me aforetime, and which was set forth in print, I therein told much of the history of King Arthur; of how he manifested his royalty in the achievement of that wonderful magic sword which he drew forth out of the anvil; of how he established his royalty; of how he found a splendid sword yclept Excalibur in a miraculously wonderful manner; of how he won the most beautiful lady in the world for his queen; and of how he established the famous Round Table of noble worthy knights, the like of whose prowess the world hath never seen, and will not be likely ever to behold again.*

*Also I told in that book the adventures of certain worthy knights and likewise how the magician Merlin was betrayed to his undoing by a sorceress hight Vivien.*

*Now, if you took any joy in reading that book, I have great hope that that which follows may be every whit as pleasing to you; for I shall hereinafter have to do with the adventures of certain other worthies with whom you may have already become acquainted through my book and otherwise; and likewise of the adventures of certain other worthies, of whom you have not yet been told by me.*

*More especially, I believe, you will find entertainment in what I shall have to tell you of the adventures of that great knight who was altogether the most noble of spirit, and the most beautiful, and the bravest of heart, of any knight who ever lived—excepting only his own son, Galahad, who was the crowning glory of his house and of his name and of the reign of King Arthur.*

*However, if Sir Launcelot of the Lake failed now and then in his behavior, who is there in the world shall say, "I never fell into error"? And if he more than once offended, who is there shall have hardihood to say, "I never committed offence"?*

*Yea, that which maketh Launcelot so singularly dear to all the world, is that he was not different from other men, but like other men, both in his virtues and his shortcomings; only that he was more strong and more brave and more untiring than those of us who are his brethren, both in our endeavors and in our failures.*

# Contents

## The Story of Launcelot

## Chapter Third

## Chapter Fourth

## Chapter Fifth

## Chapter Sixth

## Chapter Seventh

## Chapter Eighth

# The Book of Sir Tristram

## PART I

### THE STORY OF SIR TRISTRAM AND THE LADY BELLE ISOULT

## Chapter First

## Chapter Second

## Chapter Third

## Chapter Second

## Chapter Third

# PART III

## THE MADNESS OF SIR TRISTRAM

### Chapter First

### Chapter Second

### Chapter Third

## Chapter Fourth

# The Book of Sir Percival

## Chapter First

## Chapter Second

## Chapter Third

## Chapter Fourth

## Chapter Fifth

# LIST OF ILLUSTRATIONS

# The Story of
## the
# CHAMPIONS
## of the
# Round Table

# The Lady Nymue beareth away Launcelot into the Lake:

# Prologue.

IT hath already been set forth in print in a volume written by me concerning the adventures of King Arthur when he first became king, how there were certain lesser kings who favored him and were friendly allies with him, and how there were certain others of the same sort who were his enemies.

Among those who were his friends was King Ban of Benwick, who was an exceedingly noble lord of high estate and great honor, and who was of a lineage so exalted that it is not likely that there was anyone in the world who was of a higher strain.

Now, upon a certain time, King Ban of Benwick fell into great trouble; for there came against him a very powerful enemy, to wit, King Claudas of Scotland. King Claudas brought unto Benwick a huge army of knights and lords, and these sat down before the Castle of Trible with intent to take that strong fortress and destroy it. *Of King Ban and his misfortunes.*

This noble Castle of Trible was the chiefest and the strongest place of defence in all King Ban's dominions, wherefore he had intrenched himself there with all of his knights and with his Queen, hight Helen, and his youngest son, hight Launcelot.

Now this child, Launcelot, was dearer to Queen Helen than all the world besides, for he was not only large of limb but so extraordinarily beautiful of face that I do not believe an angel from Paradise could have been more beautiful than he. He had been born with a singular birth-mark upon his shoulder, which birth-mark had the appearance as of a golden star enstamped upon the skin; wherefore, because of this, the Queen would say: "Launcelot, by reason of that star upon thy shoulder I believe that thou shalt be the star of our house and that thou shalt shine with such remarkable glory that all the world shall behold thy lustre and shall marvel thereat for all time to come." So the Queen took extraordinary delight in Launcelot and loved him to the very core of her heart—albeit she knew not, at the time she spake, how that prophecy of hers concerning the star was to fall so perfectly true.

Now, though King Ban thought himself very well defended at his Castle of Trible, yet King Claudas brought so terribly big an army against that place that it covered the entire plain. A great many battles were fought under the walls of the castle, but ever King Claudas waxed greater and stronger, and King Ban's party grew weaker and more fearful.

So by and by things came to such a pass that King Ban bethought him of King Arthur, and he said to himself: "I will go to my lord the King and *King Ban* beseech help and aid from him, for he will certainly give it *bethinks him of* me. Nor will I trust any messenger in this affair other than *King Arthur.* myself; for I myself will go to King Arthur and will speak to him with my own lips."

Having thus bethought him, he sent for Queen Helen to come into his privy closet and he said to her: "My dear love, nothing remaineth for me but to go unto the court of King Arthur and beseech him to lend his powerful aid in this extremity of our misfortunes; nor will I trust any messenger in this affair but myself. Now, this castle is no place for thee, when I am away, therefore, when I go upon this business, I will take thee and Launcelot with me, and I will leave you both in safety at King Arthur's court with our other son, Sir Ector, until this war be ended and done." And to these Queen Helen lent her assent.

So King Ban summoned to him the seneschal of the castle, who was named Sir Malydor le Brun, and said to him: "Messire, I go hence to-night by a secret pass, with intent to betake me unto King Arthur, and to beseech his aid in this extremity. Moreover, I shall take with me my lady and the young child Launcelot, to place them within the care of King Arthur during these dolorous wars. But besides these, I will take no other one with me but only my favorite esquire, Foliot. Now I charge thee, sir, to hold this castle in my behalf with all thy might and main, and yield it not to our enemies upon any extremity; for I believe I shall in a little while return with sufficient aid from King Arthur to compass the relief of this place."

So when night had fallen very dark and still, King Ban, and Queen Helen, and the young child Launcelot, and the esquire Foliot left the town privily by means of a postern gate. Thence they went by a *King Ban with* secret path, known only to a very few, that led down a steep *Queen Helen* declivity of rocks, with walls of rock upon either side that were *escape from* very high indeed, and so they came out in safety beyond the *Trible.* army of King Claudas and into the forest of the valley below. And the forest lay very still and solemn and dark in the silence of the night-time.

Having thus come out in safety into the forest, that small party journeyed on with all celerity that they were able to achieve until, some little time before dawn, they came to where was a lake of water in an open meadow of the forest. Here they rested for a little while, for Queen Helen had fallen very weary with the rough and hasty journey which they had traveled.

Now whilst they sat there resting, Foliot spake of a sudden, saying unto King Ban: "Lord, what is that light that maketh the sky so bright yonder-ways?" Then King Ban looked a little and presently said: *Foliot seeth a* "Methinks it must be the dawn that is breaking." "Lord," *light.* quoth Foliot, "that cannot very well be; for that light in the sky lieth in the south, whence we have come, and not in the east, where the sun should arise."

Then King Ban's heart misgave him, and his soul was shaken with a great trouble. "Foliot," he said, "I believe that you speak sooth and that that light bodes very ill for us all." Then he said: "Stay here for a little and I will go and discover what that light may be." Therewith he mounted his horse and rode away in the darkness.

Now there was a very high hill near-by where they were, and upon the top of the hill was an open platform of rock whence a man could see a great way off in every direction. So King Ban went to this place, *King Ban* and, when he had come there, he cast his eyes in the direction *beholdeth the* of the light and he straightway beheld with a manner of terror *burning of* that the light came from Trible; and then, with that terror *Trible.* still growing greater at his heart, he beheld that the town and the castle were all in one great flame of fire.

When King Ban saw this he sat for a while upon his horse like one turned into a stone. Then, after a while, he cried out in a great voice: "Woe! Woe! Woe is me!" And then he cried out still in a very loud voice, "Certes, God hath deserted me entirely."

Therewith a great passion of grief took hold upon him and shook him like to a leaf, and immediately after that he felt that something brake within him with a very sharp and bitter pain, and he wist that it was his heart that had broken. So being all alone there upon the hilltop, and in the perfect stillness of the night, he cried out, "My heart! My heart!"

*The death of King Ban.*　And therewith, the shadows of death coming upon him, he could not sit any longer upon his horse, but fell down upon the ground. And he knew very well that death was nigh him, so, having no cross to pray upon, he took two blades of grass and twisted them into that holy sign, and he kissed it and prayed unto it that God would forgive him his sins. So he died all alone upon that hilltop.

Meanwhile, Queen Helen and Foliot sat together waiting for him to return and presently they heard the sound of his horse's hoofs coming down that rocky path. Then Queen Helen said: "Foliot, methinks my lord cometh." So in a little came the horse with the empty saddle. When Foliot beheld that he said: "Lady, here meseems is great trouble come to us, for methinks something hath befallen my lord, and that he is in sore travail, for here is his horse without him."

Then it seemed to Queen Helen as though the spirit of life suddenly went away from her, for she foresaw what had befallen. So she arose like one in a dream, and, speaking very quietly, she said: "Foliot, take me whither my lord went awhile since!" To this Foliot said: "Lady, wait until the morning, which is near at hand, for it is too dark for you to go thitherward at this present." Whereunto the Lady Helen replied: "Foliot, I cannot wait, for if I stay here and wait I believe I shall go mad." Upon this, Foliot did not try to persuade her any more but made ready to take her whither she would go.

Now the young child Launcelot was then asleep upon the Queen's knees, wherefore she took her cloak and wrapped the child in it and laid him very gently upon the ground, so that he did not wake. Then she mounted upon her palfrey and Foliot led the palfrey up the hill whither King Ban had gone a short time since.

When they came to that place of open rocks above told of, they found King Ban lying very quiet and still upon the ground and with a countenance of great peace. For I believe of a surety that God had forgiven him all

*The Lady Helen findeth the King.*　his sins, and he would now suffer no more because of the cares and the troubles of this life. Thus Queen Helen found him, and finding him she made no moan or outcry of any kind, only she looked for a long while into his dead face, which she could see very plainly now, because that the dawn had already broken. And by and by she said: "Dear Lord, thou art at this time in a happier case than I." And by and by she said to Foliot: "Go and bring his horse to this place, that we may bear him hence." "Lady," said Foliot, "it is not good for you to be left here alone." "Foliot," said the Queen, "thou dost not know how much alone I am; thy leaving me here cannot make me more alone." Therewith she fell to weeping with great passion.

Then Foliot wept also in great measure and, still weeping like rain, he went away and left her. When he came again with King Ban's horse the sun had risen and all the birds were singing with great jubilation and everything was so blithe and gay that no one could have believed that care and trouble could dwell in a world that was so beautiful.

So Queen Helen and Foliot lifted the dead king to his horse and then the Queen said: "Come thou, Foliot, at thine own gait, and I will go ahead and seek my child, for I have yet Launcelot to be my joy. *The Lady* Haply he will be needing me at this moment." So the Queen *Helen bringeth* made haste down the steep hill ahead of Foliot and by and by *her dead down* she came to the margin of that little lake where they had *Mountain.* rested awhile since.

By now the sun had risen very strong and warm so that all the lake, and the meadows circumadjacent, and the forest that stood around about that meadow were illumined with the glory of his effulgence.

Now as Queen Helen entered that meadow she beheld that a very wonderful lady was there, and this lady bare the child Launcelot in her arms. And the lady sang to Launcelot, and the young child looked up into her face and laughed and set his hand against her cheek. All this Queen Helen beheld; and she likewise beheld that the lady was of a very extraordinary appearance, being clad altogether in green that glistered and shone with a wonderful brightness. And she beheld that around the neck of the lady was a necklace of gold, inset with opal stones and emeralds; and she perceived that the lady's face was like ivory—very white and clear—and that her eyes, which were very bright, shone like jewels set into ivory. And she saw that the lady was very wonderfully beautiful, so that the beholder, looking upon her, felt a manner of fear—for that lady was Fay.

(And that lady was the Lady of the Lake, spoken of aforetime in the Book of King Arthur, wherein it is told how she aided King Arthur to obtain that wonderful, famous sword yclept Excalibur, and how she aided Sir Pellias, the Gentle Knight, in the time of his extremity, and took him into the lake with her. Also divers other things concerning her are told of therein.)

Then the Queen came near to where the lady was, and she said to her,

"Lady, I pray you give me my child again!" Upon this the Lady of the Lake smiled very strangely and said: "Thou shalt have thy child again, lady, but not now; after a little thou shalt have hīm again." Then Queen Helen cried out with great agony of passion: "Lady, would you take my child from me? Give him to me again, for he is all I have left in the world. Lo, I have lost house and lands and husband, and all the other joys that life has me to give, wherefore, I beseech you, take not my child from me." To this the Lady of the Lake said: "Thou must endure thy sorrow a while longer; for it is so ordained that I must take thy child; for I take him only that I may give him to thee again, reared in such a wise that he shall make the glory of thy house to be the glory of the world. For he shall become the greatest knight in the world, and from his loins shall spring a greater still than he, so that the glory of the House of King Ban shall be spoken of as long as mankind shall last." But Queen Helen cried out all the more in a great despair: "What care I for all this? I care only that I shall have my little child again! Give him to me!"

Therewith she would have laid hold of the garments of the Lady of the Lake in supplication, but the Lady of the Lake drew herself away from *The Lady of the Lake taketh Launcelot into the Lake.* Queen Helen's hand and said: "Touch me not, for I am not mortal, but Fay." And thereupon she and Launcelot vanished from before Queen Helen's eyes as the breath vanishes from the face of a mirror.

For when you breathe upon a mirror the breath will obscure that which lieth behind; but presently the breath will disappear and vanish, and then you shall behold all things entirely clear and bright to the sight again. So the Lady of the Lake vanished away, and everything behind her where she had stood was clear and bright, and she was gone.

Then Queen Helen fell down in a swoon, and lay beside the lake of the meadow like one that is dead; and when Foliot came he found her so and wist not what to do for her. There was his lord who was dead and his lady who was so like to death that he knew not whether she was dead or no. So he knew not what to do but sat down and made great lamentation for a long while.

What time he sat thus there came that way three nuns who dwelt in an abbey of nuns which was not a great distance away from that place. *The Lady Helen taketh to a Nunnery.* These made great pity over that sorrowful sight, and they took away from there the dead King and the woeful Queen, and the King they buried in holy ground, and the Queen they let live with them and she was thereafter known as the "Sister of Sorrows."

Now Launcelot dwelt for nigh seventeen years with the Lady Nymue of the Lake in that wonderful, beautiful valley covered over with the appearance of such a magical lake as hath been aforetime described in the Book of King Arthur.

*How Launcelot dwelt in the lake.*

And that land of the lake was of this sort that shall here be described :—

Unto anyone who could enter into the magic water of that lake (and there were very few of those who were mortal who were allowed to come to those meadows of Faery that were there concealed beneath those enchanted waters) he would behold before him a wide and radiant field of extraordinary beauty. And he would behold that that field was covered all over with such a multitude of exquisite and beautiful flowers that the heart of the beholder would be elated with pure joy to find himself in the midst of that waving sea of multitudinous and fragrant blossoms. And he would behold many fair and shady groves of trees that here and there grew up from that valley, each glade overshadowing a fountain of water as clear as crystal. And he would perhaps behold, at such pleasant places beneath the shade of those trees, some party of the fair and gentle folk of that country; and he would see them playing in sport, or he would hear them chanting to the music of shining golden harps. And he would behold in the midst of that beautiful plain a wonderful castle with towers and roofs uplifted high into the sky, and all shining in the peculiar radiance of that land, like to castles and battlements of pure gold.

Such was the land unto which Launcelot was brought, and from what I have told you you may see what a wonderful, beautiful place it was.

And the mystery of that place entered into the soul of Launcelot, so that thereafter, when he came out thence, he was never like other folk, but always appeared to be in a manner remote and distant from other of his fellow-mortals with whom he dwelt.

For though he smiled a great deal, it was not often that he laughed ; and if he did laugh, it was never in scorn, but always in loving-kindness.

It was here in this land that Sir Pellias had now dwelt for several years, with great peace and content. (For it hath been told in the Book of King Arthur how, when he was upon the edge of death, the Lady Nymue of the Lake brought him back to life again, and how, after that time, he was half fay and half mortal.)

And the reason why Launcelot was brought to that place was that Sir Pellias might teach him and train him in all the arts of chivalry. For no one in all the world was more skilful in arms than Sir Pellias, and no one could so well teach Launcelot the duties of chivalry as he.

So Sir Pellias taught Launcelot all that was best of knighthood, both as to conduct of manner, and as to the worthiness and skill at arms, wherefore it was that when Launcelot was completely taught, there was no knight in all the world who was his peer in strength of arms or in courtesy of behavior, until his own son, Sir Galahad, appeared in the courts of chivalry as shall by and by be told of.

So when Launcelot came forth into the world again he became the greatest knight in all the history of chivalry, wherefore that prophecy of his mother was fulfilled as to his being like to a bright star of exceeding lustre.

Accordingly, I have herein told you with great particularity all these circumstances of his early history so that you may know exactly how it was that he was taken away into the lake, and why it was that he was afterward known as Sir Launcelot, surnamed of the Lake.

As to how he came into the world to achieve that greatness unto which he had been preordained, and as to how King Arthur made him knight, and as to many very excellent adventures that befell him, you shall immediately read in what followeth.

# PART I

# The Story of Launcelot

*H*ERE *beginneth the story of Sir Launcelot, surnamed of the Lake, who was held by all men to be the most excellent, noble, perfect knight-champion who was ever seen in the world from the very beginning of chivalry unto the time when his son, Sir Galahad, appeared like a bright star of extraordinary splendor shining in the sky of chivalry.*

*In this Book it shall be told how he was taken into a magic lake, how he came out thence to be made knight by King Arthur, and of how he undertook several of those adventures that made him at once the wonder and the admiration of all men, and the chiefest glory of the Round Table of Arthur-Pendragon.*

**ir Launcelot greets Queen Guinevere:**

# Chapter First

*How Sir Launcelot Came Forth From the Enchanted Castle of the Lake and Entered Into the World Again, and How King Arthur Made Him Knight.*

I KNOW not any time of the year that is more full of joyfulness than the early summer season; for that time the sun is wonderfully lusty and strong, yet not so very hot; that time the trees and shrubs are very full of life and very abundant of shade and yet have not grown dry with the heats and droughts of later days; that time the grass is young and lush and green, so that when you walk athwart the meadow-lands it is as though you walked through a fair billowy lake of magical *Of the spring-* verdure, sprinkled over with a great multitude of little flowers; *time of long ago.* that time the roses are everywhere a-bloom, both the white rose and the red, and the eglantine is abundant; that time the nests are brimful of well-fledged nestlings, and the little hearts of the small parent fowls are so exalted with gladness that they sing with all their mights and mains, so that the early daytime is filled full of the sweet jargon and the jubilant medley of their voices. Yea; that is a goodly season of the year, for though, haply, the spirit may not be so hilarious as in the young and golden spring-time, yet doth the soul take to itself so great a content in the fulness of the beauty of the world, that the heart is elated with a great and abundant joy that it is not apt to feel at another season.

Now it chanced upon the day before Saint John's day in the fulness of a summer-time such as this, that King Arthur looked forth from his chamber very early in the morning and beheld how exceedingly fair and very lusty was the world out-of-doors—all in the freshness of the young daylight. For the sun had not yet risen, though he was about to rise, and the sky was like to pure gold for brightness; all the grass and leaves and flowers were drenched with sweet and fragrant dew, and the birds were singing so vehemently that the heart of any man could not but rejoice in the fulness of life that lay all around about him.

There were two knights with King Arthur at that time, one was Sir Ewain, the son of Morgana le Fay (and he was King Arthur's nephew), and the other was Sir Ector de Maris, the son of King Ban of Benwick and of Queen Helen—this latter a very noble, youthful knight, and the youngest of all the Knights of the Round Table who were at that time elected. These stood by King Arthur and looked forth out of the window with him and they also took joy with him in the sweetness of the summer season. Unto them, after a while, King Arthur spake, saying: "Messires, *King Arthur* meseems this is too fair a day to stay within doors. For, *and two knights* certes, it is a shame that I who am a king should be prisoner *ride a-hunting.* within mine own castle, whilst any ploughman may be free of the wold and the green woods and the bright sun and the blue sky and the wind that blows over hill and dale. Now, I too would fain go forth out of doors and enjoy these things; wherefore I ordain that we shall go a-hunting this day and that ye and I shall start before any others of the lords and the ladies that dwell herein are awake. So let us take our horses and our hounds and let us take certain foresters and huntsmen, and let us go forth a-hunting into the green forest; for this day shall be holiday for me and for you and we shall leave care behind us, and for a while we shall disport ourselves in pleasant places."

So they all did as King Arthur bade; they made them each man ready with his own hands, and they bade the huntsmen and the foresters to attend thereupon as the King had ordained. Then they rode forth from the castle and out into the wide world that lay beyond, and it was yet so early in the morning that none of the castle folk were astir to know of their departure.

All that day they hunted in the forest with much joy and with great sport, nor did they turn their faces toward home again until the day was so far spent that the sun had sunk behind the tops of the tall leafy trees. Then, at that time, King Arthur gave command that they should bend their ways toward Camelot once more.

Now this time, being the Eve of Saint John, fairies and those folk who are fay come forth, as is very well known, into the world from which they dwell apart at other times. So when King Arthur and those two knights and their several foresters and huntsmen came to a certain outlying part of the forest, they were suddenly aware of a damsel and a dwarf waiting where the road upon which they were travelling crossed another road, and they perceived, from her very remarkable appearance, that the damsel was very likely Fay. For both she and her dwarf sat each *King Arthur* upon a milk-white horse, very strangely still, close to where *and his* was a shrine by a hedge of hawthorne; and the damsel was so *companions find* *a strange* wonderfully fair of face that it was a marvel to behold her. *damsel and a* Moreover, she was clad all in white samite from top to toe and *dwarf.* her garments were embroidered with silver; and the trappings and garniture of her horse were of white samite studded with bright silver bosses, wherefore, because of this silver, she glistered with a sudden lustre whensoever she moved a little. When King Arthur and the two knights who were with him drew nigh this damsel, much marvelling at her appearance, she hailed him in a voice that was both high and clear, crying: "Welcome, King Arthur! Welcome, King Arthur! Welcome, King Arthur!" saying three words three times; and "Welcome, Sir Ewain!" "Welcome, Sir Ector de Maris!" addressing each of those lords by his name.

"Damsel," quoth King Arthur, "it is very singular that you should know who we are and that we should not know you. Now, will you not tell us your name and whence you come and whither you go? For of a surety I believe you are Fay."

"Lord," said the damsel, "it matters not who I am, saving that I am of the court of a wonderful lady who is your very good friend. She hath sent me here to meet you and to beseech you to come with me whither I shall lead you, and I shall lead you unto her."

"Damsel," said King Arthur, "I shall be right glad to go with you as you desire me to do. So, if you will lead me to your lady, I and my knights will gladly follow you thitherway to pay our court unto her."

Upon this the damsel waved her hand, and drawing her bridle-rein she led the way, accompanied by the dwarf, and King Arthur *King Arthur* and the two knights followed her, and all their party of forest- *and his knights* *follow the* ers and huntsmen and hounds and beagles followed them. *damsel.*

By this time the sun had set and the moon had risen very fair and round and as yellow as gold, making a great light above the silent tree-tops. Everything now was embalmed in the twilight, and all the world was enshrouded in the mystery of the midsummer eve. Yet though the sun had gone the light was wonderfully bright, wherefore all that the eye could see stood sharp-cut and very clear to the vision.

So the damsel and the dwarf led the way for somewhat of a distance, though not for so very far, until they came of a sudden to where was an open meadow in the forest, hedged all around with the trees of the woodland. And here the King and his knights were aware of a great bustle of many people, some working very busily in setting up several pavilions of white samite, and others preparing a table as for a feast, and others upon this business and others upon that; and there were various sumpter-mules and pack-horses and palfreys all about, as though belonging to a party of considerable estate.

Then King Arthur and those who were with him beheld that, at some distance away upon the other side of the meadow, there were three people sitting under a crab-apple tree upon a couch especially prepared for them, and they were aware that these people were the chief of all that company.

The first party of the three was a knight of very haughty and noble appearance, clad all in armor as white as silver; and his jupon was white embroidered with silver, and the scabbard of the sword and the sword-belt were white, and his shield hung in the crab-tree above him and that, too, was all white as of silver. This knight still wore his helmet, so that his countenance was not to be seen. The second party of the three was a lady clad all in white raiment. Her face was covered by her wimple so that her countenance also was not to be seen very clearly, but her garments were of wonderful sort, being of white sarcenet embroidered over with silver in

*King Arthur and his companions are brought to speak with strange folk.* the pattern of lily flowers. Also she wore around her breast and throat a chain of shining silver studded with bright and sparkling gems of divers sorts. The third party of the three was a youth of eighteen years, so beautiful of face that it seemed to King Arthur that he had never beheld so noble a

being. For his countenance was white and shining, and his hair was as soft as silk and as black as it was possible to be, and curled down upon his shoulders; and his eyes were large and bright and extraordinarily black, and his eyebrows arched so smoothly that if they had been painted they could not have been marked upon his forehead more evenly than they were; and his lips, which pouted a little, though not very much, were as red as coral, and his upper lip was shaded with a soft down of black. Moreover, this youth was clad altogether in white cloth of satin with no ornaments whatsoever saving only a fine chain of shining silver set with opal-stones and emeralds that hung about his neck.

Then when King Arthur approached near enough he perceived by certain signs that the lady was the chiefest of those three, wherefore he paid his court to her especially, saying to her: "Lady, it seems that I have been brought hitherward unto you and that you were aware of my name and estate when you sent for me. Now I should be exceedingly glad if you would enlighten me in the same manner as to yourself."

"Sir," she said, "that I shall be glad to do; for if I have known you aforetime, you have also seen me afore time and have known me as your friend." Therewith the lady lowered the wimple from her face and King Arthur perceived that it was the Lady of the Lake.

Upon this he kneeled down upon one knee and took her hand and set it to his lips. "Lady," quoth he, "I have indeed cause to know you very well, for you have, as you affirm, been a friend to me and to my friends upon many several occasions." Then King Arthur turned to *King Arthur findeth Sir* that knight who was with that Lady of the Lake, and he said *Pellias again.* unto him: "Messire, if I mistake not, I should know you also; and I doubt not, if you will lift the umbril of your helmet, we shall all three know your face." Upon this the knight without more ado lifted his umbril as King Arthur had desired him to do and the three beheld that it was Sir Pellias, the Gentle Knight.

Now it hath already been very fully told about Sir Pellias in the Book of King Arthur, and those of you who read of him therein will remember, no doubt, how sorely he was wounded in a combat with Sir Gawaine, who was his best friend, and of how the Lady of the Lake took him to dwell with her in that wonderful city that was hidden by the appearance as of an enchanted lake, and of how it was Sir Gawaine who last beheld him upon that occasion. But if Sir Gawaine was the dearest friend that Sir Pellias had at that time, then Sir Ewain was only less dear to him. Therefore, when Sir Ewain beheld that the strange knight was Sir Pellias, he wist not what to think for pure wonder; for no mortal eyes had ever beheld Sir Pellias since he had gone into the lake with the Lady of the Lake that time as foretold, and it was not thought that anyone would ever see him again.

So when Sir Ewain beheld that the knight was Sir Pellias he emitted a great cry of joy and ran to him and catched him in his arms, and Sir Pellias forbade him not. For though at most times those who are of Faëry do not suffer themselves to be touched by mortal hands, yet, upon the Eve of Saint John's Day, fairies and mortals may commune as though they were of the same flesh and blood. Wherefore Sir Pellias did not forbid Sir Ewain, and they embraced, as one-time brethren-in-arms should embrace. And each kissed the other upon the face, and each made great joy the one over the other. Yea, so great was their joy that all those who stood about were moved with pure happiness at beholding them.

Then Sir Pellias came to King Arthur and kneeled down before him and kissed his hand, as is the bounden duty of every knight unto his lord. "Ha, Messire," quoth King Arthur, "methought when I beheld this lady, that you would not be very far distant from her." Then he said unto the Lady of the Lake: "Lady, I prithee tell me, who is this fair youth who is with you. For methinks I never beheld before so noble and so beautiful a countenance as his. Maybe you will make us acquainted with him also."

"Lord," said the Lady Nymue, "who he is, and of what quality, shall, I hope, be made manifest in due time; just now I would not wish that he should be known even unto you. But touching him, I may say that it was for his sake that I sent my damsel to meet you at the cross-roads awhile ago. But of that, more anon; for see! the feast is now spread which we have prepared for your entertainment. So let us first eat and drink and make merry together, and then we shall speak further of this matter."

So they all six went and sat down to the table that had been spread for them in the open meadow-land. For the night was very pleasant and warm and a wonderful full moon shone down upon them with a marvellous lustre, and there was a pleasant air, soft and warm, from the forest, and, what with the scores of bright waxen tapers that stood in silver candle-sticks upon the table (each taper sparkling as bright as any star), the night was made all illuminate like to some singular mid-day. There was set before them a plenty of divers savory meats and of several excellent wines, some as yellow as gold, and some as red as carbuncle, and they ate and they drank and they made merry in the soft moonlight with talk and laughter. Somewhiles they told Sir Pellias and the lady of all that was toward at court at Camelot; otherwhiles Sir Pellias and the lady told them such marvellous things concerning the land in which they two dwelt that it would be hard to believe that the courts of Heaven could be fairer than the courts of Fairyland whence they had come.

*The Lady of the Lake prepareth a feast for King Arthur.*

Then, after the feast was ended, the Lady of the Lake said to King Arthur: "Sir, an I have won your favor in any way, there is a certain thing I would ask of you." To the which King Arthur made reply: "Ask it, Lady, and it shall be granted thee, no matter what it may be." "Sir," said the Lady of the Lake, "this is what I would ask of you. I would ask you to look upon this youth who sits beside me. He is so dear to me that I cannot very well make you know how dear he is. I have brought him hither from our dwelling-place for one certain reason; to wit, that you should make him knight. That is the great favor I would ask of you. To this intent I have brought armor and all the appurtenances of knighthood; for he is of such noble lineage that no armor in the world could be too good for him."

"Lady," quoth King Arthur, "I will do what you ask with much pleasure and gladness. But, touching that armor of which you speak, it is my custom to provide anyone whom I make a knight with armor of mine own choosing."

To this the Lady of the Lake smiled very kindly, saying, "Lord, I pray you, let be in this case, for I daresay that the armor which hath been provided for this youth shall be so altogether worthy of your nobility and of his future credit that you will be entirely contented with it." And with that, King Arthur was altogether satisfied.

And, touching that armor, the ancient history that speaketh of these matters saith that it was of such a sort as this that followeth, and that it was brought from that enchanted court of the lake in this wise; to wit, in the front came two youths, leading two white mules, and the mules bore two chests studded with silver bosses. In one chest was the hauberk of that armor and in the other were the iron boots. These were bright like to silver and were inlaid with cunningly devised figures, all of pure gold. Next to them came two esquires, clad in white robes and mounted *Of the armor,* upon white horses, bearing the one a silver shield and the *etc., of Sir* other a shining helmet, as of silver—it likewise being very *Launcelot.* wonderfully inlaid with figures of pure gold. After these came two other esquires, the one bearing a sword in a white sheath embossed with studs of silver (the belt whereof was of silver with facets of gold) and the other leading a white charger, whose coat was as soft and as shining as silk. And all the gear and furniture of this horse was of silver and of white samite embellished with silver. So from this you can see how nobly that young acolyte was provided with all that beseemed his future greatness. For, as you may have guessed, this youth was Launcelot, King Ban's son of Benwick, who shortly became the greatest knight in the world.

Now there was in that part of the forest border a small abbey of monks, and in the chapel of that abbey Launcelot watched his armor for that night and Sir Ewain was with him for all that time. Mean- *Launcelot* time King Arthur and Sir Ector de Maris slept each in a silken *guards his* pavilion provided for them by the Lady of the Lake. *armor at night.*

In the morning Sir Ewain took Launcelot to the bath and bathed him, for such was the custom of those who were being prepared for knighthood.

Now, whilst Sir Ewain was bathing the youth, he beheld that on his shoulder was a mark in the likeness of a golden star and he marvelled very much thereat; but he made no mention of it at that time, but held his peace concerning what he saw; only he marvelled very greatly thereat.

Then, after Sir Ewain had bathed Launcelot, he clothed him in raiment fitted for that ceremony unto which he was ordained, and when the youth *King Arthur creates Launcelot a Knight-Royal.* was so clothed, Sir Ewain brought him to King Arthur, and King Arthur knighted Launcelot with great ceremony, and buckled the belt around him with his own hands. After he had done this Sir Ewain and Sir Ector de Maris set the golden spurs to his heels, and Sir Ector wist not that he was performing such office for his own brother.

So Sir Launcelot was made knight with great estate and ceremony, whereof I have told you all, unto every particular. For it is fitting that all things should be so told concerning that most great and famous knight.

After King Arthur had so dubbed Sir Launcelot knight, it was time that those two parties should part company—to wit, the party of the Lady of the Lake and the party of King Arthur. But when they were about to leave one another the Lady of the Lake took Sir Launcelot aside, and she spake to him after this manner:

"Launcelot, forget not that you are a king's son, and that your lineage is as noble as that of anyone upon earth—for so I have often told you aforetime. Wherefore, see to it that your worthiness shall be as great as your beauty, and that your courtesy and gentleness shall be as great as your prowess. To-day you shall go unto Camelot with King Arthur to make yourself known unto that famous Court of Chivalry. But do not tarry there, but, ere the night cometh, depart and go forth into the world to prove your knighthood as worthily as God shall give you grace to do. For I would not have you declare yourself to the world until you have *The Lady of the Lake gives Sir Launcelot good advice.* proved your worthiness by your deeds. Wherefore, do not yourself proclaim your name, but wait until the world proclaimeth it; for it is better for the world to proclaim the worthiness of a man than that the man should proclaim his own worthiness. So hold yourself ready to undertake any adventure whatsoever that God sendeth to you to do, but never let any other man complete a task unto which you yourself have set your hand." Then, after the Lady of the Lake had so advised Sir Launcelot, she kissed him upon the face, and therewith gave him a ring curiously wrought and set with a wonderful purple stone, which ring had such power that it would dissolve every enchantment. Then she said: "Launcelot, wear this ring and never let it be from off your finger." And Launcelot said: "I will do so." So Sir

Launcelot set the ring upon his finger and it was so that it never left his finger whilst he drew the breath of life.

Then King Arthur and Sir Ewain and Sir Ector de Maris and the young Sir Launcelot laid their ways toward Camelot. And, as they journeyed so together, Sir Ewain communicated privily to Sir Ector de Maris how that the youth had a mark as of a golden star upon the skin of his shoulder, and upon this news Sir Ector fell very silent. For Sir Ector knew that that sign was upon his own brother's shoulder, and he did not know how it could be upon the shoulder of any other man. Wherefore, he wist not what to think that it should be upon the shoulder of this youth. But he said naught of these thoughts to Sir Ewain, but held his peace.

So they reached Camelot whilst it was still quite early in the morning and all they who were there made great joy at the coming of so wonderfully fair and noble a young knight as Sir Launcelot appeared to be. Wherefore, there was great sound of rejoicing at his coming.

*Sir Launcelot cometh to Camelot.*

Then, after a while, King Arthur said: "Let us go and see if, haply, this youth's name is marked upon any of the seats of the Round Table, for I think it should be there." So all they of the court went to that pavilion afore described, where the Round Table was established, and they looked; and lo! upon the seat that King Pellinore had one time occupied was this name:

## THE KNIGHT OF THE LAKE.

So the name stood at first, nor did it change until the name of Sir Launcelot of the Lake became so famous in all the world. Then it became changed to this:

## SIR LAUNCELOT OF THE LAKE.

So Sir Launcelot remained at Camelot for that entire day and was made acquainted with a great many of the lords and ladies and knights and dames of King Arthur's court. And all that while he was like one that walked in a dream, for he had never before beheld anything of the world of mankind since he had been carried away into the lake, wherefore he wist not very well whether what he saw was real or whether he beheld it in a vision of enchantment. For it was all very new and wonderful to him and he took great delight in it because that he was a man and because this world was the world of mankind. Wherefore, though that Castle of the Lake was so

*Sir Launcelot becometh knight of the Round Table.*

beautiful, yet he felt his heart go forth to this other and less beautiful land as it did not go forth to that, because he was human and this was human.

Nevertheless, though that was so joyful a day for him, yet Sir Launcelot did not forget what the Lady of the Lake had said concerning the time he was to abide there! Wherefore, when it drew toward evening he besought leave of King Arthur to depart from that place in search of adventures, and King Arthur gave him leave to do as he desired.

So Sir Launcelot prepared to depart, and whilst he was in his chamber making ready there came in unto him Sir Ector de Maris. And Sir Ector said unto him: "Sir, I prithee tell me—is it true that you bear upon your right shoulder a mark like unto a golden star?" And Sir Launcelot made reply: "Yea, that is true." Then Sir Ector said: "I beseech you to tell me if your name is Launcelot." And Sir Launcelot said: "Yea, that is my name."

Upon this Sir Ector broke out into great weeping and he catched Sir Launcelot in his arms and he cried out: "Launcelot, thou art mine own brother! For thy father was my father, and my mother was thy mother! For we are both sons unto King Ban of Benwick, and Queen Helen was our

*Of the brother-hood of Sir Ector and Sir Launcelot.*

mother." Therewith he kissed Sir Launcelot with great passion upon the face. And Sir Launcelot upon his part kissed Sir Ector with a great passion of joy that he had found a brother in this strange world into which he had so newly come. But Sir Launcelot charged Sir Ector that he should say nothing of this to any man; and Sir Ector pledged his knightly word to that effect. (Nor did he ever tell anyone who Sir Launcelot was until Sir Launcelot had performed such deeds that all the world spake his name.)

For when Sir Launcelot went out into the world in that wise he undertook several very weighty achievements and brought them all to a successful issue, so that his name very quickly became known in every court of chivalry.

First he removed an enchantment that overhung a castle, hight Dolorous Gard; and he freed that castle and liberated all the sad, sorry captives that lay therein. (And this castle he held for his own and changed the name from Dolorous Gard to Joyous Gard and the castle became very famous afterward as his best-loved possession. For this was the first of all his possessions that he won by the prowess of his arms and he loved it best of all and considered it always his home.) After that Sir Launcelot, at the

*Of sundry adventures of Sir Launcelot.*

bidding of Queen Guinevere, took the part of the Lady of Nohan against the King of Northumberland, and he overcame the King of Northumberland and made him subject unto King Arthur. Then he overcame Sir Gallehaut, King of the Marches, and sent him captive to the court of King Arthur (and afterward Sir Gallehaut and Sir Launcelot became great friends for aye). So in a little while all the world spoke of Sir Launcelot, for it was said of him, and truly, that he had never been overcome by any other knight, whether upon horseback or upon foot, and that he always succeeded in every adventure which he undertook, whether that adventure were great or whether it were small. So it was as the Lady of the Lake desired it to be, for Sir Launcelot's name became famous, not because he was his father's son, but because of the deeds which he performed upon his own account.

So Sir Launcelot performed all these famous adventures, and after that he returned again to the court of King Arthur crowned with the glory of his successful knighthood, and there he was received with joy and acclaim and was duly installed in that seat of the Round Table that was his. And in that court he was held in the greatest honor and esteem of all the knights who were there. For King Arthur spake many times concerning him to this effect: that he knew not any honor or glory that could belong to a king greater than having such a knight for to serve him as was Sir Launcelot of the Lake. For a knight like Sir Launcelot came hardly ever into the world, and when he did come his glory must needs illuminate with its effulgence the entire reign of that king whose servant he was.

So it was that Sir Launcelot was greatly honored by everybody at the court of King Arthur, and he thereafter abided at that court for the most part of his life.

And now I must needs make mention of that friendship that existed betwixt Sir Launcelot and Queen Guinevere, for after he thus returned to the court of the king, they two became such friends that no two people could be greater friends than they were.

Now I am aware that there have been many scandalous things said concerning that friendship, but I do not choose to believe any such evil sayings. For there are always those who love to think and say evil things of others. Yet though it is not to be denied that Sir Launcelot never had for his lady any other dame than the Lady Guinevere, still no one hath ever said with truth that she regarded Sir Launcelot otherwise than as her very dear friend. For Sir Launcelot always avouched with *Of Sir* his knightly word, unto the last day of his life, that the Lady *Launcelot and* Guinevere was noble and worthy in all ways, wherefore I *Queen* *Guinevere.* choose to believe his knightly word and to hold that what he said was true. For did not he become an hermit, and did not she become a nun in their latter days, and were they not both broken of heart when

King Arthur departed from this life in so singular a manner as he did? Wherefore I choose to believe good of such noble souls as they, and not evil of them.

Yet, though Sir Launcelot thus abided at the court of the King, he ever loved the open world and a life of adventure above all things else.   For he *How Sir* had lived so long in the Lake that these things of the sturdy life *Launcelot dwelt* of out-of-doors never lost their charm for him.   So, though he *at Camelot.* found, for a while, great joy in being at the court of the King (for there were many jousts held in his honor, and, whithersoever he rode forth, men would say to one another: "Yonder goeth that great knight, Sir Launcelot, who is the greatest knight in the world"), yet he longed ever to be abroad in the wide world again.   So one day he besought King Arthur for leave to depart thence and to go forth for a while in search of adventures; and King Arthur gave him leave to do as he desired.

So now shall be told of several excellent adventures that Sir Launcelot undertook, and which he carried through with entire success, and to the great glory and renown of the Round Table, of which he was the foremost knight.

 ir Lionel of Britain.

# Chapter Second

*How Sir Launcelot and Sir Lionel Rode Forth Errant Together and How Sir Lionel Met Sir Turquine to His Great Dole. Also How Sir Ector Grieved for the Departure of His Brother Launcelot and So, Following Him, Fell into a Very Sorry Adventure.*

NOW after King Arthur had thus given Sir Launcelot leave to go errant and whilst Sir Launcelot was making himself ready to depart there came to him Sir Lionel, who was his cousin germain, and Sir Lionel besought leave to go with him as his knight-companion, and Sir Launcelot gave him that leave.

So when King Arthur confirmed Sir Launcelot's permission Sir Lionel also made himself ready very joyfully, and early of the morning of the next day they two took their leave of the court and rode away together; the day being very fair and gracious and all the air full of the joy of that season—which was in the flower of the spring-time.

*Sir Launcelot and Sir Lionel depart in search of adventure.*

So, about noon-tide, they came to a certain place where a great apple-tree stood by a hedge, and by that time they had grown an-hungered. So they tied their horses near-by in a cool and shady place and straightway sat them down under the apple-tree in the soft tall grass, which was yet fresh with the coolness of the morning.

Then when they had ended their meal Sir Launcelot said: "Brother, I have a great lust to sleep for a little space, for I find myself so drowsy that mine eyelids are like scales of lead." Unto which Sir Lionel made reply: "Very well; sleep thou for a while, and I will keep watch, and after that thou shalt watch, and I will sleep for a little space."

*Sir Launcelot sleepeth beneath an apple-tree.*

So Sir Launcelot put his helmet beneath his head and turned upon his side, and in a little had fallen into a sleep which had neither dream nor thought of any kind, but which was deep and pure like to a clear well of water in the forest.

And, whilst he slept thus, Sir Lionel kept watch, walking up and down in the shade of a hedge near-by.

Where they were was upon the side of a hill, and beneath them was a little valley; and a road ran through the valley, very white and shining in the sunlight, like a silken ribbon, and the road lay between growing fields of corn and pasture-land. Now as Sir Lionel walked beside the hedge he beheld three knights come riding into that valley and along that road with very great speed and in several clouds of dust; and behind them came a fourth knight, who was very huge of frame and who was clad altogether in black armor. Moreover, this knight rode upon a black horse and his shield was black and his spear was black and the furniture of his horse was black, so that everything appertaining to that knight was as black as any raven.

*Sir Lionel perceives how one knight pursues three knights.*

And Sir Lionel beheld that this one knight pursued those other three knights and that his horse went with greater speed than theirs, so that by and by he overtook the hindermost knight. And Sir Lionel beheld that the sable knight smote the fleeing knight a great buffet with his sword, so that that knight fell headlong from his horse and rolled over two or three times upon the ground and then lay as though he were dead. Then the black knight catched the second of the three, and served him as he had served his fellow. Then the third of the three, finding that there was no escape for him, turned as if to defend himself; but the black knight drave at him, and smote him so terrible a blow that I believe had a thunderbolt smitten him he would not have fallen from his horse more suddenly than he did. For, though that combat was full three furlongs away, yet Sir Lionel heard the sound of that blow as clearly as though it had been close by.

Then after the black knight had thus struck down those three knights he went to each in turn and tied his hands behind his back. Then, lifting each man with extraordinary ease, he laid him across the saddle of that horse from which he had fallen, as though he were a sack of grain. And all this Sir Lionel beheld with very great wonder, marvelling much at the strength and prowess of that black knight. "Ha," quoth he to himself, "I will go and inquire into this business, for it may haply be that yonder black knight shall not find it to be so easy to deal with a knight of the Round Table as with those other three knights."

So, with this, Sir Lionel loosed his horse very quietly and went his way so softly that Sir Launcelot was not awakened. And after he had gone some way, he mounted his steed and rode off at a fast gallop down into that valley.

When Sir Lionel had come to that place where the knight was, he found that he had just bound the last of the three knights upon the saddle of his horse as aforetold. So Sir Lionel spoke to the sable knight in this wise: "Sir, I pray you tell me your name and degree and why you treat those knights in so shameful a fashion as I behold you to do." *Sir Lionel addresses the sable knight.*

"Messire," said the black knight very fiercely, "this matter concerns you not at all; yet I may tell you that those knights whom I have overthrown are knights of King Arthur's court, and so I serve all such as come into this place. So will I serve you, too, if you be a knight of King Arthur's."

"Well," said Sir Lionel, "that is a very ungracious thing for you to say. And as for that, I too am a knight of King Arthur's court, but I do not believe that you will serve me as you have served those three. Instead of that, I have great hope that I shall serve you in such a fashion that I shall be able to set these knights free from your hands."

Thereupon, without more ado, he made him ready with spear and shield, and the black knight, perceiving his design, also made him ready. Then they rode a little distance apart so as to have a fair course for a tilt upon the roadway. Then each set spur to his horse and the two drave together with such violence that the earth shook beneath them. So they met fair in the middle of the course, but lo! in that encounter the spear of Sir Lionel broke into as many as thirty or forty pieces, but the spear of the black knight held, so that Sir Lionel was lifted clean out from his saddle and over the crupper of his horse with such violence that when he smote the ground he rolled three times over ere he ceased to fall. And because of that fierce, terrible blow he swooned away entirely, and all was black before his eyes, and he knew nothing. *The sable knight overcomes Sir Lionel*

Therewith the black knight dismounted and tied Sir Lionel's arms behind his back and he laid him across the saddle of his horse as he had laid those others across the saddles of their horses; and he tied him there very securely with strong cords so that Sir Lionel could not move.

And all this while Sir Launcelot slept beneath the apple-tree upon the hillside, for he was greatly soothed by the melodious humming of the bees in the blossoms above where he lay.

Now you are to know that he who had thus taken Sir Lionel and those three knights prisoner was one Sir Turquine, a very cruel, haughty knight, who had a great and strong castle out beyond the mouth of that valley in which these knights took combat as aforetold. Moreover, it was the

custom of Sir Turquine to make prisoner all the knights and ladies who
came that way; and all the knights and ladies who were not of King Arthur's
*Of Sir* court he set free when they had paid a sufficient ransom
*Turquine the* unto him; but the knights who were of King Arthur's court,
*sable knight.* and especially those who were of the Round Table, he held
prisoner for aye within his castle. The dungeon of that castle was a very
cold, dismal, and unlovely place, and it was to this prison that he proposed
to take those four knights whom he had overcome, with intent to hold them
prisoner as aforetold.

And now turn we to King Arthur's court and consider what befell there
after Sir Launcelot and Sir Lionel had left it in search of adventures.

When Sir Ector found that Sir Launcelot and Sir Lionel had gone away
in that fashion he was very much grieved in spirit; wherefore he said to
himself, "Meseems my brother might have taken me with him as well as
our cousin." So he went to King Arthur and besought his leave to quit
the court and to ride after those other two and to join in their
adventures, and King Arthur very cheerfully gave him that *Sir Ector*
leave. So Sir Ector made him ready with all despatch, and *follows Sir*
*Launcelot and*
rode away at a great gait after Sir Launcelot and Sir Lionel. *Sir Lionel.*
And ever as Sir Ector rode he made diligent inquiry and he found that
those two knights had ridden before him, so he said to himself: "By and by
I shall overtake them—if not to-day, at least by night, or by to-morrow
day."

But after a while he came to a cross-roads, and there he took a way that
Sir Launcelot and Sir Lionel had not taken; so that, after he had gone a
distance, he found that he had missed them by taking that road. Nevertheless, he went on until about the prime of the day, what time he met a
forester, to whom he said: "Sirrah, saw you two knights ride this way—
one knight clad in white armor with a white shield upon which was depicted
the figure of a lady, and the other knight clad in red armor with the figure
of a red gryphon upon his shield?" "Nay," said the forester, "I saw not
such folk." Then said Sir Ector, "Is there any adventure to be found
hereabouts?" Upon this the forester fell to laughing in great measure.
"Yea," he said, "there is an adventure to be found hard by *Sir Ector seeks*
and it is one that many have undertaken and not one yet hath *adventure.*
ever fulfilled." Then Sir Ector said, "Tell me what that adventure is and
I will undertake it."

"Sir," said the forester, "if you will follow along yonder road for a distance, you will find a very large, strong castle surrounded by a broad

moat.   In front of that castle is a stream of water with a fair, shallow ford, where the roadway crosses the water.   Upon this side of that ford there groweth a thorn-tree, very large and sturdy, and upon it hangs a basin of brass.   Strike upon that basin with the butt of your spear, and you shall presently meet with that adventure concerning which I have just now spoken."   "Fellow," said Sir Ector, "grammercy for your news."   And, therewith, straightway he rode off in search of that adventure.

He rode a great distance at a very fast gait and by and by he came to the top of a hill and therewith he saw before him the mouth of a fair valley. Across from where he stood was another hill not very large or high, but exceedingly steep and rocky.   Upon this farther hill was builded a tall, noble castle of gray stone with many towers and spires and tall chimneys and with several score of windows, all shining bright in the clear weather. A fair river ran down into the mouth of that valley and it was as bright and as smooth as silver, and on each side of it were smooth level meadow-lands —very green—and here and there shady groves of trees and plantations of fruit-trees.   And Sir Ector perceived that the road upon which he travelled crossed the aforesaid river by a shallow ford, and he wist that this must be the ford whereof the forester had spoken.   So he rode down unto that ford, and when he had come nigh he perceived the thorn-tree of which the forester had told him, and he saw that a great basin of brass hung to the thorn-tree, just as the forester had said.

Then Sir Ector rode to that thorn-tree and he smote upon that basin of brass with the butt of his spear, so that the basin rang with a noise like *Sir Ector* thunder; and he smote it again and again, several times over. *smites upon the* But though he was aware of a great commotion within that *brazen basin.* fair castle, yet no adventure befell him, although he smote the brazen basin several times.

Now, his horse being athirst, Sir Ector drove him into the ford that he might drink, and whilst he was there he was suddenly aware where, on the other side of the stream, was a singular party coming along the roadway. For first of all there rode a knight entirely clad in black, riding upon a black horse, and all the harness and furniture of that horse entirely of black. Behind him, that knight led four horses as though they were pack-horses, and across each one of those four horses was a knight in full armor, bound fast to the saddle like to a sack of grain, whereat Sir Ector was very greatly astonished.

As soon as that sable knight approached the castle, several came running forth and relieved him of those horses he led and took them into the castle, and as soon as he had been thus relieved the sable knight rode very violently up to where Sir Ector was. As soon as he had come to the water's edge he cried out: "Sir Knight, come forth from out of that water and do me battle."

"Very well," said Sir Ector, "I will do so, though it will, I think, be to thy very great discomfort."

With that he came quickly out from the ford, the water whereof was all broken and churned into foam at his passing, and straight-

*Sir Ector essays battle with the sable knight.*

way he cast aside his spear and drew his sword and, driving against that sable knight, he smote him such a buffet that his horse turned twice about.

"Ha," said the black knight, "that is the best blow that ever I had struck me in all of my life." Therewith he rushed upon Sir Ector, and without using a weapon of any sort he catched him about the body, under-neath the arms, and dragged him clean out of his saddle, and flung him across the horn of his own saddle. Thereupon, having accomplished this marvellous feat, and with Sir Ector still across his saddle-bow, he rode up unto his castle, nor stopped until he had reached the court-yard of the keep. There he set Sir Ector down upon the stone pavement. Then he said: "Messire, thou hast done to me this day what no other knight hath ever done to me before, wherefore, if thou wilt promise to be my man from henceforth, I will let thee go free and give thee great rewards for thy services as well."

But Sir Ector was filled very full of shame, wherefore he cried out fiercely, "Rather would I lie within a prison all my life than serve so catiff a knight as thou, who darest to treat other knights as thou hast just now treated me."

"Well," said the black knight very grimly, "thou shalt have thy choice." Therewith he gave certain orders, whereupon a great many fierce fellows set upon Sir Ector and stripped him of all his armor, and immediately haled him off, half-naked, to that dungeon aforementioned.

There he found many knights of King Arthur's court, and several of the Round Table, all of whom he knew, and when they beheld Sir Ector flung in unto them in that fashion they lifted up their voices in great lamentation that he should have been added to their number, instead of freeing them from their dolorous and pitiable case. "Alas," said they, "there is no knight alive may free us from this dungeon, unless it be Sir Launcelot. For this Sir Turquine is, certes, the greatest knight in all the world, unless it be Sir Launcelot."

*The sable knight makes prisoner of Sir Ector.*

Queen Morgana appears unto
Sir Launcelot.

# Chapter Third

*How Sir Launcelot was Found in a Sleep by Queen Morgana le Fay and Three Other Queens who were with Her, and How He was Taken to a Castle of Queen Morgana's and of What Befell Him There.*

SO Sir Launcelot lay in deep slumber under that apple-tree, and knew neither that Sir Lionel had left him nor what ill-fortune had befallen that good knight. Whilst he lay there sleeping in that wise there came by, along the road, and at a little distance from him, a very fair procession of lordly people, making a noble parade upon the highway. The chiefest of this company were four ladies, who were four *Four Queens and* queens. With them rode four knights, and, because the day *their courts pass* was warm, the four knights bore a canopy of green silk by the *by where Sir* *Launcelot lies* four corners upon the points of their lances in such wise as to *sleeping.* shelter those queens from the strong heat of the sun. And those four knights rode all armed cap-a-pie on four noble war-horses, and the four queens, bedight in great estate, rode on four white mules richly caparisoned with furniture of divers colors embroidered with gold. After these lordly folk there followed a very excellent court of esquires and demoiselles to the number of a score or more; some riding upon horses and some upon mules that ambled very easily.

Now all these folk of greater or lesser degree were entirely unaware that Sir Launcelot lay sleeping so nigh to them as they rode by chattering very gayly together in the spring-time weather, taking great pleasure in the warm air, and in growing things, and the green fields, and the bright sky; and they would have had no knowledge that the knight was there, had not Sir Launcelot's horse neighed very lustily. Thereupon, they were aware of the horse, and then they were aware of Sir Launcelot where he lay asleep under the apple-tree, with his head lying upon his helmet.

Now foremost of all those queens was Queen Morgana le Fay (who was King Arthur's sister, and a potent, wicked enchantress, of whom much hath been told in the Book of King Arthur), and besides Queen Morgana there was the Queen of North Wales, and the Queen of Eastland, and the Queen of the Outer Isles.

Now when this party of queens, knights, esquires, and ladies heard the war-horse neigh, and when they beheld Sir Launcelot where he lay, they drew rein and marvelled very greatly to see a knight sleeping so soundly at that place, maugre all the noise and tumult of their passing. So Queen Morgana called to her one of the esquires who followed after them, and she said to him: "Go softly and see if thou knowest who is yonder knight; but do not wake him."

So the esquire did as she commanded; he went unto that apple-tree and he looked into Sir Launcelot's face, and by hap he knew who it was because he had been to Camelot erstwhiles and he had seen Sir Launcelot at that

*An esquire knoweth Sir Launcelot.* place. So he hastened back to Queen Morgana and he said to her: "Lady, I believe that yonder knight is none other than the great Sir Launcelot of the Lake, concerning whom there is now such report; for he is reputed to be the most powerful of all the knights of King Arthur's Round Table, and the greatest knight in the world, so that King Arthur loves him and favors him above all other knights."

Now when Queen Morgana le Fay was aware that the knight who was asleep there was Sir Launcelot, it immediately entered her mind for to lay some powerful, malignant enchantment upon him to despite King Arthur. For she too knew how dear Sir Launcelot was to King Arthur, and so she had a mind to do him mischief for King Arthur's sake. So she went softly to where Sir Launcelot lay with intent to work some such spell upon him. But when she had come to Sir Launcelot she was aware that this purpose of mischief was not possible whilst he wore that ring upon his finger which the Lady of the Lake had given him; wherefore she had to put by her evil design for a while.

But though she was unable to work any malign spell upon him, she was

*Queen Morgana le Fay sets a mild enchantment upon Sir Launcelot.* able to cause it by her magic that that sleep in which he lay should remain unbroken for three or four hours. So she made certain movements of her hands above his face and by that means she wove the threads of his slumber so closely together that he could not break through them to awake.

After she had done this she called to her several of the esquires who were of her party, and these at her command fetched the shield of Sir Launcelot and laid him upon it. Then they lifted him and bore him away, carrying him in that manner to a certain castle in the forest that was no great distance away. And the name of that castle was Chateaubras and it was one of Queen Morgana's castles.

And all that while Sir Launcelot wist nothing, but lay in a profound sleep, so that when he awoke and looked about him he was so greatly astonished that he knew not whether he was in a vision or whether he was awake. For whilst he had gone asleep beneath that apple-tree, here he now lay in a fair chamber upon a couch spread with a coverlet of flame-colored linen. *Sir Launcelot awakens in a fair chamber.*

Then he perceived that it was a very fair room in which he lay, for it was hung all about with tapestry hangings representing fair ladies at court and knights at battle. And there were woven carpets upon the floor, and the couch whereon he lay was of carved wood, richly gilt. There were two windows to that chamber, and when he looked forth he perceived that the chamber where he was was very high from the ground, being built so loftily upon the rugged rocks at its foot that the forest lay far away beneath him like a sea of green. And he perceived that there was but one door to this chamber and that the door was bound with iron and studded with great bosses of wrought iron, and when he tried that door he found that it was locked.

So Sir Launcelot was aware from these things that he was a prisoner—though not a prisoner in a hard case—and he wist not how he had come thither nor what had happened to him.

Now when the twilight of the evening had fallen, a porter, huge of frame and very forbidding of aspect, came and opened the door of the chamber where Sir Launcelot lay, and when he had done so there entered a fair damsel, bearing a very good supper upon a silver tray. Moreover, she bore upon the tray three tapers of perfumed wax set in three silver candlesticks, and these gave a fair light to the entire room. But, when Sir Launcelot saw the maiden coming thus with intent to serve him, he arose and took the tray from her and set it himself upon the table; and for this civility the damsel made acknowledgement to him. *A fair damsel beareth light and food unto Sir Launcelot.* Then she said to him: "Sir Knight, what cheer do you have?" "Ha, damsel," said Sir Launcelot, "I do not know how to answer you that, for I wist not what cheer to have until I know whether I be with friends or with enemies. For though this chamber wherein I lie is very fair and well-bedight, yet meseems I must have been brought here by some enchantment, and that I am a prisoner in this place; wherefore I know not what cheer to take."

Then the damsel looked upon Sir Launcelot, and she was very sorry for him. "Sir," quoth she, "I take great pity to see you in this pass, for I hear tell you are the best knight in the world and, of a surety, you are of a very noble appearance. I must tell *The damsel has pity for Sir Launcelot.*

you that this castle wherein you lie is a castle of enchantment, and they who dwell here mean you no good; wherefore I would advise you to be upon your guard against them."

"Maiden," said Sir Launcelot, "I give you grammercy for your kind words, and I will be upon my guard as you advise me."

Then the damsel would have said more, but she durst not for fear that she should be overheard and that evil should befall her, for the porter was still without the door. So in a little she went away and Sir Launcelot was left alone.

But though the damsel bade Sir Launcelot have good cheer, yet he had no very good cheer for that night, as anyone may well suppose, for he wist not what was to befall him upon the morrow.

Now when the morning had come Sir Launcelot was aware of someone at his chamber door, and when that one entered it was Queen Morgana le Fay.

She was clad in all the glory at her command, and her appearance was so shining and radiant that when she came into that room Sir Launcelot knew not whether it was a vision his eyes beheld or whether she was a creature of flesh and blood. For she came with her golden crown upon her head, and her hair, which was as red as gold, was bound around with ribbons *How Queen* of gold; and she was clad all in cloth of gold; and she wore *Morgana* golden rings with jewels upon her fingers and golden bracelets *cometh to Sir* upon her arms and a golden collar around her shoulders; *Launcelot.* wherefore, when she came into the room she shone with an extraordinary splendor, as if she were a marvellous statue made all of pure gold—only that her face was very soft and beautiful, and her eyes shone exceedingly bright, and her lips, which were as red as coral, smiled, and her countenance moved and changed with all the wiles of fascination that she could cause it to assume.

When Sir Launcelot beheld her come thus gloriously into his room he rose and greeted her with a very profound salutation, for he was astonished beyond measure at beholding that shining vision. Then Queen Morgana gave him her hand, and he kneeled, and took her jewelled fingers in his and set her hand to his lips. "Welcome, Sir Launcelot!" quoth she; "welcome to this place! For it is indeed a great honor to have here so noble and famous a knight as you!"

"Ha, Lady," said Sir Launcelot, "you are gracious to me beyond measure! But I pray you tell me how I came to this place and by what means? For when I fell asleep yesterday at noon I lay beneath an apple-tree upon a hillside; and when I awoke—lo! I found myself in this fair chamber."

To this Queen Morgana le Fay made smiling reply as follows: "Sir, I am Queen Morgana le Fay, of whom you may have heard tell, for I am the sister of King Arthur, whose particular knight you are. Yesterday, at noon, riding with certain other queens and a small court of knights, esquires, and demoiselles, we went by where you lay sleeping. Finding you lying so, alone and without any companion, I was able, by certain *Queen Morgana* arts which I possess, to lay a gentle enchantment upon you so *seeks to beguile* that the sleep wherein you lay should remain unbroken for *Sir Launcelot.* three or four hours. So we brought you to this place in hopes that you would stay with us for two or three days or more, and give us the pleasure of your company. For your fame, which is very great, hath reached even as far as this place, wherefore we have made a gentle prisoner of you for this time being."

"Lady," said Sir Launcelot, "such constraint as that would be very pleasing to me at another time. But when I fell asleep I was with my cousin, Sir Lionel, and I know not what hath become of him, and haply he will not know what hath become of me should he seek me. Now I pray you let me go forth and find my cousin, and when I have done so I will return to you again at this place with an easy spirit."

"Well, Messire," said Queen Morgana, "it shall be as you desire, only I require of you some pledge of your return." (Herewith she drew from her finger a golden ring set very richly with several jewels.) "Now take this ring," she said, "and give me that ring which I see upon your finger, and when you shall return hither each shall have his ring again from the other."

"Lady," said Sir Launcelot, "that may not be. For this ring was placed upon my finger with such a pledge that it may never leave where it is whilst my soul abideth in my body. Ask of me any other pledge and you shall have it; but I cannot give this ring to you."

Upon this Queen Morgana's cheeks grew very red, and her eyes shone like sparks of fire. "Ha, Sir Knight," she said, "I do not think you are very courteous to refuse a lady and a queen so small a pledge as *Queen Morgana* that. I am much affronted with you that you should have *hath anger for* done so. Wherefore, I now demand of you, as the sister of *Sir Launcelot.* King Arthur whom you serve, that you give me that ring."

"Lady," said Sir Launcelot, "I may not do that, though it grieveth me much to refuse you."

Then Queen Morgana looked at Sir Launcelot awhile with a very angry countenance, but she perceived that she was not to have her will with him, wherefore she presently turned very quickly and went out of the room, leaving Sir Launcelot much perturbed in spirit. For he knew how great were the arts of Queen Morgana le Fay, and he could not tell what harm she might seek to work upon him by those arts. But he ever bore in mind how that the ring which he wore was sovereign against such malignant arts as she practised, wherefore he took what comfort he could from that circumstance.

Nevertheless, he abode in that chamber in great uncertainty for all that day, and when night came he was afraid to let himself slumber, lest they of the castle should come whilst he slept and work him some secret ill; wherefore he remained awake whilst all the rest of the castle slept. Now at the middle of the night, and about the time of the first cock-crow, he was aware of a sound without and a light that fell through the crack of the door. Then, in a little, the door was opened and there entered that young damsel who had served him with his supper the night before, and she bare a lighted taper in her hand.

When Sir Launcelot perceived that damsel he said: "Maiden, do you come hither with good intent or with evil intent?" "Sir," she said, "I come with good intent, for I take great pity to see you in such a sorry case as this. I am a King's daughter in attendance upon Queen Morgana le *The damsel* Fay, but she is so powerful an enchantress that, in good sooth, *cometh again to* I am in great fear lest she some time do me an ill-hap. So *Sir Launcelot.* to-morrow I leave her service and return unto my father's castle. Meantime, I am of a mind to help you in your adversity. For Queen Morgana trusts me, and I have knowledge of this castle and I have all the keys thereof, wherefore I can set you free. And I will set you free if you will, upon your part, serve me in a way that you can very easily do."

"Well," said Sir Launcelot, "provided I may serve you in a way fitting my knightly honor, I shall be glad to do so under any condition. Now I pray you tell me what it is you would have of me."

"Sir," said the damsel, "my father hath made a tournament betwixt him and the King of North Wales upon Tuesday next, and that is just a fortnight from this day. Now, already my father hath lost one such a tournament, for he hath no very great array of knights upon his side, and *The damsel* the King of North Wales hath three knights of King Arthur's *speaketh to Sir* Round Table to aid his party. Because of the great help of *Launcelot of her* *father, King* these knights of the Round Table, the King of North Wales *Bagdemagus.* won the last tournament and my father lost it, and now he feareth to lose the tournament that is to be. Now if you will enter upon my father's side upon the day of the tournament, I doubt not that he shall win that tournament; for all men say that you are the greatest knight in the world at this time. So if you will promise to help my father and will seal that promise with your knightly word, then will I set you free of this castle of enchantment."

"Fair maiden," said Sir Launcelot, "tell me your name and your father's name, for I cannot give you my promise until I know who ye be."

"Sir," said the demoiselle, "I am called Elouise the Fair, and my father is King Bagdemagus." "Ha!" quoth Sir Launcelot, "I know your father, and I know that he is a good king and a very worthy knight besides. If you did me no service whatsoever, I would, at your simple asking, were I free of this place, lend him such aid as it is in my power to give."

*Sir Launcelot promises to aid King Bagdemagus.*

At this the damsel took great joy and gave Sir Launcelot thanks beyond measure. So they spoke together as to how that matter might be brought about so that Sir Launcelot should be brought to talk to King Bagdemagus. And the damsel Elouise said: "Let it be this way, Sir Launcelot. Imprimis—thou art to know that somewhat of a long distance to the westward of that place where thou didst fall asleep yesterday, there standeth a very large, fair abbey known as the Abbey of Saint James the Lesser. This abbey is surrounded by an exceedingly noble estate that lieth all around about it so that no man that haps in that part of the country can miss it if he make inquiry for it. Now I will go and take lodging at that abbey a little while after I leave this place. So when it suits thee to do so, come thou thither and thou wilt find me there and I will bring thee to my father."

"Very well," said Sir Launcelot, "let it be that way. I will come to that place in good time for the tournament. Meantime, I prithee, rest in the assurance that I shall never forgot thy kindness to me this day, nor thy gracious behavior and speech unto me. Wherefore I shall deem it not a duty but a pleasure to serve thee."

So, having arranged all these matters, the damsel Elouise opened the door of that room and led Sir Launcelot out thence; and she led him through various passages and down several long flights of steps, and so brought him at last unto a certain chamber, where was his armor. Then the damsel helped Sir Launcelot to encase him in his armor, so that in a little while he was altogether armed as he had been when he fell asleep under that apple-tree. Thereafter the damsel brought him out past the court-yard and unto the stable where was Sir Launcelot's horse, and the horse knew him when he came. So he saddled the horse by the light of a half-moon which sailed like a boat high up in the sky through the silver, floating clouds, and therewith he was ready to

*The damsel bringeth Sir Launcelot to freedom.*

depart.    Then the damsel opened the gate and he rode out into the night, which was now drawing near the dawning of the day.

Thus Elouise the Fair aided Sir Launcelot to escape from that castle of enchantment, where else great ill might have befallen him.

And now it shall be told how Sir Launcelot did battle with Sir Turquine and of what happened thereat.

ir Launcelot doeth battle
with Sir Turquine.

# Chapter Fourth

*How Sir Launcelot Sought Sir Lionel and How a Young Damsel Brought Him to the Greatest Battle that Ever He Had in All His Life.*

SO Sir Launcelot rode through the forest, and whilst he rode the day began to break. About sunrise he came out into an open clearing where certain charcoal-burners were plying their trade.

To these rude fellows he appeared out of the dark forest like some bright and shining vision; and they made him welcome and offered him to eat of their food, and he dismounted and sat down with them and brake his fast with them. And when he had satisfied his hunger, he gave them grammercy for their entertainment, and took horse and rode away. *Sir Launcelot breaks his fast in the forest.*

He made forward until about the middle of the morning, what time he came suddenly upon that place where, two days before, he had fallen asleep beneath the blooming apple-tree. Here he drew rein and looked about him for a considerable while; for he thought that haply he might find some trace of Sir Lionel thereabouts. But there was no trace of him, and Sir Launcelot wist not what had become of him. *Sir Launcelot cometh again to the place of the apple-tree.*

Now whilst Sir Launcelot was still there, not knowing what to do to find Sir Lionel, there passed that way a damsel riding upon a white palfrey. Unto her Sir Launcelot made salutation, and she made salutation to him and asked him what cheer. "Maiden," said Sir Launcelot, "the cheer that I have is not very good, seeing that I have lost my companion-at-arms and know not where he is." Then he said: "Did you haply meet anywhere with a knight with the figure of a red gryphon upon his shield?" whereunto the damsel answered: "Nay, I saw none such." Then Sir Launcelot said: "Tell me, fair damsel, dost thou know of any adventure hereabouts that I may undertake? For, as thou seest, I am errant, and in search of such." *Sir Launcelot perceives a damsel upon a palfrey.*

Upon this the damsel fell a-laughing: "Yea, Sir Knight," said she, "I know of an adventure not far away, but it is an adventure that no knight yet that ever I heard tell of hath accomplished. I can take thee to that adventure if thou hast a desire to pursue it."

"Why should I not pursue it," said Sir Launcelot, "seeing that I am here for that very cause—to pursue adventure?"

"Well," said the damsel, "then come with me, Sir Knight, I will take thee to an adventure that shall satisfy thee."

So Sir Launcelot and that damsel rode away from that place together; he upon his great war-horse and she upon her ambling palfrey beside him. And the sun shone down upon them, very pleasant and warm, and all who passed them turned to look after them; for the maiden was very fair and slender, and Sir Launcelot was of so noble and stately a mien that few could behold him even from a distance without looking twice or three times upon him. And as they travelled in that way together they fell into converse, and the damsel said to Sir Launcelot: "Sir, thou appearest to be a very good knight, and of such a sort as may well undertake any adventure with great hope of success. Now I prithee to tell me thy name and what knight thou art."

*The damsel leads Sir Launcelot to an adventure.*

"Fair maiden," said Sir Launcelot, "as for telling you my name, that I will gladly do. I am called Sir Launcelot of the Lake, and I am a knight of King Arthur's court and of his Round Table."

At this the damsel was very greatly astonished and filled with admiration. "Hah!" quoth she, "it is a great pleasure to me to fall in with you, Sir Launcelot, for all the world now bespeaketh your fame. Little did I ever think to behold your person, much less speak with you, and ride in this way with you. Now I will tell you what this adventure is on which we are set; it is this—there is, some small distance from this, a castle of a knight hight Sir Turquine, who hath in his prison a great many knights of King Arthur's court, and several knights of his Round Table. These knights he keepeth there in great dole and misery, for it is said that their groans may be heard by the passers along the high-road below the castle. This Sir Turquine is held to be the greatest knight in the world (unless it be thou) for he hath never yet been overcome in battle, whether a-horseback or a-foot. But, indeed, I think it to be altogether likely that thou wilt overcome him."

*Sir Launcelot and the maiden discourse together.*

"Fair damsel," quoth Sir Launcelot, "I too have hope that I shall hold mine own with him, when I meet him, and to that I shall do my best endeavor. Yet this and all other matters are entirely in the hands of God."

Then the damsel said, "If you should overcome this Sir Turquine, I know of still another adventure which, if you do not undertake it, I know of no one else who may undertake to bring it to a successful issue."

Quoth Sir Launcelot, "I am glad to hear of that or of any other adventure, for I take great joy in such adventuring. Now, tell me, what is this other adventure?"

"Sir," said the damsel, "a long distance to the west of this there is a knight who hath a castle in the woods and he is the evilest disposed knight that ever I heard tell of. For he lurks continually in the outskirts of the woods, whence he rushes forth at times upon those who pass by. Especially he is an enemy to all ladies of that country, *The maiden tells Sir Launcelot* for he hath taken many of them prisoners to his castle and *of the savage forest knight.* hath held them in the dungeon thereof for ransom; and sometimes he hath held them for a long while. Now I am fain that thou undertake that adventure for my sake."

"Well," said Sir Launcelot, "I believe it would be a good thing for any knight to do to rid the world of such an evil-disposed knight as that, so if I have the good fortune to overcome this Sir Turquine, I give my knightly word that I will undertake this adventure for thy sake, if so be thou wilt go with me for to show me the way to his castle."

"That I will do with all gladness," said the damsel, "for it is great pride for any lady to ride with you upon such an adventure."

Thus they talked, and all was arranged betwixt them. And thus they rode very pleasantly through that valley for the distance of two leagues or a little more, until they came to that place where the road crossed the smooth stream of water afore told of; and there was the castle of Sir Turquine as afore told of; and there was the thorn-bush and the basin hanging upon the thorn-bush as afore told of. Then the maiden said: "Sir Launcelot, beat upon that basin and so thou shalt summon Sir Turquine to battle with thee."

So Sir Launcelot rode to that basin where it hung and he smote upon *Sir Launcelot* it very violently with the butt of his spear. And he smote *smites upon the* upon that basin again and again until he smote the bottom *basin.* from out it; but at that time immediately no one came.

Then, after a while, he was ware of one who came riding toward him, and he beheld that he who came riding was a knight very huge of frame, and long and strong of limb. And he beheld that the knight was clad entirely in black, and that the horse upon which he rode and all the furniture of the horse was black. And he beheld that this knight drave before him another horse, and that across the saddle of that other horse there lay an armed knight, bound hand and foot; and Sir Launcelot wist that the sable knight who came riding was that Sir Turquine whom he sought.

So Sir Turquine came very rapidly along the highway toward where Sir Launcelot sat, driving that other horse and the captive knight before him all the while. And as they came nearer and nearer Sir Launce-

*The sable knight bringeth Sir Gaheris captive.* lot thought that he should know who the wounded knight was; and when they came right close, so that he could see the markings of the shield of that captive knight, he wist that it was Sir Gaheris, the brother of Sir Gawaine, and the nephew of King Arthur, whom Sir Turquine brought thither in that wise.

At this Sir Launcelot was very wroth; for he could not abide seeing a fellow-knight of the Round Table treated with such disregard as that which Sir Gaheris suffered at the hands of Sir Turquine; wherefore Sir Launcelot rode to meet Sir Turquine, and he cried out: "Sir Knight! put that wounded man down from his horse, and let him rest for a while, and we two will prove our strength, the one against the other! For it is a shame for thee to treat a noble knight of the Round Table with such despite as thou art treating that knight."

"Sir," said Sir Turquine, "as I treat that knight, so treat I all knights of the Round Table—and so will I treat thee if thou be of the Round Table."

"Well," said Sir Launcelot, "as for that, I am indeed of the Round Table, and I have come hither for no other reason than for to do battle with thee."

"Sir Knight," said Sir Turquine, "thou speakest very boldly; now I pray thee to tell me what knight thou art and what is thy name."

"Messire," said Sir Launcelot, "I have no fear to do that. I am called Sir Launcelot of the Lake, and I am a knight of King Arthur's, who made me knight with his own hand."

"Ha!" said Sir Turquine, "that is very good news to me, for of all knights in the world thou art the one I most desire to meet, for I have looked for thee for a long while with intent to do battle with thee. For it was thou who didst slay my brother Sir Caradus at Dolorous Gard, who was held to be the best knight in all the world. Wherefore, because of this, I have the greatest despite against thee of any man in the world, and it was because of that despite that I waged particular battle against all the knights of King Arthur's court. And in despite of thee I now hold five score and eight knights, who are thy fellows, in the dismallest dungeon of my castle. Also I have to tell thee that among those knights is thine own brother, Sir Ector, and thy kinsman, Sir Lionel. For I overthrew Sir Ector and Sir Lionel only a day or two ago, and now they lie almost naked in the lower parts of that castle yonder. I will put down this knight as thou biddst me, and when I have done battle with thee I hope to tie thee on his saddle-horn in his place."

So Sir Turquine loosed the cords that bound Sir Gaheris and set him from off the horse's back, and Sir Gaheris, who was sorely wounded and very weak, sat him down upon a slab of stone near-by.

Then Sir Launcelot and Sir Turquine made themselves ready at all points, and each took such stand as seemed to him to be best; and when each was ready for the assault, each set spurs to his horse and rushed the one against the other with such terrible violence that they smote together like a clap of thunder. *Sir Launcelot and Sir Turquine do battle together.*

So fierce was that onset that each horse fell back upon the ground and only by great skill and address did the knight who rode him void his saddle, so as to save himself from a fall. And in that meeting the horse of Sir Turquine was killed outright and the back of Sir Launcelot's horse was broken and he could not rise, but lay like dead upon the ground.

Then each knight drew his sword and set his shield before him and they came together with such wrath that it appeared as though their fierce eyes shot sparks of fire through the occulariums of their helmets. So they met and struck; and they struck many scores of times, and their blows were so violent that neither shield nor armor could withstand the strokes they gave. For their shields were cleft and many pieces of armor were hewn from their limbs, so that the ground was littered with them. And each knight gave the other so many grim wounds that the ground presently was all sprinkled with red where they stood.

Now that time the day had waxed very hot, for it was come high noontide, so presently Sir Turquine cried out: "Stay thee, Sir Launcelot, for I have a boon to ask!" At this Sir Launcelot stayed his hand and said: "What is it thou hast to ask, Sir Knight?" Sir Turquine said: "Messire, I am athirst —let me drink." And Sir Launcelot said: "Go and drink."

So Sir Turquine went to that river and entered into that water, which was presently stained with red all about him. And he stooped where he stood and drank his fill, and presently came forth again altogether refreshed.

Therewith he took up his sword once more and rushed at Sir Launcelot and smote with double strength, so that Sir Launcelot bent before him and had much ado to defend himself from these blows.

Then by and by Sir Launcelot waxed faint upon his part and was athirst, and he cried out: "I crave of thee a boon, Sir Knight!" "What wouldst thou have?" said Sir Turquine. "Sir Knight," said Sir Launcelot, "bide while I drink, for I am athirst." "Nay," said Sir Turquine, "thou shalt not drink until thou quenchest thy thirst in Paradise." "Ha!" cried Sir Launcelot, "thou art a foul churl and no true knight. For when thou wert athirst, I let thee drink; and now that I am athirst, thou deniest me to quench my thirst."

Therewith he was filled with such anger that he was like one gone wode; wherefore he flung aside his shield and took his sword in both hands and rushed upon Sir Turquine and smote him again and again; and the blows he gave were so fierce that Sir Turquine waxed somewhat bewildered and bore aback, and held his shield low for faintness.

Then when Sir Launcelot beheld that Sir Turquine was faint in that wise, he rushed upon him and catched him by the beaver of his helmet and pulled *Sir Launcelot* him down upon his knees. And Sir Launcelot rushed Sir *overcometh Sir* Turquine's helmet from off his head. And he lifted his sword *Turquine.* and smote Sir Turquine's head from off his shoulders, so that it rolled down upon the ground.

Then for a while Sir Launcelot stood there panting for to catch his breath after that sore battle, for he was nearly stifled with the heat and fury thereof. Then he went down into the water, and he staggered like a drunken man as he went, and the water ran all red at his coming. And Sir Launcelot stooped and slaked his thirst, which was very furious and hot.

Thereafter he came up out of the water again, all dripping, and he went to where the damsel was and he said to her: "Damsel, lo, I have overcome Sir Turquine; now I am ready to go with thee upon that other adventure, as I promised thee I would."

At this the damsel was astonished beyond measure, wherefore she cried: "Sir, thou art sorely hurt, and in need of rest for two or three days, and maybe a long time more, until thy wounds are healed."

"Nay," said Sir Launcelot, "no need to wait; I will go with thee now."

Then Sir Launcelot went to Sir Gaheris—for Sir Gaheris had been sitting for all that while upon that slab of stone. Sir Launcelot said to Sir Gaheris: "Fair Lord, be not angry if I take your horse, for I must presently go with this damsel, and you see mine own horse hath broke his back."

"Sir Knight," said Sir Gaheris, "this day you have saved both me and my horse, wherefore it is altogether fitting that my horse or anything that is mine should be yours to do with as you please. So I pray you take my horse, only tell me your name and what knight you are; for I swear by my sword that I never saw any knight in all the world do battle so wonderfully as you have done to-day."

"Sir," said Sir Launcelot, "I am called Sir Launcelot of the Lake, and I am a knight of King Arthur's. So it is altogether fitting that I should do such service unto you as this, seeing that you are the brother of that dear knight, Sir Gawaine. For if I should not do this battle that I have done for your sake, I should yet do it for the sake of my lord, King Arthur, who is your uncle and Sir Gawaine's uncle." *Sir Launcelot makes himself known to Sir Gaheris.*

Now when Sir Gaheris heard who Sir Launcelot was, he made great exclamation of amazement. "Ha, Sir Launcelot!" he cried, "and is it thou! Often have I heard of thee and of thy prowess at arms! I have desired to meet thee more than any knight in the world; but never did I think to meet thee in such a case as this." Therewith Sir Gaheris arose, and went to Sir Launcelot, and Sir Launcelot came to him and they met and embraced and kissed one another upon the face; and from that time forth they were as brethren together.

Then Sir Launcelot said to Sir Gaheris: "I pray you, Lord, for to go up unto yonder castle, and bring succor to those unfortunates who lie therein. For I think you will find there many fellow-knights of the Round Table. And I believe that you will find therein my brother, Sir Ector, and my cousin, Sir Lionel. And if you find any other of my kindred I pray you to set them free and to do what you can for to comfort them and to put them at their ease. And if there is any treasure in that castle, I bid you give it unto those knights who are prisoners there, for to compensate them for the pains they have endured. Moveover, I pray you tell Sir Ector and Sir Lionel not to follow after me, but to return to court and wait for me there, for I have two adventures to undertake and I must essay them alone." *Sir Launcelot bids Sir Gaheris to free the castle captives.*

Then Sir Gaheris was very much astonished, and he cried out upon Sir Launcelot: "Sir! Sir! Surely you will not go forth upon another adventure at this time, seeing that you are so sorely wounded."

But Sir Launcelot said: "Yea, I shall go now; for I do not think that my wounds are so deep that I shall not be able to do my devoirs when my time cometh to do them."

At this Sir Gaheris was amazed beyond measure, for Sir Launcelot was very sorely wounded, and his armor was much broken in that battle, wherefore Sir Gaheris had never beheld a person who was so steadfast of purpose as to do battle in such a case.

So Sir Launcelot mounted Sir Gaheris' horse and rode away with that young damsel, and Sir Gaheris went to the castle as Sir Launcelot had bidden him to do. *Sir Launcelot departs with the damsel.*

In that castle he found five score and eight prisoners in dreadful case, for some who were there had been there for a long time, so that the hair of them had grown down upon their shoulders, and their beards had grown down upon their breasts. And some had been there but a short time, as *Sir Gaheris frees the castle captives.* was the case of Sir Lionel and Sir Ector. But all were in a miserable sorry plight; and all of those sad prisoners but two were knights of King Arthur's court, and eight of them were knights of the Round Table. All these crowded around Sir Gaheris, for they saw that he was wounded and they deemed that it was he had set them free, wherefore they gave him thanks beyond measure.

"Not so," said Sir Gaheris, "it was not I who set you free; it was Sir Launcelot of the Lake. He overcame Sir Turquine in such a battle as I never before beheld. For I saw that battle with mine own eyes, being at a little distance seated upon a stone slab and wounded as you see. And I make my oath that I never beheld so fierce and manful a combat in all of my life. But now your troubles are over and done, and Sir Launcelot greets you all with words of good cheer and bids me tell you to take all ease and comfort that you can in being free, and in especial he bids me greet you, Sir Ector, and you, Sir Lionel, and to tell you that you are to follow him no farther, but to return to court and bide there until he cometh; for he goeth upon an adventure which he must undertake by himself."

"Not so," said Sir Lionel, "I will follow after him, and find him." And *Sir Lionel and Sir Ector and Sir Kay follow after Sir Launcelot.* so said Sir Ector likewise, that he would go and find Sir Launcelot. Then Sir Kay the Seneschal said that he would ride with those two; so the three took horse and rode away together to find Sir Launcelot.

As for those others, they ransacked throughout the castle of Sir Turquine, and they found twelve treasure-chests full of treasure, both of silver and of gold, together with many precious jewels; and they found many bales of cloth of silk and of cloth of gold. So, as Sir Launcelot had bid them do so, they divided the treasure among themselves, setting aside a part for Sir Ector and a part for Sir Lionel and a part for Sir Kay. Then, whereas before they had been mournful, now they were joyful at having been made so rich with those precious things.

Thus happily ended that great battle with Sir Turquine which was very likely the fiercest and most dolorous fight that ever Sir Launcelot had in all of his life. For, unless it was Sir Tristram, he never found any other knight so big as Sir Turquine except Sir Galahad, who was his own son.

And now it shall be told how Sir Launcelot fared upon that adventure which he had promised the young damsel to undertake.

ir Launcelot sits with Sir
Hilaire and Croisette.

# Chapter Fifth

*How Sir Launcelot Went Upon an Adventure with the Damsel Croisette as Companion, and How He Overcame Sir Peris of the Forest Sauvage.*

NOW after Sir Launcelot had finished that battle with Sir Turquine as aforetold, and when he had borrowed the horse of Sir Gaheris, he rode away from that place of combat with the young damsel, with intent to carry out the other adventure which he had promised her to undertake.

But though he rode with her, yet, for a while, he said very little to her, for his wounds ached him sorely and he was in a great deal of pain. So, because of this, he had small mind to talk, but only to endure what he had to endure with as much patience as he might command. And *How Sir Launcelot was suffering and she was right sorry for him, pain him.* the damsel upon her part was somewhat aware of what Sir *celot's wounds* Launcelot was suffering and she was right sorry for him, wherefore she did not trouble him with idle discourse at that moment, but waited for a while before she spake.

Then by and by she said to him: "Messire, I would that thou wouldst rest for some days, and take thine ease, and have thy wounds searched and dressed, and have thy armor looked to and redded. Now there is a castle at some distance from this, and it is my brother's castle, and thither we may go in a little pass. There thou mayst rest for this night and take thine ease. For I know that my brother will be wonderfully glad to see thee because thou art so famous."

Then Sir Launcelot turned his eyes upon the damsel: "Fair maiden," quoth he, "I make confession that I do in sooth ache a very great deal, and that I am somewhat aweary with the battle I have endured this day. Wherefore I am very well content to follow thy commands in this matter. But I prithee, damsel, tell me what is thy name, for I know not yet how thou art called."

"Sir," she said, "I am called Croisette of the Dale, and my brother is called Sir Hilaire of the Dale, and it is to his castle that I am about to take thee to rest for this time."

Then Sir Launcelot said: "I go with thee, damsel, wherever it is thy will to take me."

So they two rode through that valley at a slow pace and very easily. And toward the waning of the afternoon they left the valley by a narrow

*Of how Sir Launcelot and the damsel ride together.*

side way, and so in a little while came into a shallow dale, very fertile and smiling, but of no great size. For the more part that dale was all spread over with fields and meadow-lands, with here and there a plantation of trees in full blossom and here and there a farm croft. A winding river flowed down through the midst of this valley, very quiet and smooth, and brimming its grassy banks, where were alder and sedge and long rows of pollard willows overreaching the water.

At the farther end of the valley was a castle of very comely of appear-

*Sir Launcelot and Croisette come to a fair valley.*

ance, being built part of stone and part of bright red bricks; and the castle had many windows of glass and tall chimneys, some a-smoke. About the castle and nigh to it was a little village of thatched cottages, with many trees in blossom and some without blossom shading the gables of the small houses that took shelter beneath them.

Now when Sir Launcelot and Croisette came into that little valley it was at the declining of the day and the sky was all alight with the slanting sun, and the swallows were flying above the smooth shining surface of the river in such multitudes that it was wonderful to behold them. And the lowing herds were winding slowly along by the river in their homeward way, and all was so peaceful and quiet that Sir Launcelot drew rein for pure pleasure, and sat for some while looking down upon that fair, happy dale. Then by and by he said: "Croisette, meseems I have never beheld so sweet and fair a country as this, nor one in which it would be so pleasant to live."

Upon this Croisette was very much pleased, and she smiled upon Sir Launcelot. "Think you so, Sir Launcelot?" quoth she. "Well, in sooth, I am very glad that this valley pleasures you; for I love it beyond any other place in all the world. For here was I born and here was I raised in that castle yonder. For that is my brother's castle and it was my father's castle before his time; wherefore meseems that no place in all the world can ever be so dear to my heart as this dale."

Thereupon they went forward up that little valley, and along by the smoothly flowing river, and the farther they went the more Sir Launcelot took pleasure in all that he beheld. Thus they came through the pretty village where the folk stood and watched with great admiration how that noble knight rode that way; and so they came to the castle and rode into the court-yard thereof. Then presently there came the lord of that castle, who was Sir Hilaire of the Dale. And Sir Hilaire greeted Sir Launcelot, saying: "Welcome, Sir Knight. This is great honor you do me to come into this quiet dale with my sister, for we do not often have with us travellers of such quality as you."

*Croisette bringeth Sir Launcelot to her brother's house.*

"Brother," said Croisette, "you may well say that it is an honor to have this knight with us, for this is none other knight than the great Sir Launcelot of the Lake. This day I beheld him overcome Sir Turquine in fair and honorable battle. So he doth indeed do great honor for to visit us in this wise."

Then Sir Hilaire looked at Sir Launcelot very steadily, and he said: "Sir Launcelot, your fame is so great that it hath reached even unto this peaceful outland place; wherefore it shall not soon be forgotten here how you came hither. Now, I pray you, come in and refresh yourself, for I see that you are wounded and I doubt not you are weary."

Upon this several attendants came, and they took Sir Launcelot and led him to a pleasant chamber. There they unarmed him and gave him a bath in tepid water, and there came a leech and searched his wounds and dressed them. Then those in attendance upon him gave him a soft robe of cloth of velvet, and when Sir Launcelot had put it on he felt much at ease, and in great comfort of body.

*Sir Launcelot is made at ease.*

By and by, when evening had fallen, a very good, excellent feast was spread in the hall of the castle, and there sat down thereto Sir Launcelot and Sir Hilaire and the damsel Croisette. As they ate they discoursed of various things, and Sir Launcelot told many things concerning his adventures, so that all who were there were very quiet, listening to what he said. For it was as though he were a visitor come to them from some other world, very strange and distant, of which they had no knowledge, wherefore they all listened so as not to lose a single word of what he told them. So that evening passed very pleasantly, and Sir Launcelot went to his bed with great content of spirit.

So Sir Launcelot abided for several days in that place until his wounds were healed. Then one morning, after they had all broken their fast, he made request that he and the damsel might be allowed to depart upon that adventure which he had promised her to undertake, and unto this Sir Hilaire gave his consent.

*How Sir Launcelot abides at the castle of Sir Hilaire.*

Now, during this while, Sir Launcelot's armor had been so pieced and mended by the armor-smiths of that castle that when he donned it it was, in a measure, as sound as it had ever been, and of that Sir Launcelot was very glad.   So having made ready in all ways he and Croisette took leave of that place, and all they who were there bade them adieu and gave Sir Launcelot God-speed upon that adventure.

Now some while after they left that dale they rode through a very ancient forest, where the sod was exceedingly soft underfoot and silent to the tread of the horses, and where it was very full of bursting foliage overhead. And as they rode at an easy pace through that woodland place they talked of many things in a very pleasant and merry discourse.

Quoth the damsel unto Sir Launcelot: "Messire, I take very great wonder that thou hast not some special lady for to serve in all ways as a knight should serve a lady."

"Ha, damsel," said Sir Launcelot, "I do serve a lady in that manner and she is peerless above all other ladies; for that lady is the Lady Guinevere,

*Sir Launcelot and Croisette discourse to- gether.* who is King Arthur's queen.   Yet though I am her servant I serve her from a very great distance.   For in serving her I am like one who standeth upon the earth, yet looketh upward ever toward the bright and morning star.   For though such an one may delight in that star from a distance, yet may he never hope to reach an altitude whereon that star standeth."

"Heyday!" quoth Croisette, "for that matter, there are other ways of serving a lady than that wise.   Were I a knight meseems I would rather serve a lady nearer at hand than at so great distance as that of which thou speakest.   For in most cases a knight would rather serve a lady who may smile upon him nigh at hand, and not stand so far off from him as a star in the sky."   But to this Sir Launcelot made no reply but only smiled.   Then in a little Croisette said: "Dost thou never think of a lady in that wise, Sir Launcelot?"

"Nay," said Sir Launcelot, "and neither do I desire so to serve any lady. For it is thus with me, Croisette—for all that while of my life until I was eighteen years of age I lived in a very wonderful land beneath a magical lake, of which I may not tell thee.   Then I came out of that lake and into this world and King Arthur made me a knight.   Now because I was so long absent from this world of mankind and never saw aught of it until I was grown into a man, meseems I love that world so greatly that I cannot

*Sir Launcelot speaketh of the Lady Guinevere.* tell thee how beautiful and wonderful it seems to me.   For it is so wonderful and so beautiful that methinks my soul can never drink its fill of the pleasures thereof.   Yea; methinks I love every blade of grass upon the fields, and every leaf upon every tree; and that I love everything that creepeth or that flyeth, so that when I am abroad under the sky and behold those things about me I am whiles like to weep for very joy of them. Wherefore it is, Croisette, that I would rather be a knight-errant in this world which I love so greatly than to be a king seated upon a throne with a golden crown upon my head and all men kneeling unto me. Yea; meseems that because of my joy in these things I have no room in my heart for such a love of lady as thou speakest of, but only for the love of knight-errantry, and a great wish for to make this world in which I now live the better and the happier for my dwelling in it. Thus it is, Croisette, that I have no lady for to serve in the manner thou speakest of. Nor will I ever have such, saving only the Lady Guinevere, the thought of whom standeth above me like that bright star afore spoken of."

"Ha," quoth Croisette, "then am I sad for the sake of some lady, I know not who. For if thou wert of another mind thou mightest make some lady very glad to have so great a knight as thou art to serve her." Upon this Sir Launcelot laughed with a very cheerful spirit, for he and the damsel were grown to be exceedingly good friends, as you may suppose from such discourse as this.

So they wended their way in this fashion until somewhat after the prime of day, and by that time they had come out of that forest and into a very rugged country. For this place into which they were now come was a sort of rocky valley, rough and bare and in no wise beautiful. *Sir Launcelot* When they had entered into it they perceived, a great way *perceives the* off, a castle built up upon the rocks. And that castle was *Castle of Sir* built very high, so that the roofs and the chimneys thereof *Peris.* stood wonderfully sharp and clear against the sky; yet the castle was so distant that it looked like a toy which you might easily take into your hand and hold betwixt your fingers.

Then Croisette said to Sir Launcelot: "Yonder is the castle of that evil-minded knight of whom I spake to thee yesterday, and his name is Sir Peris of the Forest Sauvage. Below that castle, where the road leads into that woodland, there doth he lurk to seize upon wayfarers who come thitherward. And indeed he is a very catiff knight, for, though he is strong and powerful, he doth not often attack other knights, but only ladies and demoiselles who come hither. For these he may take captive without danger to himself. For I believe that though he is so big of frame yet is he a coward in his heart."

Then Sir Launcelot sat for a while and regarded that castle, and fell into thought; and he said, "Damsel, if so be this knight is such a coward as thou sayest, meseems that if I travel with thee I shall have some ado to come

upon him; because, if he sees me with thee, he may keep himself hidden
in the thicket of the forest from my sight.   Now I will have it
this way; do thou ride along the highway in plain sight of
the castle, and I will keep within the woodland skirts, where I
may have thee in sight and still be hidden from the sight of others.   Then
if this knight assail thee, as I think it likely he may do, I will come out and
do battle with him ere he escapes.''

*Sir Launcelot advises Croisette what to do.*

So it was arranged as Sir Launcelot said and they rode in that wise:
Croisette rode along the highway, and Sir Launcelot rode under the trees
in the outskirts of the forest, where he was hidden from the eyes of anyone
who might be looking that way.   So they went on for a long pass until
they came pretty nigh to where the castle was.

Then, as they came to a certain part of the road that dipped down
toward a small valley, they were suddenly aware of a great noise, and
immediately there issued out from the forest a knight, large and strong of
frame, and followed close behind by a squire dressed altogether in scarlet
from head to foot.   This knight bore down with great speed
upon where Croisette was, and the esquire followed close
behind him.   When these two had come near to Croisette, the esquire leaped
from off his horse and caught her palfrey by the bridle, and the knight came
close to her and catched her as though to drag her off from her horse.

*Sir Peris attacks Croisette.*

With that Croisette shrieked very loud, and immediately Sir Launcelot
broke out from the woods and rode down upon where all this was toward
with a noise like to thunder.   As he came he cried aloud in a great and
terrible voice: "Sir Knight, let go that lady, and turn thou to me and
defend thyself!''

Then Sir Peris of the Forest Sauvage looked this way and that with intent
to escape, but he was aware that he could not escape from Sir Launcelot,
wherefore he took his shield in hand and drew his sword and put himself
into a position of defence; for, whereas he could not escape, he was, perforce, minded to do battle.   Then Sir Launcelot threw aside his spear, and
he set his shield before him and he took his sword in his hand,
and he drave his horse against Sir Peris.   And when he had
come nigh to Sir Peris he raised himself in his stirrups and
struck him such a buffet that I believe nothing in the world could withstand its force.   For though Sir Peris raised his shield against that blow,
yet the sword of Sir Launcelot smote through the shield and it smote down
the arm that held the shield, and it smote with such a terrible force upon the
helm of Sir Peris that Sir Peris fell down from his horse and lay in a swoon
without any motion at all.

*Sir Launcelot overthrows Sir Peris.*

Then Sir Launcelot leaped down from his horse and rushed off the helm of Sir Peris, and lifted his sword with intent to strike off his head.

Upon that the senses of Sir Peris came somewhat back to him, and he set his palms together and he cried out, though in a very weak voice: "Spare me, Sir Knight! I yield myself to thee!"

"Why should I spare thee?" said Sir Launcelot.

"Sir," said Sir Peris, "I beseech thee, by thy knighthood, to spare me."

"Well," said Sir Launcelot, "since thou hast besought me upon my knighthood I cannot do else than spare thee. But if I do spare thee, thou shalt have to endure such shame that any true knight in thy stead would rather die than be spared in such a manner."

"Sir Knight," said Sir Peris, "I am content with anything thou mayst do, so be that thou wilt spare my life."

Upon this Sir Launcelot bade Sir Peris rise. And he took the halter of Sir Peris's horse, and he bound Sir Peris's arms behind his back, and when he had done this he drove him up to his castle at the point of his lance. And when they came to the castle he bade Sir Peris have open the castle; and Sir Peris did so; and thereupon Sir Launcelot and Sir Peris entered the castle and the damsel and the squire followed after them.

In that castle were fourteen ladies of high degree held captive for ransom; and some of these had been there for a considerable time, to their great discomfort. All these were filled with joy when they were *Sir Launcelot* aware that Sir Launcelot had set them free. So they came *liberates the* to Sir Launcelot and paid their court to him and gave him *captive ladies.* great thanks beyond measure.

Sir Launcelot and Croisette abode in that castle all that night, and when the next morning had come Sir Launcelot made search all over that castle, and he found a considerable treasure of silver and gold, which had been gathered there by the ransom of the ladies and the *Sir Launcelot* damsels of degree whom Sir Peris had made prisoner afore- *gives the castle* time. All this treasure Sir Launcelot divided among those *treasure to the* ladies who were prisoners, and a share of the treasure he gave *captive ladies.* to the damsel Croisette, because that they two were such good friends and because Croisette had brought him thither to that adventure, and thereof Croisette was very glad. But Sir Launcelot kept none of that treasure for himself.

Then Croisette said: "How is this, Sir Launcelot? You have not kept any of this treasure for yourself, yet you won it by your own force of arms, wherefore it is altogether yours to keep if you will to do so."

"Croisette," said Sir Launcelot, "I do not care for such things as this treasure; for when I lived within that lake of which I have spoken to thee, such things as this treasure were there as cheap as pebbles which you may gather up at any river-bed, wherefore it has come to pass that such things have no value to me."

Now, after all this had been settled, Sir Launcelot had Sir Peris of the Forest Sauvage haled before him, and Sir Launcelot said: "Catiff Knight, now is it time for thy shame to come upon thee." Therewith he had Sir

*Sir Launcelot makes Sir Peris a dis- honored captive.* Peris stripped of all armor and raiment, even to his jerkin and his hose, and he had his arms tied behind his back, and he had a halter set about his neck; and Sir Launcelot tied the halter that was about the neck of Sir Peris to the horn of the saddle of his own horse, so that when he rode away with Croisette Sir Peris must needs follow behind him at whatever gait the horse of Sir Launcelot might take.

So Sir Launcelot and Croisette rode back to the manor of Sir Hilaire of the Dale with Sir Peris running behind them, and when they had come there Sir Launcelot delivered Sir Peris unto Sir Hilaire, and Sir Hilaire had Sir Peris bound upon a horse's back with his *Sir Hilaire sendeth Sir Peris to King Arthur.* feet underneath the belly of the horse; and sent him to Camelot for King Arthur to deal with him as might seem to the King to be fit.

But Sir Launcelot remained with Sir Hilaire of the Dale all the next day and he was very well content to be in that pleasant place. And upon the day after that, which was Sunday, he set forth at about the prime of the day to go to that abbey of monks where he had appointed to meet the damsel Elouise the Fair, as aforetold.

And now you shall hear how Sir Launcelot behaved at the tournament of King Bagdemagus, if it please you to read that which herewith immediately followeth.

ir Launcelot and Elouise the Fair.

# Chapter Sixth

*How Sir Launcelot Took Part in the Tournament Between King Bagdemagus and the King of North Wales, and How He Won that Battle for King Bagdemagus.*

SIR LAUNCELOT rode by many highways and many byways at a very slow pace, stopping now and then when it pleased him to do so, for he took great joy in being free in the open air again. For the day was warm and that time the clouds were very thick, drifting in great abundance across the sky. And anon there would fall a sudden shower of rain, and anon the sun would shine forth again, very warm and strong, so that all the world sparkled as with incredible myriads of jewels. Then the cock crowed lustily because the shower was past, and another cock answered him far away, and all the world suddenly smiled, and the water trickled everywhere, and the *How Sir Launcelot rode to find Elouise the Fair.* little hills clapped their hands for joy. So Sir Launcelot took great pleasure in the day and he went his way at so easy a pace that it was night-time ere he reached that abbey of monks where he was to meet Elouise the Fair.

Now that evening Elouise was sitting in a certain apartment of the abbey overlooking the court-yard, and a maiden was reading to her by the light of several waxen tapers from a book of painted pictures. And the maiden read in a voice that was both high and clear; meanwhile, Elouise sat very still and listened to what she read. Now while Elouise the Fair sat so, there was of a sudden the sound of a great horse coming on the stone pavement of the court below. Therewith Elouise arose hastily and ran to the window and looked down into that court-yard. Then she saw who he was that came, and that it was Sir Launcelot of the Lake. For the light was not yet altogether gone from the sky, which was all shining with gray, so that she could see who it was who came there.

Then Elouise gave great exclamation of joy, and clapped her hands. And she ran down to the court where Sir Launcelot was, and several of her maidens went with her.

When she had come to the court she gave great welcome to Sir Launcelot, and she summoned many attendants and she bade them look to Sir Launce-
*Elouise the Fair gives welcome to Sir Launcelot.* lot. So some of them aided Sir Launcelot to dismount and some took his horse, and some brought him up to a chamber that had been set apart for him, and there unarmed and served him, and set him at his ease.

Then Elouise sent to him a soft robe of purple cloth of velvet, lined with fur, and Sir Launcelot put it upon him and took great comfort in it.

After that Sir Launcelot descended to where Elouise was, and he found that a fair supper had been set for his refreshment. So he sat and ate, and Elouise the Fair herself served him.

Meanwhile she had sent for her father, King Bagdemagus, who was at
*Elouise sends for King Bagdemagus.* that time no great distance away, and a little after Sir Launce-lot had finished his supper King Bagdemagus came to that place, much wondering why Elouise had sent for him.

When King Bagdemagus came, Elouise took him by the hand and led him to Sir Launcelot, and she said: "Sire, here is a knight who, for my sake, is come to help you in this tournament upon Tuesday."

Now King Bagdemagus had never before seen Sir Launcelot, so he knew not who that knight was. Wherefore he said to him: "Messire, I am much beholden to you for coming to my aid in this battle. Now I pray you that you tell me your name and what knight you are."

"Lord," said Sir Launcelot, "I am hight Launcelot, and am surnamed 'He of the Lake.'"

Now when King Bagdemagus heard this he was astonished beyond measure, wherefore he cried out, "This is wonderful, that you who are the very flower of knighthood should be here, and that you should come to aid me in my battle!"

"Sire," said Sir Launcelot, "I know not how much aid I may be to thee until that matter is proven. But of a surety I owe it to this damsel to do what I am able at her request, in return for all that she hath done for
*Sir Launcelot talks with King Bagdemagus.* me to aid me in my time of great peril. So it is a very small repayment for me to aid thee, her father, in thy time of difficulties. Wherefore if, by good hap, I may be of use to thee in this battle which is nigh at hand, then I shall be glad beyond measure that I have paid some part of that debt which I owe to this lady."

"Messire," said King Bagdemagus, "I give thee grammercy for thy good will in this matter. I am sure that, with thy aid, I shall be successful in this battle, and that it will always be most renowned in the history of chivalry because thou hast taken part in it."

So spake they with great courtesy to one another. Then, by and by, Sir Launcelot said: "Sir, I pray you tell me who are those knights of King Arthur's court who are upon the part of the King of North Wales? For I would fain know against whom I am to do battle." To which King Bagdemagus said: "Messire, those three knights of the Round Table are as follows—there is Sir Mordred, nephew unto King Arthur, and there is Sir Galahantine, and there is Sir Mador de la Porte."

"Ha," quoth Sir Launcelot, "these are three very good knights indeed, and I am not at all astonished that the King of North Wales should have had such good fortune aforetime in that other tournament with you, seeing that he had three such knights as they to do battle upon his side."

After this they fell into discourse as to the manner in which they should do battle upon the morrow, and Sir Launcelot advised in this wise: "Lord, let me take three knights of yours, such as you trust, and such as you hold to be the strongest knights of your party. Let these three knights paint their shields altogether white and I will paint mine white, *Sir Launcelot* and then no man will know who we are. For I would have it *arranges the* so that I should not be known to be in this battle until I shall *order of battle* have approved myself in it. Now, when you have chosen those *Bagdemagus.* three knights, we four will take hiding in some wood or glade nigh to the place of combat, and when you are most busily engaged, and when you begin to be hard-pressed, then we will come forth and fall upon the flank of the party of the King of North Wales with intent to throw them into confusion. Then you will push your assault very hard, and I doubt not by the grace of God that we shall betwixt us be able to bear back their array in confusion."

This advice seemed very good to King Bagdemagus, and so he did as Sir Launcelot said. He chose him three very strong, worthy, honorable knights, and these made their shields white as Sir Launcelot directed.

Thus, all things being arranged as Sir Launcelot willed, it came to be the eve before the battle. So a little after sunset Sir Launcelot and those three knights whom King Bagdemagus had chosen rode over toward the place of tourney (which was some twelve miles from the abbey where the damsel Elouise was lodged). There they found a little woodland of tall, leafy trees fit for Sir Launcelot's purpose, and that wood stood to one side of the meadow of battle and at about the distance of three furlongs from it. In this little wood Sir Launcelot and the three knights-companion whom King Bagdemagus had chosen laid themselves down upon the ground and wrapped, each man, his cloak about him. So they slept there until the morrow, when the battle was ordained to be.

Now there had been very great preparation made for this tournament, for on three sides of the meadow of battle scaffolds had been built and rows of seats had been placed. These were covered over with tapestries and hangings of divers colors—some of figured and some of plain weaving—so that the green and level meadow-land was hung all about with these gay and gaudy colors.

Now when the morning had come, the folk who came to witness that tournament began to assemble from all directions—lords and ladies of high degree, esquires and damsels of lesser rank, burghers and craftsmen with their wives, townspeople from the town, yeomen from the woodlands, and freeholders from the farm crofts. With these came many knights of the two parties in contest, and with the knights came their esquires in attendance. Now these knights were all in full armor, shining very bright, and the esquires were clad in raiment of many textures and various colors, so that they were very gay and debonair. So, with all this throng moving along the highway toward the meadow of battle, it seemed as though the entire world was alive with gay and moving figures.

Now the place where Sir Launcelot and those three knights who were with him lay hidden was not far from the highway, so, whence they lay, they could see all that goodly procession of folk taking their way toward *Sir Launcelot* the lists, and they could look down upon the meadow of *and his com-* battle, which, as hath been said, was not more than three *panions lie near* *the place of* furlongs distant, and they could see the crowds of people of *tournament.* high and low degree taking their places upon those seats according to their rank and station. And they could see how the knights-contestant arrayed themselves upon this side of the field and upon that, and how the esquires and attendants hurried hither and thither, busying themselves in making their lords ready for the encounter that was soon to befall. Yea, all this could they see as plainly as though it lay upon the palm of a hand.

So they saw that about noontide all those who had come thither had taken their places, and that the field was clean, and that the two parties of combat were arrayed in order for battle.

Then Sir Launcelot perceived that the party of the King of North Wales was very much greater than the party of King Bagdemagus; for while the party of the King of North Wales had nigh eight score of helms, the party of King Bagdemagus had hardly four score of helms. So Sir Launcelot perceived that that party of King Bagdemagus would have much labor to do if it was to win in the battle.

Now, all being prepared, the marshal stood forth and blew upon his trumpet, and therewith those two parties of knights rushed the one against the other, each in so great a cloud of dust that one could hardly see *How the* the knights in their passage. Therewith they met in the midst *battle began.* of the meadow of battle, with such a crash and uproar of splintered lances as was terrible to hear.

And for a while no man could see what was toward, so great was the dust and the tumult. But by and by the dust raised itself a little and then Sir Launcelot perceived that the party of King Bagdemagus had been pushed back by that other party, as might have been supposed in such a case.

So Sir Launcelot looked upon the battle for some while and he saw that the party of King Bagdemagus was pushed farther and farther back. Then by and by Sir Launcelot said to his knights-companion: "Messires, methinks now is our time to enter this engagement."

Therewith he and they rode forth out of that woods, and they rode down the hill and across the fields and so came into that meadow-of-battle.

At that time the party of the King of North Wales was so busily engaged in its assault upon the party of King Bagdemagus that very few of those knights engaged were aware of those four knights coming, and those who were aware of them thought but very little of the coming of *Sir Launcelot* so small a number. So no one interfered with their coming, *and his com-* wherefore they were able to bear down with great speed upon *panions enter* the flank of the party of the King of North Wales. Therewith *the battle.* they struck that flank with such force that both horses and horsemen were overturned by their assault.

In that encounter Sir Launcelot carried a spear that was wonderfully strong and tough. With it he ran with great fierceness into the very thickest of the press, and before he was checked he struck down five knights with that one spear. And likewise those three knights that were with him did such good service that all that flank of the party of the King of North Wales was thrown into great confusion and wist not what to do for to guard themselves against that fierce, furious onset.

Then Sir Launcelot and his three companions bore back a little, and when they got their distance they ran again into the press, and this time Sir Launcelot overthrew the King of North Wales himself, and that with such violence that the bone of his thigh was broken, and he had to be carried away out of that field by his attendants. And in this second assault Sir Launcelot and the three knights who were with him overthrew eleven knights besides the King of North Wales, wherefore all that part of the press began to break away from them and to seek some place where they could defend themselves from such another assault.

Now when the party of King Bagdemagus saw into what confusion the other party were thrown by these four knights-champion, they began a very fierce and furious attack, and with such vehemence that in a little the party of the King of North Wales began to bear back before them. So, what with those who withdrew before Sir Launcelot's assault, and what with those who withdrew from the assault of King Bagdemagus, there was a great deal of confusion in the ranks of the party of the King of North Wales.

Now those three knights who were of King Arthur's court perceived how Sir Launcelot and his knights-companion were throwing the ranks of the party of the King of North Wales into confusion, and they knew that unless the onset of Sir Launcelot was checked, the day would of a surety be lost unto them. Wherefore said Sir Mador de la Porte: "Yonder is a very strong and fierce-fighting knight; if we do not check his onset we will very likely be brought to shame in this battle." "Yea," said Sir Mordred, "that is so. Now I will take it upon me to joust with that knight and to

*Sir Launcelot overthrows Sir Mordred.* overthrow him." Upon that those other two knights bade him go and do as he said. So Sir Mordred made way to where Sir Launcelot was, coming forward very fiercely and with great violence, and Sir Launcelot was aware of Sir Mordred's coming and made him ready for that assault. So the two came together with terrible violence and Sir Launcelot struck Sir Mordred such a buffet that the breast-band of Sir Mordred's saddle brake, and both the saddle and Sir Mordred flew over his horse's tail. Therewith Sir Mordred fell upon his head and struck with such violence upon the ground that his neck was nigh broken, and he lay altogether in a dead swoon and had to be carried out of the lists by his attendants.

This saw Sir Mador de la Porte, and he cried out: "Ha! see what hath befallen Sir Mordred!" And therewith he also bare down upon Sir Launcelot with all his might and main with intent to overthrow him. And Sir

*Sir Launcelot overthrows Sir Mador.* Launcelot ran against him, and they struck together so fiercely that it was terrible to behold. But the spear of Sir Mador de la Porte burst into pieces, whilst the spear of Sir Launcelot held, so that both Sir Mador and his horse were overthown, the horse rolling upon the man. And in that encounter Sir Mador's shoulder went out of place, and he also had to be borne away by his attendants.

Then Sir Galahantine took a great spear from his esquire, who was nigh him, and he also ran against Sir Launcelot with all his might; and Sir Launcelot met him in full course and that onset was more terrible than either of the other two. For the spear of each knight was burst into splinters, even to the butt thereof. Then each threw away the butt of his spear and drew out his sword, and Sir Galahantine struck Sir Launcelot such a blow that the legs of Sir Launcelot's horse trembled under him because of the weight of that stroke. At this Sir Launcelot waxed wroth beyond measure and he rose in his stirrups and he smote Sir Galahantine such a buffet that the blood burst out from his nose and his ears, and all his senses so went away from him that he might hardly behold the light of day because of the swimming of his sight.

*Sir Launcelot strikes Sir Galahantine a sad blow.*

Therewith Sir Galahantine's head hung down upon his breast and he had no power to guide his horse, wherefore his horse made way out of the press and galloped off, bearing Sir Galahantine away, whether he would or no. And after the horse had galloped a little distance Sir Galahantine could not any longer sit upon his saddle, but he fell off of his horse and rolled over upon the ground and had not strength to rise therefrom.

Then Sir Launcelot catched another spear, great and strong, from the esquire who followed him, and before ever that spear broke he overthrew sixteen knights therewith. Wherefore all who beheld him were amazed and terrified at what he did.

By now the party of the King of North Wales began to bear more and more aback and in a little they broke, and then the party of King Bagdemagus pursued them hither and thither, and those who did not surrender were overthrown so that it was not possible for them to make any new order of battle. Then that party surrendered itself as conquered, one and all, and so King Bagdemagus won that tournament with the greatest glory that it was possible for him to have. For it had never been heard of before that a party of four-score knights should overcome in that way a party of eightscore knights, with three knights of the Round Table to champion them. Nor would such a victory have been possible only for what Sir Launcelot did in that battle.

*Sir Launcelot wins the battle for King Bagdemagus.*

So Sir Launcelot won that tournament for King Bagdemagus, and after the battle was over and done King Bagdemagus came to Sir Launcelot and said to him: "Messire, thou hast brought to me the greatest glory this day that ever fell to my lot in all of my life. Now I prithee come with me and refresh thyself with me, so that I may give thee fitting thanks for all thou hast done, and so that I may reward thee in such a way as is fit for a king to reward a knight-champion such as thou art."

Unto this Sir Launcelot made reply: "Lord, I give you thanks for *Sir Launcelot departs without reward.* your courtesy, but I need no reward; for it is meet that I should have done what I could for the sake of the demoiselle Elouise the Fair, seeing that she rescued me from the mischiefs that Queen Morgana had intent to do me."

Then King Bagdemagus besought Sir Launcelot that he would tarry awhile and rest, but Sir Launcelot would not do so, but would be going upon his way without any tarrying. But he said to King Bagdemagus: "I prithee greet your daughter for me, and say to her that if ever she hath need of my services again let her send to me, and I will come to her even if it be to the end of the earth. For I have not yet repaid her for what she hath done for me."

Therewith Sir Launcelot went his way from that meadow of battle, and, coming to the skirts of the forest he entered therein, and those who were there at the meadow of battle did not see him any more.

So endeth the history of that famous tournament betwixt King Bagdemagus and the King of North Wales.

# Sir Launcelot climbs to catch the lady's falcon

# Chapter Seventh

*How Sir Launcelot Fell Into the Greatest Peril that Ever He Encountered in all His Life. Also How He Freed a Misfortunate Castle and Town From the Giants Who Held Them, and How He Released the Lord Thereof From a Dungeon.*

NOW Sir Launcelot wandered errant for many days, meeting no adventure of any moment, but taking great joy in all that he beheld of the wide world about him, and in that time he found lodging wheresoever he chanced to be (if not in house, then beneath the skies), and he endured all sorts of weather, both wet and dry.

Upon a certain day, in the prime of the morning, he came across a hilltop, and beheld beneath him a valley, very fertile and well-tilled, with fields and meadow-lands spread all over it like to a fair green carpet woven in divers patterns. And in the midst of the *Sir Launcelot cometh to a fair* valley was a very large and noble castle, with many towers, *valley with* and tall, steep roofs, and clustering chimneys. So Sir Launce- *a castle.* lot descended into that valley, and the road which he took ended in front of the castle and under the shade of the tall gray walls thereof. But he did not stop at that castle but went on by it.

Now after Sir Launcelot had passed by that castle it seemed to him that he heard very delicate silver bells ringing sweetly in the air above him, and when he looked up he beheld that a falcon was flying over his head toward a high elm tree that stood at a little distance, and he wist that it was the bells upon the cap of the falcon that rang so sweetly. And Sir Launcelot beheld that long lunes hung from the feet of the falcon as she flew, wherefore he was aware that the falcon had slipped her lunes and had flown from her owner.

So Sir Launcelot watched the falcon, and he beheld that she lit in a tall elm tree, where she took her perch and rested, balancing with her wings part spread. Then by and by she would have taken her *Sir Launcelot* flight again, but the lunes about her feet had become en- *beholdeth a fal-* tangled around the bough on which she sat, so that when she *con entangled.* would have flown she could not do so. Now Sir Launcelot was very sorry to see the falcon beating herself in that wise, straining to escape from where she was prisoner, but he knew not what to do to aid her, for the tree was very high, and he was no good climber of trees.

While he stood there watching that falcon he heard the portcullis of the castle lifted, with a great noise, and the drawbridge let fall, and therewith there came a lady riding out of the castle very rapidly upon a white mule, and she rode toward where Sir Launcelot watched the falcon upon the tree. When that lady had come nigh to Sir Launcelot, she cried out to him: "Sir Knight, didst thou see a falcon fly this way?" Sir Launcelot said: "Yea, Lady, and there she hangs, caught by her lunes in yonder elm-tree."

Then when that lady beheld how that her falcon hung there she smote her hands together, crying out: "Alas, alas! what shall I do? That falcon is my lord's favorite hawk! While I was playing with her a while since, she slipped from me and took flight, and has sped as thou dost see. Now when my lord findeth that I have lost his hawk in that wise he will be very angry with me, and will haply do me some grievous hurt."

Quoth Sir Launcelot: "Lady, I am very sorry for you." "Sir," she said, "it boots nothing for you to be sorry for me unless you can aid me." "How *The Lady* may I aid you in this?" said Sir Launcelot. "Messire," quoth *beseeches Sir* she, "how otherwise could you aid me than by climbing up *Launcelot to get* into this tree for my hawk? For if you aid me not in such a *her the falcon* *again.* fashion, I know not what I shall do, for my lord hath a very hot and violent temper, and he is not likely to brook having his favorite hawk lost to him, as it is like to be."

Upon this Sir Launcelot was put to a great pass and knew not what to do, for he had no good mind to climb that tree. "Lady," quoth he, "I prithee tell me what is thy lord's name." "Messire," she replied, "he is hight Sir Phelot, and is a knight of the court of the King of North Wales."

"Well, Lady," said Sir Launcelot, "thou dost put upon me a very sore task in this, for God knoweth I am no climber of trees. Yea, I would rather do battle with twenty knights than to climb one such tree as this. Ne'ertheless, I cannot find it in me to refuse the asking of any lady, if so be it lieth at all in my power to perform her will. Now if you will aid me to unarm myself, I will endeavor to climb this tree and get your hawk."

So the lady dismounted from her mule, and Sir Launcelot dismounted from his horse, and the lady aided Sir Launcelot to unarm himself. And *Sir Launcelot* when he had unarmed himself he took off all his clothes sav- *climbs the tree.* ing only his hosen and his doublet. Then he climbed that tree, though with great labor and pain to himself, and with much dread lest he should fall. So he, at last, reached the falcon where it was, and he loosened the lunes from where they were entangled about the branch, and he freed the bird. Then he brake off a great piece of rotten bough of the tree and he tied the lunes of the falcon to it and he tossed the falcon down to where the lady was; and the lady ran with great joy and caught the falcon and loosed it from the piece of branch and tied the lunes to her wrist, so that it could not escape again.

Then Sir Launcelot began to descend the tree with as great labor and pain as he had climbed into it.

But he had not come very far down when he perceived a knight who came riding very rapidly toward that tree, and he saw that the knight was in full armor. When this knight came to the tree he drew rein and bespoke the lady who was there, though Sir Launcelot could not hear what he said. So, after he had spoken for a little, the knight dismounted from his horse and went to Sir Launcelot's shield and looked upon *Sir Phelot* the face of it very carefully. Then presently he looked up- *threatens Sir* ward toward Sir Launcelot, and he said: "Art thou Sir Laun- *Launcelot's life.* celot of the Lake?" And Sir Launcelot said: "Yea." "Very well," said the knight, "I am pleased beyond measure at that. For I am Sir Phelot, the lord of this castle, and the brother of that Sir Peris of the Forest Sauvage, whom thou didst treat so shamefully after thou hadst overcome him in battle."

"Sir," said Sir Launcelot, "I treated him nowise differently from what he deserved." "No matter for that," said Sir Phelot, "he was my brother, and thou didst put great despite and shame upon him. So now I will be revenged upon thee, for now I have thee where I would have thee, and I will slay thee as shamefully as thou didst put shame upon him. So say thy prayers where thou art, for thou shalt never go away from this place alive."

"Sir Knight," said Sir Launcelot, "I do not believe that thou wouldst really assault a naked and harmless man, for it would certainly be a great shame to thee to do me a harm in that wise. For lo! thou art armed in full, and I am a naked man, and to slay me as I am would be both murder and treason."

"No matter for that," said Sir Phelot; "as for the shame of it, I take no thought of it. I tell thee thou shalt have no grace nor mercy from me. Wherefore make thy peace with Heaven, for thine hour is come."

"Sir Knight," said Sir Launcelot, "I ask only one boon of thee; if thou art of a mind to take so much shame upon thee, as appears to be the case, let me not, at least, die like a felon without any weapon. Let me have my sword in my hand, even if I have no other defence. For if a knight must die, it is a shame for him to die without weapons. So hang my sword upon yonder bough, where I may reach it, and then thou mayst slay me."

"Nay," said Sir Phelot, "I will not do that, for I know very well how wonderful is thy prowess. Wherefore I believe that even if thou wert otherwise unarmed thou mightst overcome me if thou hadst thy sword. So I will give thee no such chance, but will have my will of thee as thou art."

Then Sir Launcelot was put to a great pass of anxiety, for he wist not what to do to escape from that danger in which he lay. Wherefore he looked all about him and above him and below him, and at last he beheld *Sir Launcelot is put to a sad pass to escape.* a great branch of the elm tree just above his head, very straight and tough. So he catched this branch and broke it off from the tree and shaped it to a club of some sort. Then he came lower, and the knight waited to strike him with his sword, when he was low enough; but Sir Launcelot did not come low enough for that.

Then Sir Launcelot perceived that his horse stood below him and a little to one side, so of a sudden he ran out along the branch whereon he stood and he leaped quickly down to the earth upon the farther side of his horse from where the knight stood.

At this Sir Phelot ran at him and lashed at him with his sword, thinking to slay him before he had recovered from his leap. But Sir Launcelot was quicker than he, for he recovered his feet and put away the blow of Sir Phelot with his club which he held. Then he ran in upon Sir Phelot under his sword arm, and before he could use his sword he struck Sir Phelot with *Sir Launcelot overcomes Sir Phelot with a strange weapon.* all his might upon the side of his head. And he struck him very quickly again, and he struck him the third time, all in the space whilst one might count two. And those blows he struck were so direful that Sir Phelot fell down upon his knees, all stunned and bedazed, and the strength went out of his thews because of faintness. Then Sir Launcelot took the sword out of the hand of Sir Phelot and Sir Phelot did not have strength to deny him. And Sir Launcelot plucked off Sir Phelot's helm and catched him by the hair and dragged his neck forward so as to have ease to strike his head from off his body.

Now all this while the lady had been weeping and watching what befell. But when she saw the great danger Sir Phelot was in, she ran and clasped her arms about him, and cried out in a very loud and piercing voice upon Sir Launcelot to spare Sir Phelot and to slay him not. But Sir Launcelot, still holding him by the hair of the head, said: "Lady, I cannot spare him, for he has treated me more treacherously than any other knight with whom I ever had dealings." But the lady cried out all the more vehemently, "Sir Launcelot, thou good knight, I beseech thee, of thy knighthood, to spare him."

"Well," said Sir Launcelot, "it hath yet to be said of me that I have denied anything that I was able to grant unto any lady that hath asked it of me upon my knighthood. And yet I know not how to trust either of ye. For thou didst not say one word in my behalf when I was in danger of being slain so treacherously just now. As for this knight, I perceive that he is every whit as great a traitor and a coward as was his brother Sir Peris of the Sauvage Forest. So I will spare him, but I will not trust him, lest he turn against me ere I arm myself again. Where- *Sir Launcelot* fore give me hither the halter rein of your mule." So the *spares Sir* lady gave Sir Launcelot the halter rein, weeping amain as she *Phelot's life.* did so. And Sir Launcelot took the halter rein and he tied the arms of Sir Phelot behind him. Then he bade the lady of Sir Phelot to help him arm himself from head to foot, and she did so, trembling a very great deal. Then, when she had done so, quoth Sir Launcelot: "Now I fear the treachery of no man." Therewith he mounted his horse and rode away from that place. And he looked not behind him at all, but rode away as though he held too much scorn of that knight and of that lady to give any more thought to them

So after that Sir Launcelot travelled for a while through the green fields of that valley, till by and by he passed out of that valley, and came into a forest through which he travelled for a very long time.

For it was about the slanting of the afternoon ere he came forth out of that forest and under the open sky again. And when he came out of the forest he beheld before him a country of perfectly level marish, very lush and green, with many ponds of water and sluggish streams bordered by rushes and sedge, and with pollard willows standing in rows beside the waters. In the midst of this level plain of green (which was like to the surface of a table for flatness) there stood a noble castle, part *Sir Launcelot* built of brick and part of stone, and a town of no great size *cometh to a* and a wall about the town. And this castle and town stood *marish country.* upon an island surrounded by a lake of water, and a long bridge, built upon stone buttresses, reached from the mainland to the island. And this castle and town were a very long distance away, though they appeared very clear and distinct to the sight across the level marish, like, as it were, to a fine bit of very small and cunning carving.

Now the way that Sir Launcelot travelled, led somewhat toward that town, wherefore he went along that way with intent to view the place more near by. So he conveyed by that road for some time without meeting any soul upon the way. But at last he came of a sudden upon an archer hiding behind an osier tree with intent to shoot the water-fowl that came to a pond that was there—for he had several such fowl hanging at

his girdle.  To him Sir Launcelot said: "Good fellow, what town is that
yonderway?"  "Sir," said the yeoman, "that is called the Town of the
Marish because it stands in these Fenlands.  And that castle is called the
Castle of the Fenlands for the same reason."

Quoth Sir Launcelot: "What manner of place is that?  Is it a good place,
or is it otherwise?"  "Sir," said the archer, "that place was one while a
very good, happy place; for in times gone by there was a lord who dwelt
there who was both just and noble, and kind to all folk, wherefore he was
loved by all the people.  But one night there came two very grim and
*Sir Launcelot* horrible giants thither from the Welsh Mountains and these
*talks with a* entered into the castle by treachery and made prisoner of the
*yeoman.* lord of the castle.  Him they cast into the dungeon of the
castle, where they held him prisoner as an hostage.  For they threaten that
if friends of that lord's should send force against them to dispossess them,
they will slay him.  As for any other rescue, there is no knight who dareth
to go against them because of their terrible size, and their strength, and their
dreadful, horrible countenances."

"Well," said Sir Launcelot, "that is a pity and I am sorry for that noble
lordling.  Now, since there is no other single knight who dareth to undertake this adventure, I myself will go and encounter these giants."

"Nay, Sir Knight," said the yeoman, "do not do so, for they are not like
mortal men, but rather like monsters that are neither beast nor man.
Wherefore anyone who beholdeth them, feareth them."

"Grammercy for thy thought of me, good fellow," quoth Sir Launcelot,
" but if I shall refuse an adventure because I find it perilous, then I am not
like to undertake any adventure at all."

Therewith he bade good den to that yeoman and rode upon his way,
directing his course toward that town at an easy pass.

So he came at last to the long bridge that reached from the land to the
island, and he saw that at the farther end of the bridge was the gateway
of the town and through the arch thereof he could perceive a street of the
town, and the houses upon either side of the street, and the people thereof
coming and going.

So he rode forth upon the bridge and at the noise of his coming (for the
*Sir Launcelot* hoofs of his horse sounded like thunder upon the floor of the
*crosses the bridge* bridge) the people of the town came running to see who it
*to the town.* was that dared to come so boldly into their town.

These, when Sir Launcelot came nigh, began to call to him on high,
crying: "Turn back, Sir Knight!  Turn back!  Else you will meet your
death at this place."

But Sir Launcelot would not turn back, but advanced very steadfastly upon his way.

Now somewhat nigh the farther end of that bridge there stood a little lodge of stone, built to shelter the warden of the bridge from stress of weather. When Sir Launcelot came nigh to this lodge there started suddenly out from it a great churl, above seven feet high, who bore in his hand a huge club, shod with iron and with great spikes of iron at the top. This churl ran to Sir Launcelot and catched his horse by the bridle-rein and thrust it back upon its haunches, crying out in a great hoarse voice: "Whither goest thou, Sir Knight, for to cross this bridge?" Sir Launcelot said: "Let go my horse's rein, Sir Churl." Whereunto the churl made answer: "I will not let go thy horse's rein, and thou shalt not cross this bridge."

At this Sir Launcelot waxed very angry, and he drew his sword and struck the churl a blow with the flat thereof upon the shoulder, so that he dropped the rein very quickly. Therewith that churl drew back and took his great iron-shod club in both hands and struck at Sir Launcelot a blow that would have split a millstone. But Sir Launcelot put by *Sir Launcelot* the blow with his sword so that it did him no harm. But there- *slays the huge* with he waxed so wroth that he ground his teeth together with *churl.* anger, and, rising in his stirrups, he lashed that churl so woeful a blow that he cleft through his iron cap and his head and his breast even to the paps.

Now when the people of the town beheld that terrible blow they lifted up their voices in a great outcry, crying out: "Turn back, Sir Knight! Turn back! For this is a very woful thing for thee that thou hast done!" and some cried out: "Thou hast killed the giants' warder of the bridge!" And others cried: "Thou art a dead man unless thou make haste away from this." But to all this Sir Launcelot paid no heed, but wiped his sword and thrust it back into its sheath. Then he went forward *The folk warn* upon his way across the bridge as though nothing had befallen, *Sir Launcelot.* and so came to the farther side. Then, without paying any heed to all the people who were there, he rode straight to the castle and into the gate of the castle and into the court-yard thereof.

Now by this time all the castle was astir, and in great tumult, and many people came running to the windows and looked down upon Sir Launcelot. And Sir Launcelot sat his horse and looked all about him. So he perceived that beyond the court-yard was a fair space of grass, very smooth and green, well fitted for battle, wherefore he dismounted from his horse and tied it to a ring in the wall, and then he went to that green field and made him ready for whatever might befall.

Meantime all those people who were at the windows of the castle cried out to him, as the people of the town had done: "Go away, Sir Knight! Go away whilst there is still time for you to escape, or else you are a dead man!"

But Sir Launcelot replied not, but stood there and waited very steadfastly. Then the great door of the castle hall opened, and there came forth therefrom those two giants of whom he had heard tell.

And in truth Sir Launcelot had never beheld such horrible beings as they; for they were above ten feet high, and very huge of body and long of limb. *Two giants attack Sir Launcelot.* And they were clad in armor of bull-hide with iron rings upon it, and each was armed with a great club, huge and thick, and shod with iron, and studded with spikes. These came toward Sir Launcelot swinging their clubs and laughing very hideously and gnashing their long white teeth, for they thought to make easy work of him.

Then Sir Launcelot, seeing them coming thus, set his shield before him, and made ready for that assault with great calmness of demeanor. Then the giants rushed suddenly upon him and struck at him, the both of them together; for they deemed that by so doing the enemy could not escape both blows, but if one failed the other would slay him. But Sir Launcelot put aside the blow of one giant with his sword and of the other with his *How Sir Launcelot slays the first giant.* shield, with marvellous dexterity. Thereupon, ere they could recover themselves, he turned upon that giant who was upon his left hand and he struck him so terrible a blow upon the shoulder that he cut through the armor and through the shoulder and half-way through the body, so that the head and one arm of the giant leaned toward one way, and the other arm and the shoulder leaned toward the other way. Therewith the giant fell down upon the ground bellowing, so that it was most terrible to hear; and in a little he had died where he had fallen.

Now when the fellow of that giant beheld that dreadful, horrible stroke, he was so possessed with terror that he stood for a while trembling and like one in a maze. But when he saw Sir Launcelot turn upon him with intent to make at him also, he let fall his club and ran away with great and fearful outcry. Therewith he ran toward the castle and would have entered therein, but those within the castle had closed the doors and the gates against him, so that he could not escape in that way. So the giant ran around and around the court with great outcry, seeking for some escape *How Sir Launcelot slays the second giant.* from his pursuer, and Sir Launcelot ran after him. And Sir Launcelot struck him several times with his sword, so that at last, what with terror and pain and weariness, that giant stumbled and fell upon the ground. Therewith Sir Launcelot ran at him, and, ere he could rise, he took his sword in both hands and smote off his head so that it rolled down upon the ground like a ball. Then Sir Launcelot stood there panting for breath, for he had raced very hard after the giant, and could hardly catch his breath again. As he stood so, many of those of the castle and many of those who were of the town came to him from all sides; and they crowded around him and gave him great acclaim for ridding that place of those giants.

Then Sir Launcelot said to them: "Where is your lord?" Whereunto they made reply: "Sir, he lieth in the dungeon of the castle under the ground chained to the walls thereof, and there he hath been for three years or more, and no one hath dared to bring him succor until you came hither." "Go find him," said Sir Launcelot, "and set him free, and lose no time in doing so. And put him at all ease that you can."

They say: "Will you not stay and see him, Messire, and receive his acknowledgements for what you have done?" But Sir Launcelot replied: "Nay, not so." Then they say: "Will you not have some refreshment after this battle?" Whereunto Sir Launcelot said: "I do not need such refreshment." Then they say: "But will you not rest a little?" *Sir Launcelot* "Nay," said Sir Launcelot: "I may not tarry, for I have far *departs without* to go and several things to do, so that I do not care to stay." *refreshment.* So he loosed his horse from the ring in the wall, and mounted upon it and rode away from that castle and from that town and across the bridge whence he had come. And all the people followed after him, giving him great acclaim.

So Sir Launcelot left the castle, not because he needed no rest, but because he could not endure to receive the thanks of those whom he benefited. For though he loved to bring aid to the needy, yet he did not love to receive their thanks and their praise. Wherefore, having freed the lord of that castle from that brood of giants, he was content therewith and went his way without resting or waiting for thanks.

For so it was with those noble gallant knights of those days; that whilst they would perform signal service for mankind, yet they were not pleased to receive thanks or reward for the same, but took the utmost satisfaction, not in what they gained by their acts, but in the doing of knightly deeds, for they found all their reward in their deeds, because that thereby they made the world in which they lived better; and because they made the glory of the King, whose servants they were, the more glorious.

And I hold that such behavior upon the part of anyone makes him the peer of Sir Launcelot or Sir Tristram or Sir Lamorack or Sir Percival; yea, of Sir Galahad himself. For it does not need either the accolade or the bath to cause a man to be a true knight of God's making; nor does it need that a mortal King should lay sword upon shoulder to constitute a man the fellow of such knightly company as that whose history I am herewith writing; it needs only that he should prove himself at all times worthy in the performance of his duty, and that he shall not consider the hope of reward, or of praise of others in the performance of that duty.

So look to it that in all your services you take example of the noble Sir Launcelot of the Lake, and that you do your uttermost with might and main, and that you therewith rest content with having done your best, maugre any praise.　So you shall become a worthy fellow of Sir Launcelot and of his fellows.

ir Launcelot takes the
armor of Sir Kay

# Chapter Eighth

*How Sir Launcelot Rescued Sir Kay From a Perilous Pass. Also How He Changed Armor with Sir Kay and what Befell.*

ONE day Sir Launcelot came at early nightfall to a goodly manor-house and there he besought lodging for the night, and lodging was granted to him very willingly.

Now there was no lord of that manor, but only an old gentlewoman of very good breeding and address. She made Sir Launcelot right welcome and gave such cheer as she could, setting before him a very good supper, hot and savory, and a great beaker of humming mead wherewith to wash it down. Whilst Sir Launcelot ate, the gentlewoman inquired of him his name and he told her it was Sir Launcelot of the Lake. "Ha!" quoth she, "I never heard that name before, but it is a very good name." *The old gentle-woman makes Sir Launcelot welcome.*

At this Sir Launcelot laughed: "I am glad," said he, "that my name belikes thee. As for thy not having heard of it—well, I am a young knight as yet, having had but three years of service. Yet I have hopes that by and by it may be better known than it is at this present."

"Thou sayest well," quoth she, "for thou art very young yet, wherefore thou mayst not know what thou canst do till thou hast tried." And therewith Sir Launcelot laughed again, and said: "Yea, that is very true."

Now after Sir Launcelot had supped, his hostess showed him to the lodging she had provided for him wherein to sleep, and the lodging was in a fair garret over the gateway of the court. So Sir Launcelot went to his bed and, being weary with journeying, he presently fell into a deep and gentle sleep.

Now about the middle of the night there fell of a sudden the noise of some-one beating upon the gate and calling in a loud voice and de-manding immediate admittance thereat. This noise awoke Sir Launcelot, and he arose from his couch and went to the window and looked out to see who it was that shouted so loudly and made such uproar. *Sir Launcelot is aroused from sleep.*

The moon was shining at that time, very bright and still, and by the light thereof Sir Launcelot beheld that there was a knight in full armor seated upon horseback without the gate, and that the knight beat upon the gate with the pommel of his sword, and shouted that they should let him in.

But ere anyone could run to answer his call there came a great noise of horses upon the highroad, and immediately after there appeared three knights riding very fiercely that way, and these three knights were plainly pursuing that one knight. For, when they perceived him, they rode very violently to where he was, and fell upon him fiercely, all three at one time; wherefore, though that one knight defended himself as well as he could, yet was he in a very sorry way, and altogether likely to be overborne. For those three surrounded him so close to the gate that he could do little to shift himself away from their assaults.

Now when Sir Launcelot beheld how those three knights attacked that one knight, he said to himself: "Of a surety, yonder knight is in a very sorry way. I will do what I can to help him; for it is a shame to behold three knights attack one knight in that way. And if he be slain in this assault, meseems I shall be a party to his death."

Therewith he ran and put his armor upon him, and made ready for battle. Then he drew the sheet from his bed, and he tied the sheet to the

*Sir Launcelot goeth to the rescue of the knight assaulted.* bar of the window and by it he let himself quickly down to the ground not far from where those knights were doing battle. So being safely arrived in that way he cried out in a very loud voice: "Messires, leave that knight whom ye assail, and turn to me, for I have a mind to do battle with you myself."

Then one of those knights, speaking very fiercely, said: "Who are you, and what business have you here?"

"It matters not who I am," said Sir Launcelot, "but I will not have it that you three shall attack that one without first having had to do with me."

"Very well," said that knight who had spoken, "you shall presently have your will of that."

Therewith he and his fellows immediately descended from their horses, and drew their swords and came at Sir Launcelot upon three sides at once. Then Sir Launcelot set his back against the gate and prepared to defend himself.

Therewith that knight whom he would defend immediately got down from his horse with intent to come to the aid of Sir Launcelot, but Sir Launcelot forbade him very fiercely, saying: "Let be, Sir Knight, this is my quarrel, and you shall not meddle in it."

Upon this, those three knights rushed upon him very furiously, and they struck at him all at once, smiting at him wherever they could and with all their might and main. So Sir Launcelot had much ado to defend himself from their assault. But he made shift that they should not all rush in upon him at once, and by and by he found his chance with one of them. Whereupon he turned suddenly upon that one, and suddenly he lashed so terrible a buffet at him that the knight fell down and lay as though he had been struck dead with the force thereof.

*Sir Launcelot does battle with three knights.*

Then, ere those other two had recovered themselves, he ran at a second and struck him so fierce a blow that his wits left him, and he staggered like a drunken man and ran around and around in a circle, not knowing whither he went. Then he rushed upon the third and thrust him back with great violence, and as he went back Sir Launcelot struck him, too, as he had struck his companions and therewith that knight dropped his sword and fell down upon his knees and had not power to raise himself up.

Then Sir Launcelot ran to him and snatched off his helmet, and catched him by the hair with intent to cut off his head. But at that the fallen knight embraced Sir Launcelot about the knees, crying out: "Spare my life!"

"Why should I spare you?" said Sir Launcelot. "Sir," cried the knight, "I beseech you of your knighthood to spare me."

"What claim have you upon knighthood," said Sir Launcelot, "who would attack a single knight, three men against one man?"

Then the other of those knights who had been staggered by Sir Launcelot's blow, but who had by now somewhat recovered himself, came and kneeled to Sir Launcelot, and said: "Sir, spare his life, for we all yield ourselves unto you, for certes, you are the greatest champion in all the world."

Then Sir Launcelot was appeased, but he said: "Nay, I will not take your yielding unto me. For as you three assaulted this single knight, so shall you all three yield to him."

"Messire," said the knight who kneeled: "I am very loth to yield us to that knight, for we chased him hither, and he fled from us, and we would have overcome him had you not come to his aid."

"Well," said Sir Launcelot, "I care nothing for all that, but only that you do as I will. And if ye do not do it, then I must perforce slay your companions and you two. Wherefore you may take your choice."

Then said that knight who kneeled: "Messire, I see no other thing to do than to yield us as you would have, wherefore we submit ourselves unto this knight whom you have rescued from us."

*The three knights must yield to the one knight.*

Then Sir Launcelot turned to that knight to whom he had brought aid in that matter, and he said: "Sir Knight, these knights yield themselves unto you to do as you command them. Now I pray you of your courtesy to tell me your name and who you are."

"Sir," said that knight, "I am Sir Kay the Seneschal, and am King Arthur's foster-brother, and a knight of the Round Table. I have been errant now for some time in search of Sir Launcelot of the Lake. Now, I deem either that you are Sir Launcelot, or else that you are the peer of Sir Launcelot."

"Thou art right, Sir Kay," said Sir Launcelot, "and I am Sir Launcelot of the Lake." So thereat they two made great joy over one another, and embraced one another as brothers-in-arms should do.

Then Sir Kay told Sir Launcelot how it was with those three knights who had assailed him; that they were three brethren, and that he had over-thrown the fourth brother in an adventure at arms and had hurt him very sorely thereby. So those three had been pursuing him for three days with intent to do him a harm.

Now Sir Kay was very loath to take submission of those three knights, but Sir Launcelot would have it so and no other way. So Sir Kay consented *Sir Kay* to let it be as Sir Launcelot willed. Thereupon those three *taketh sub-* knights came and submitted themselves to Sir Kay, and Sir *mission of the* Kay ordained that they should go to Camelot and lay their *three knights.* case before King Arthur, and that King Arthur should adjudge their case according to what he considered to be right and fitting.

Then those three knights mounted upon their horses and rode away, and when they had done so the gates of the manor were opened, and Sir Launce-lot and Sir Kay entered in. But when the old lady who was his hostess beheld Sir Launcelot come in, she was very greatly astonished, for she wist he was still asleep in his bed-chamber. Wherefore she said: "Sir, methought you were in bed and asleep." "So indeed I was," said Sir Launcelot, "but when I saw this knight in peril of his life against three knights, I leaped out of my window and went to his aid." "Well," said his hostess, "meseems that you will sometime be a very good knight, if you have so much courage whilst you are so young." And at that both Sir Launcelot and Sir Kay laughed a great deal.

Then the chatelaine set bread and wine before Sir Kay, and he ate and refreshed himself, and thereafter he and Sir Launcelot went to that garret above the gate, and there fell asleep with great ease of body.

Now before the sun arose Sir Launcelot awoke but Sir Kay still slept very soundly. Then Sir Launcelot beheld how Sir Kay slept, and he had a mind for a jest. So he clad himself in Sir Kay's armor altogether from head to foot, and he took Sir Kay's shield and spear, and he *Sir Launcelot* left his armor and shield and spear for Sir Kay to use. Then *takes Sir Kay's* he went very softly from that room, and left Sir Kay still sleep- *armor.* ing. And he took Sir Kay's horse and mounted upon it and rode away; and all that while Sir Kay knew not what had befallen, but slept very deeply.

Now after a while Sir Kay awoke, and he found that Sir Launcelot was gone, and when he looked he found that his own armor was gone and that Sir Launcelot's armor was left. Then he wist what Sir Launcelot had done, and he said: "Ha! what a noble, courteous knight is the gentleman. For he hath left me his armor for my protection, and whilst I wear it and carry his shield and ride his horse, it is not likely that anyone will assail me upon my way. As for those who assail him, I do not believe that they will be likely to find great pleasure in their battle."

Therewith he arose and clad himself in Sir Launcelot's armor, and after he had broken his fast he thanked his hostess for what she had given him, and rode upon his way with great content of spirit.

(And it was as Sir Kay had said, for when he met other knights upon the road, and when they beheld the figure upon his shield, they all said: "It is not well to meddle with that knight, for that is Sir Launcelot." And so he came to Camelot without having to do battle with any man.)

As for Sir Launcelot, he rode upon his way with great cheerfulness of spirit, taking no heed at all of any trouble in the world, but chanting to himself as he rode in the pleasant weather. But ever he made *How Sir* his way toward Camelot, for he said: "I will return to Camelot *Launcelot travels* for a little, and see how it fares with my friends at the court *toward Camelot.* of the King."

So by and by he entered into the country around about Camelot, which is a very smooth and fertile country, full of fair rivers and meadows with many cots and hamlets, and with fair hedge-bordered highways, wonderfully pleasant to journey in. So travelling he came to a very large meadow where were several groves of trees standing here and there along by a river. And as he went through this meadow he saw before him a long bridge, and at the farther side of the bridge were three pavilions of silk of divers colors, *Sir Launcelot* which pavilions had been cast in the shade of a grove of beech- *perceives three* trees. In front of each pavilion stood a great spear thrust in *knights at feast.* the earth, and from the spear hung the shield of the knight to whom the pavilion belonged. These shields Sir Launcelot read very easily, and so knew the knights who were there. To wit: that they were Sir

Gunther, Sir Gylmere, and Sir Raynold, who were three brothers of the Court of King Arthur. As Sir Launcelot passed their pavilions, he saw that the three knights sat at feast in the midmost pavilion of the three, and that a number of esquires and pages waited upon them and served them, for those knights were of very high estate, and so they were established as high lords should be.

Now when those knights perceived Sir Launcelot they thought it was Sir Kay because of the armor he wore, and Sir Gunther, who was the eldest of the three brothers, cried out: "Come hither, Sir Kay, and eat with us!"

*The three knights bid Sir Launcelot come to feast with them.* But to this Sir Launcelot made no reply, but rode on his way. Then said Sir Gunther: "Meseems Sir Kay hath grown very proud this morning. Now I will go and bring him back with me, or else I will bring down his pride to earth." So he made haste and donned his helmet and ran and took his shield and his spear, and mounted his horse and rode after Sir Launcelot at a hard gallop. As he drew nigh to Sir Launcelot he cried out: "Stay, Sir Knight! Turn again, and go with me!" "Why should I go with you?" said Sir Launcelot. Quoth Sir Gunther: "Because you must either return with me or do battle with me." "Well," said Sir Launcelot, "I would rather do battle than return against my will." And at that Sir Gunther was astonished, for Sir Kay was not wont to be so ready for a battle. So Sir Launcelot set his shield and spear and took his stand, and Sir Gunther took his stand. Then, when they were in all ways prepared, each set spur to his horse and

*Sir Launcelot overthrows Sir Gunther.* rushed together with terrible speed. So each knight struck the other in the midst of his shield, but the onset of Sir Launce- lot was so terrible that it was not to be withstood, wherefore both Sir Gunther and his horse were overthrown in such a cloud of dust that nothing at all was to be seen of them until that cloud lifted.

At this both Sir Raynold and Sir Gylmere were astonished beyond measure, for Sir Gunther was reckoned to be a much better knight than Sir Kay, wherefore they wist not how it was that Sir Kay should have overthrown him in that fashion.

So straightway Sir Gylmere, who was the second of those brothers, called out to Sir Launcelot to tarry and do battle. "Very well," said Sir Launce- lot, "if I cannot escape thee I must needs do battle. Only make haste, for I would fain be going upon my way."

So Sir Gylmere donned his helm in haste and ran and took his shield and spear and mounted upon his horse. So when he had made himself ready in all ways he rushed upon Sir Launcelot with all his might and Sir Launce- lot rushed against him.

In that encounter each knight struck the other in the midst of his shield, and the spear of Sir Gylmere burst into pieces, but Sir Laun- *Sir Launcelot* celot's spear held, so the breast-strap of Sir Gylmere's saddle *overthrows Sir* bursting, both saddle and knight were swept entirely off the *Gylmere.* horse and to the earth, where Sir Gylmere lay altogether stunned.

Then Sir Raynold came against Sir Launcelot in like manner as the others had done, and in that encounter Sir Launcelot over- *Sir Launcelot* threw both horse and man so that, had not Sir Raynold *wins from Sir* voided his horse, he would likely have been very sadly *Raynold.* hurt.

Then Sir Raynold drew his sword and cried out in a loud voice: "Come, Sir Knight, and do me battle afoot!" But Sir Launcelot said: "Why will you have it so, Sir Knight? I have no such quarrel with you as to do battle with swords." "Ha!" said Sir Raynold, "you shall fight with me. For though you wear Sir Kay's armor, I wot very well that you are not Sir Kay, but a great deal bigger man than ever Sir Kay is like to be."

"Nay," said Sir Launcelot, "I will not do any more battle with you." And therewith he drew rein and rode away, leaving Sir Raynold standing very angry in the middle of the highway.

After that Sir Launcelot rode very easily at a quiet gait, with no great thought whither he rode, until after a while he came to a place where a road went across a level field with two rows of tall poplar trees, one upon either side of the highway. Then Sir Launcelot perceived where, beneath the shade of these poplar trees, were four knights standing *Sir Launcelot* each by his horse. And these four knights were conversing *meets four noble* very pleasantly together. Now as Sir Launcelot drew nigh *knights.* he perceived that those were four very famous noble knights of the Round Table; to wit: one of those knights was his own brother, Sir Ector de Maris, another was Sir Gawain, another was Sir Ewain, and the fourth was Sir Sagramore le Desirous.

Now as Sir Launcelot drew nigh Sir Gawain said: "Look, yonder cometh Sir Kay the Seneschal." Unto this Sir Sagramore le Desirous said: "Yea, this is he; now bide you here for a little while, and I will go and take a fall of him."

So straightway he mounted upon his horse, and he rode toward Sir Launcelot, and he cried out: "Stay, Sir Knight, you cannot go farther until you have had to do with me." "What would you have of me?" quoth Sir Launcelot. "Sir," said Sir Sagramore, "I will have a fall of you." "Well," said Sir Launcelot, "I suppose I must pleasure you, since it cannot be otherwise."

Therewith he dressed his shield and his spear and Sir Sagramore dressed

*Sir Launcelot overthrows Sir Sagramore.* his shield and his spear, and when they were in all ways prepared they ran together at full tilt. In that encounter Sir Sagramore's spear broke, but Sir Launcelot struck so powerful a blow that he overthrew both horse and man into a ditch of water that was near-by.

Then Sir Ector de Maris said: "Ha, surely some very ill chance has befallen Sir Sagramore for to be overthrown by Sir Kay. Now I will go and have ado with him, for if the matter rests here there will be no living at court with the jests which will be made upon us."

So he took horse and rode to where Sir Launcelot was, and he went at a very fast gallop. When he had come near to Sir Launcelot he cried out: "Have at thee, Sir Kay, for it is my turn next!" "Why should I have at thee?" said Sir Launcelot, "I have done thee no harm." "No matter," said Sir Ector, "you can go no farther until you have had to do with me." "Well," said Sir Launcelot, "if that is so, the sooner I have to do with thee, the sooner shall I be able to go upon my way."

Therewith each knight made himself ready and when they were in all

*Sir Launcelot overthrows Sir Ector.* ways prepared they came together with such force that Sir Launcelot's spear went through Sir Ector's shield and smote him upon the shoulder, and Sir Ector was thrown down upon the ground with such violence that he lay where he had fallen, without power to move.

Then said Sir Ewain to Sir Gawain where they stood together: "That is the most wonderful thing that ever I beheld, for never did I think to behold Sir Kay bear himself in battle in such a fashion as that. Now bide thee here and let me have a try at him." Therewith Sir Ewain mounted his horse and rode at Sir Launcelot, and there were no words spoken this time, but each knight immediately took his stand to do battle. Then they ran their horses together, and Sir Launcelot gave Sir Ewain such a buffet that he was astonished, and for a little he knew not where he was, for his spear fell down out of his hand, and he bore his shield so low that Sir Launcelot might have slain him where he stood if he had been minded to do so.

Then Sir Launcelot said: "Sir Knight, I bid thee yield to me." And Sir Ewain said: "I yield me. For I do not believe that thou art Sir Kay

*Sir Ewain yields to Sir Launcelot.* but a bigger man than he shall ever be. Wherefore I yield me." "Then that is well," said Sir Launcelot. "Now stand thou a little aside where thou mayst bring succor unto these other two knights, for I see that Sir Gawain has a mind to tilt with me."

And it was as Sir Launcelot said, for Sir Gawain also had mounted his horse and had made himself ready for that encounter. So Sir Gawain and Sir Launcelot took stand at such place as suited them. Then each knight set spurs to his horse and rushed together like thunder, and each knight smote the other knight in the midst of his shield; and in that encounter the spear of Sir Gawain brake in twain but the spear of Sir Launcelot held, and therewith he gave Sir Gawain *Sir Gawain jails with Sir Launcelot.* such a buffet that Sir Gawain's horse reared up into the air, and it was with much ado that he was able to void his saddle ere his horse fell over backward. For if he had not leaped to earth the horse would have fallen upon him.

Then Sir Gawain drew his sword and cried very fiercely: "Come down and fight me, Sir Knight! For thou art not Sir Kay!"

"Nay, I will not fight thee that way," said Sir Launcelot, and therewith he passed on his way without tarrying further.

But he laughed to himself behind his helmet as he rode, and he said: "God give Sir Kay joy of such a spear as this, for I believe there came never so good a spear as this into my hand. For with it I have overthrown seven famous knights in this hour.

As for those four knights of the Round Table, they comforted one another as best they could, for they knew not what to think of that which had befallen them. Only Sir Ector said: "That was never Sir Kay who served us in this wise, but such a man as is better than ten Sir Kays, or twice ten Sir Kays, for the matter of that."

Now Sir Launcelot came to Camelot about eventide, what time King Arthur and his court were assembled at their supper. Then there was great joy when news was brought of his coming and they brought him in to the court and set him beside the King and the Lady Guinevere all armed as he was. Then King Arthur said: "Sir Launcelot, how is it with thee?" and Sir Launcelot said: "It is well." Then King Arthur said: "Tell us what hath befallen thee." *How Sir Launcelot returned to Camelot.* And Sir Launcelot told all that had happened in that month since he had left court. And all they who were there listened, and were much astonished.

But when Sir Launcelot told how he had encountered those seven knights, in the armor of Sir Kay, all laughed beyond measure excepting those of the seven who were there, for they took no very good grace to be laughed at in that wise.

So now I hope I have made you acquainted with Sir Launcelot of the Lake, who was the greatest knight in the world. For not only have I told you how he was created a knight at the hands of King Arthur, but I have also led you errant along with him, so that you might see for yourself how he adventured his life for other folk and what a noble and generous gentleman he was; and how pitiful to the weak and suffering, and how terrible to the evil-doer. But now I shall have to leave him for a while (but after a while in another book that shall follow this, I shall return to him to tell you a great many things concerning other adventures of his), for meantime it is necessary that I should recount the history of another knight, who was held by many to be nearly as excellent a knight as Sir Launcelot was himself.

## CONCLUSION

*Here endeth the story of Sir Launcelot.   That which followeth is the story of Sir Tristram of Lyonesse, who was knit with Sir Launcelot into such close ties of friendship that if they had been brothers of the same blood, with the same father and mother, they could not have loved one another more than they did.*

*For indeed it would not be possible to tell any history of Sir Launcelot of the Lake without telling that of Sir Tristram of Lyonesse as well, for as the web of a fair fabric is woven in with the woof thereof, so were the lives of Sir Launcelot and Sir Tristram woven closely together.*

*Wherefore you shall now hear tell of the goodly adventures of Sir Tristram of Lyonesse; and God grant that you may have the same joy in reading thereof that I shall have in telling of them to you.*

# The Book of Sir Tristram

# Sir Tristram of Lyonesses

# Prologue.

THERE was a certain kingdom called Lyonesse, and the King of that country was hight Meliadus, and the Queen thereof who was hight the Lady Elizabeth, was sister to King Mark of Cornwall.

In the country of Lyonesse, there was a very beautiful lady, who was a cunning and wicked sorceress. This lady took great love for King Meliadus, who was of an exceedingly noble appearance, and she meditated continually how she might bring him to her castle so as to have him near her.

Now King Meliadus was a very famous huntsman, and he loved the chase above all things in the world, excepting the joy he took in the love of his Queen, the Lady Elizabeth. So, upon a certain day, in *King Meliadus* the late autumn season he was minded to go forth a-hunting, *rides a-hunting.* although the day was very cold and bleak.

About the prime of the day the hounds started, of a sudden, a very wonderful stag. For it was white and its horns were gilded very bright, shining like pure gold, so that the creature itself appeared like a living miracle in the forest. When this stag broke cover, the hounds immediately set chase to it with a great outcry of yelling, as though they were suddenly gone frantic, and when the King beheld the creature, he also was immediately seized as with a great fury for chasing it. For, beholding it, he shouted aloud and drove spurs into his horse, and rushed away at such a pass that

101

his court was, in a little while, left altogether behind him, and he and the chase were entirely alone in the forest.

The stag, with the hounds close behind it, ran at a great rate through the passes of the woodlands, and King Meliadus pursued it with might and main until the chase burst out of the forest into an open plain beyond the *King Meliadus* woodland.   Then King Meliadus beheld that in the midst of *chases the stag.* the plain was a considerable lake of water; and that in the midst of the water was an island; and that upon the island was a very tall and stately castle.   Toward this castle the stag ran with great speed, and so, coming to the lake, it leaped into the water and swam across to the island—and there was a thin sheet of clear ice upon the water close to either bank.

But when the hounds that pursued the stag came to that frozen water, they stinted their pursuit and stood whimpering upon the brink, for the ice and the water repelled them.   But King Meliadus made no such pause, but immediately leaped off from his horse, and plunged into the water and swam across in pursuit of the stag.   And when he reached the other side, he chased the stag afoot with great speed, and therewith the stag ran to the castle and into the court-yard thereof, and King Meliadus ran after it.   Then, immediately he had entered in, the gates of the castle were shut and King Meliadus was a prisoner.

(Now you are to know that that castle was the abode of the beautiful enchantress afore spoken of, and you are to know that she had sent that *King Meliadus* enchanted stag to beguile King Meliadus to her court, and *is made prisoner* so she made King Meliadus her captive.   Further, it is to be *at an enchanted* told that when she had him there within her castle, she wove *castle.* a web of enchantment all about him so that he forgot the Lady Elizabeth and his court and his kingdom and thought of nothing but that beautiful sorceress who had thus beguiled him into her power.)

Now, when those who were with the King returned to the castle of Lyonesse without him, and when the King did not return that day nor the *The Lady Eliza-* next day nor at any time, the Lady Elizabeth grew more *beth grieves to* and more distracted in her anxiety because of him.   And when *distraction.* a fortnight had gone by and still there was no news of the King, her grief and apprehension became so great that she turned distracted and they had to set watch and ward upon her lest she do herself a harm in her madness.

So for a long time they kept her within the castle; but upon a certain day she broke away from her keepers and ran out from the castle and into the forest ere those in attendance upon her knew she had gone.   Only one gentlewoman saw her, and she called upon a young page to follow her, and thereupon ran after the Queen whither she went, with intent to bring her back again.

But the Lady Elizabeth ran very deep into the forest, and the gentlewoman and the page ran after her; and the Queen thought that she was going to find her lord in the forest. So she ran very rapidly *The Lady Eliza-* for a great distance, until by and by she waxed faint with *beth escapes into* weariness from running and sank down upon the ground; *the forest.* and there they that followed her found her lying. And they found that the Queen was in a great passion of pain and sick to death. For the day was very wintry, with a fine powder of snow all over the ground, so that the cold of the weather pierced through the garments of the Lady Elizabeth and entered into her body and chilled her to the heart.

Now the gentlewoman, seeing how it was with the Queen, called the page to her and said: "Make haste! Go back to the castle of Lyonesse, and bring some of the knights of the castle with all speed, else the Queen will die at this place." And upon that the page ran off with great speed to do her bidding and the Queen was left alone with her gentlewoman.

Then the gentlewoman said, "Lady, what cheer?" And the Queen said, "Alas, I am sick to death." The gentlewoman said, "Lady, cannot you bear up a little until help cometh?" Thereupon the Lady Elizabeth fell to weeping very piteously, and said, "Nay, I cannot bear up any longer, for the cold hath entered into my heart." (Yea, even at that time death was upon her because of the cold at her heart.)

Then by and by in the midst of her tears and in very sore travail a man-child was born to the Queen, and when that came to pass a great peace fell suddenly upon her.

Then she said, speaking to the nurse like one in great weariness, "What child is it that I have given unto the world?" The nurse said, "It is a man-child." The Queen said to her, "Hold him up until I see him." *How Tristram* Thereupon the nurse held the child up and the Queen looked *is born in the* at him, though she could hardly see him because it was as though *forest.* a mist lay upon her eyes which she could not clear away from her sight; for at that time she was drawing deep draughts of death. Then, when she had seen the child and had beheld that he was very strong and lusty and exceedingly comely, she said: "Behold, this is my child, born in the midst of sore travail and great sorrow; wherefore his name shall be called Tristram because he hath caused so many tears to be shed."

Then in a little while the Lady died, and the gentlewoman stood weeping beside her, making great outcry in that cold and lonely forest.

Anon there came those knights who were sent from the castle to find the Queen; and when they came to that place, they beheld that she lay upon the ground all cold and white like to a statue of marble stone. So they lifted her up and bare her away upon a litter, and the gentlewoman followed weeping and wailing in great measure, and bearing the child wrapped in a mantle.

So Tristram was born in that wise, and so his name was given to him because of the tears that were shed at his birth.

And now it is to be told how King Meliadus returned from that castle of enchantment where he was held prisoner.

At this time Merlin was still living in the world, for Vivien had not yet bewitched him, as hath been told in the Book of King Arthur. So by and *King Meliadus* by it came to pass that he discovered where King Meliadus *is released from* was imprisoned and how it fared with him in the castle of that *durance.* enchantress. So he made greater spells than those that enmeshed King Meliadus, and he brought King Meliadus back into his memory of the Queen and his kingdom. Then straightway the King broke out from the castle of the enchantress and returned to his kingdom. But when he came there it was to find everything in great sorrow and dole; for the Lady Elizabeth was no longer upon this earth to bring joy to the heart of the King. So for a long while after his return King Meliadus lay altogether stricken down with the grief of that bereavement.

Here followeth the story of Tristram, how he passed his youth, and how he became a knight of Cornwall of King Mark's making.

# PART I

# The Story of Sir Tristram and the Lady Belle Isoult

*HERE followeth the story of Sir Tristram of Lyonesse, who, with Sir Launce-*
*lot of the Lake, was deemed to be one of the two most worthy and perfect*
*knights champion of his day.*

*Likewise herein shall be told the story of the Lady Belle Isoult, who next*
*to Queen Guinevere, was reckoned to be the most fair, gentle lady in all of the*
*world.*

Tristram succors the Lady Moeya.

# Chapter First

*How the new Queen of Lyonesse sought Tristram's life; how he went to*
*France, and how he returned again to Lyonesse and was received*
*with love at that place.*

SO King Meliadus grieved very bitterly for the Lady Elizabeth for the
space of seven years, and in that time he took but little pleasure in
life, and still less pleasure in that son who had been born to him in
that wise. Then one day a certain counsellor who was in great favor with
the King came to him and said: "Lord, it is not fitting that you should
live in this wise and without a mate; for you should have a queen, and you
should have other children besides Tristram, else all the fate of this kingdom shall depend upon the life of that one small child."

And King Meliadus took this counsel to heart, and after a while he said:
*King Meliadus* "What you tell me is true, and so I shall take another Queen,
*taketh the Lady* even though it is not in me to love any other woman in all of
*Moeya to second* the world but that dear one who is dead and gone." So
*wife.* a while after that he took to wife the Lady Moeya, who was
the daughter of King Howell of Britain.

Now Queen Moeya had been married to an Earl of Britain, and by him
she had a son who was about the age of Tristram. So she brought this son
to Lyonesse with her, and he and Tristram were very good companions.

But the Lady Moeya took great hatred of Tristram, for she said in her heart: "Except for this Tristram, mayhap my son might be King and overlord of this land." And these thoughts brooded with her, so that after a while she began to meditate how she might make away with Tristram so that her own son might come into his inheritance.

Now at that time Tristram was about thirteen years of age and very large and robust of form and of extraordinary strength of body and beauty of countenance. But the son of Queen Moeya was not of such a sort, so the more beautiful and noble Tristram was the more the Queen hated him. So one day she called to her a very cunning chemist and she said to him: "Give me a drink of such and such a sort, so that he who drinks thereof shall certainly die, maugre help of any kind." And the chemist gave her what she desired, and it was in a phial and was of a golden color.

Now Tristram and the son of the Lady Moeya were wont to play ball in a certain court of the castle, and when they would play there they would wax all of a heat with their sport. This the Lady Moeya was

*The Lady Moeya devises mischief against Tristram.* well aware of; so one day she took that phial of poison and she poured a part of it into a chalice and she filled the chalice with clear water and she set the chalice upon a bench where those

two would play at ball. For she said to herself: "When they grow warm with their play, Tristram will certainly drink of this water to quench his thirst, and then my son will maybe enter into his inheritance."

So the two youths played very fiercely at their game, and they waxed exceedingly hot and presently were both very violently athirst. Then Tristram said, "I would I had somewhat to drink," and his stepbrother said, "Look, yonder is a chalice of water; drink! and when thou hast quenched thy thirst, then I will drink also." But Tristram said: "Nay, brother, drink thou first, for thou art more athirst than I." Then at first the son of the Lady Moeya would not have it so, but would have Tristram drink; but afterward he did as Tristram bade him, and, taking the chalice in both hands, he drank freely of that poison which his own mother had

*The son of the Queen drinks of the poison.* prepared. Then when he had drunk his fill, Tristram took the chalice and would have drunk too; but the other said, "Stay, Tristram, there is great bitterness in that chalice"; and then

he said, "Methinks I feel a very bitter pang within my vitals," and then he cried out, "Woe is me! I am in great pain!" Therewith he fell down upon the ground and lay there in a great passion of agony. Then Tristram cried aloud for help in a piercing voice; but when help came thither it was too late, for the son of the Lady Moeya was dead.

Then the Lady Moeya was in great torment of soul, and beat her breast and tore her hair and King Meliadus had much ado for to comfort her. And after this she hated Tristram worse than ever before, for she would say to herself: "Except for this Tristram, my own son would yet be alive!"

So she brooded upon these things until she could not rest, whether by day or night. Then one day she took the rest of the poison that was in the phial and poured it into a goblet of yellow wine. This goblet she gave to one of her pages, saying: "Take this to Tristram, and offer it to him when I shall tell you to do so!"

Therewith she went down to the hall where Tristram was, and she said, "Tristram, let there be peace betwixt us." And Tristram said: "Lady, that meets my wishes, for I have never had in my heart aught but loving-kindness toward you, and so I would have it in your heart toward me." With this the page came in the hall with that goblet of yellow wine. Then the Lady Moeya took the *The Lady Moeya seeks Tristram's life a second time.* goblet and said: "Take this cup, and drink of the wine that is in it, and so there shall be peace betwixt us forever." And as she said that she looked very strangely upon Tristram, but Tristram was altogether innocent of any evil against him. So he reached out his hand to take the cup which the page brought to him.

Now at that moment King Meliadus came into the hall fresh from the chase, and he was much heated and greatly athirst, wherefore, when he saw that cup of wine he said: "Stay, Tristram, let me drink, for I am greatly athirst. After I have quenched my thirst, then thou shalt drink."

Therewith he took the goblet of wine and made to lift it to his lips. But at that the Lady Moeya cried out, in a very loud and piercing voice, "Do not drink of that wine!" The King said, "Why should I not drink of it?" "No matter," said the Lady Moeya, "thou shalt not drink of it, for there is death in it."

Therewith she ran to the King and catched him by the hand, and she plucked away the goblet so that the wine was spilled out of it upon the ground.

Then King Meliadus gazed at the Lady Moeya, and he thought of many things in very little time. Thereupon he seized her by the hair and dragged her forward, so that she fell down upon her hands and knees to the pavement of the hall. And King Meliadus drew his great sword so that it flashed like lightning, and he cried: "Tell *King Meliadus threatens to slay the Queen.* me what thou hast done, and tell me quickly, or thou shalt not be able to tell me at all!" Then the Lady Moeya clutched King Meliadus about the thighs, and she cried out: "Do not slay me with thine own hand, or else my blood will stain thee with dishonor! I will tell thee all, and then thou

mayst deal with me according to the law, for indeed I am not fit to live."
So therewithal the Lady Moeya confessed everything to the King.

Then King Meliadus shouted aloud and called the attendants and said:
"Take this woman and cast her into prison, and see that no harm befall
her there; for the lords of this country shall adjudge her, and not I." And
therewith he turned away and left her.

And thereafter, in due season, the Lady Moeya was brought to trial and
was condemned to be burned at the stake.

Now when the day came that she was to be burnt, Tristram was very
sorry for her. So when he beheld her tied fast to the stake he came to
*Tristram begs* where King Meliadus was and he kneeled before him, and he
*mercy for the* said, "Father, I crave a boon of thee." Thereupon King
*Queen.* Meliadus looked upon Tristram, and he loved him very tenderly
and he said: "My son, ask what thou wilt, and it shall be thine." Then
Tristram said: "Father, I pray thee, spare the life of this lady, for methinks
she hath repented her of her evil, and surely God hath punished her very
sorely for the wickedness she hath tried to do."

Then King Meliadus was very wroth that Tristram should interfere with
the law; but yet he had granted that boon to his son and could not withdraw.
So after a while of thought he said: "Well, I have promised, and so I will
perform my promise. Her life is thine; go to the stake and take her. But
when thou hast done so I bid thee go forth from this place and show
thy face here no more. For thou hast interfered with the law, and hast
done ill that thou, the son of the King, should save this murderess. So
thou shalt leave this place, for I mistrust that between you two some murder
will befall in this country."

So Tristram went weeping to where the Queen was bound to the stake;
and he cut her bonds with his dagger and set her free. And he said: "Lady,
thou art free; now go thy way, and may God forgive thee as I do." Then
the Queen wept also, and said, "Tristram, thou art very good to me." And
because she was barefoot and in her shift, Tristram took his cloak and
wrapped it about her.

After that, Tristram straightway left Lyonesse, and King Meliadus appointed that a noble and honorable lord of the court, hight Gouvernail,
*Tristram departs* should go with him. They two went to France, and there
*from Lyonesse.* they were made very welcome at the court of the King. So
Tristram dwelt in France till he was eighteen years old, and everyone at
the court of the King of France loved him and honored him so that he
dwelt there as though he were of the blood of France.

During the time that he was in France he became the greatest hunter in the world, and he wrote many books on venery that were read and studied long after he had ceased to live. Also he became so skilful with the harp that no minstrel in the world was his equal. And ever he waxed more sturdy of frame and more beautiful of countenance, and more well-taught in all the worship of knighthood. For during that time he became so wonderfully excellent in arms that there was no one in France who was his equal.

Thus Tristram dwelt at peace in that land for five years, but even he longed for his own home with all the might and main of his heart. So one day he said to Gouvernail: "Gouvernail, I cannot deny myself any longer from seeing my father and my own country, for I feel that I must see them or else my heart will certainly break because of its great longing." Nor would he listen to anything that Gouvernail might say contrary to this. So they two took their departure from France, and Tristram travelled as a harper and Gouvernail as his attendant. Thus they came to Lyonesse in that wise.

One day whilst King Meliadus sat at meat, they two came into the hall, and Gouvernail wore a long white beard which altogether disguised him so that no one knew him. But Tristram shone with such a *How Tristram returns to Lyonesse.* great radiance of beauty and of youth that all who looked upon him marvelled at him. And the heart of King Meliadus went out to Tristram very strongly, and he said before all of his court, "Who art thou, fair youth? And whence comest thou?" To which Tristram made reply: "Lord, I am a harper, and this is my man, and we have come from France." Then King Meliadus said to Tristram: "Sir, have you seen a youth in France whom men call Tristram?" And Tristram replied, "Yea, I have seen him several times." King Meliadus said, "Doth he do well?" "Yea," said Tristram, "he doeth very well, though at times he is sore oppressed with a great desire for his own country." At this King Meliadus turned away his face, for his heart went very strongly out at the thought of his son. Then by and by he said to Tristram, "Wilt thou play upon thy harp?" And Tristram said, "Yea, if it will please thee to hear me." Therewith he took his harp and he set it before him, and he struck the strings and played upon it, and he sang in such a wise that no one who was there had ever heard the like thereof.

Then King Meliadus' heart was melted at Tristram's minstrelsy, and he said: "That is wonderful harping. Now ask what thou wilt of me, and it shall be thine, whatever it may be."

To this Tristram said, "Lord, that is a great thing that thou sayest." "Nevertheless," said King Meliadus, "it shall be as I say." Then Tristram left his harp and he came to where King Meliadus sat, and he kneeled down before him and he said: "Lord, if so be that is the case, then that which I ask of thee is this: that thou wilt forgive me and bring me back into thy favor again."

At that King Meliadus was filled with a great wonder, and he said: "Fair youth, who art thou, and what have I to forgive thee?" "Lord," said *King Meliadus* Tristram, "I am thy son, and ask thee to forgive me that I *is reconciled to* should have saved the life of that lady who is thy Queen." *Tristram.* At this King Meliadus cried out with joy, and he came down from where he sat and he took Tristram into his arms and kissed him upon the face, and Tristram wept and kissed his father upon the face.

So they were reconciled.

After that, Tristram abode in peace in Lyonesse for some while, and during that time he made peace betwixt King Meliadus and Queen Moeya, and the Queen loved him because he was so good to her.

Now after the return of Tristram as aforesaid, King Meliadus would have made him a knight, but Tristram would not suffer the honor of *Tristram refuses* knighthood to be bestowed upon him at that time, but always *knighthood.* said: "Lord, think not ill of me if I do not accept knighthood at this time. For I would fain wait until the chance for some large adventure cometh; then I would be made a knight for to meet that adventure, so that I might immediately win renown. For what credit could there be to our house if I should be made knight, only that I might sit in hall and feast and drink and make merry?"

So spoke Sir Tristram, and his words sounded well to King Meliadus, wherefore from thenceforth King Meliadus refrained from urging knighthood upon him.

Now the way that Sir Tristram achieved knighthood shall be told in that which followeth, and also it shall then be told how he fought his first battle, which was one of the most famous that ever he fought in all of his life.

# ing Mark of Cornwall :·

# Chapter Second

*How Sir Tristram was made knight by the King of Cornwall, and how he fought a battle with a famous champion.*

NOW first of all it is to be here said that at that time there was great trouble come to King Mark of Cornwall (who, as aforesaid, was uncle to Sir Tristram) and the trouble was this:

The King of Cornwall and the King of Ireland had great debate concerning an island that lay in the sea betwixt Cornwall and Ireland. For though that island was held by Cornwall, yet the King of Ireland laid claim to it and demanded that the King of Cornwall should pay him truage for the same. This King Mark *The King of Ireland claims truage of Cornwall.* refused to do, and there was great contention betwixt Cornwall and Ireland, so that each country made ready for war.

But the King of Ireland said: "Let there not be war betwixt Ireland and Cornwall concerning this disagreement, but let us settle this affair in some other way. Let us each choose a champion and let those two champions decide the rights of this case by a combat at arms. For so the truth shall be made manifest."

Now you are to know that at that time the knights of Cornwall were held in great disregard by all courts of chivalry; for there was not in those days any knight of repute in all the court of Cornwall. Wherefore King Mark knew not where he should find him a champion to meet that challenge from the King of Ireland. Yet he must needs meet it, for he was ashamed to refuse such a challenge as that, and so to acknowledge that Cornwall had no knight-champion to defend it. So he said it should be as the King of Ireland would have it, and that if the King of Ireland would choose a champion, he also would do the same.

Thereupon the King of Ireland chose for his champion Sir Marhaus of Ireland, who was one of the greatest knights in the world. For in the Book of King Arthur (which I wrote aforetime) you may *The King of Ireland chooses Sir Marhaus for his champion.* there read in the story of Sir Pellias how great and puissant a champion Sir Marhaus was, and how he overthrew Sir Gawaine and others with the greatest ease. Wherefore at that time he was believed by many to be the greatest knight in the world (it being before the days of Sir Launcelot of the Lake), and even in the days of Sir Launcelot it was doubted whether he or Sir Launcelot were the greater champion.

So King Mark could not find any knight in Cornwall to stand against Sir Marhaus. Nor could he easily find any knight outside of Cornwall to do battle with him. For Sir Marhaus, being a knight of the Round Table, no other knight of the Round Table would fight against him—and there were no other knights so great as that famous brotherhood of the Table Round.

Accordingly, King Mark knew not where to turn to find him a champion to do battle in his behalf.

In this strait, King Mark sent a letter by a messenger to Lyonesse, asking if there was any knight at Lyonesse who would stand his champion against Sir Marhaus, and he offered great reward if such a champion would undertake his cause against Ireland.

Now when young Tristram heard this letter of his uncle King Mark, he straightway went to his father and said: "Sire, some whiles ago you desired that I should become a knight. Now I would that you *Tristram asks* would let me go to Cornwall upon this occasion. For when I *leave to go to* come there I will beseech my uncle King Mark to make me a *Cornwall.* knight, and then I will go out against Sir Marhaus. For I have a great mind to undertake this adventure in behalf of King Mark, and to stand his champion against Sir Marhaus. For though Sir Marhaus is so great a knight and so famous a hero, yet if I should have the good fortune to overcome him in battle, there would, certes, be great glory to our house through my knighthood."

Then King Meliadus looked upon Tristram and loved him very dearly, and he said: "Tristram, thou hast assuredly a very great heart to undertake this adventure, which no one else will essay. So I bid thee go, in God's name, if so be thy heart bids thee to go. For maybe God will lend the strength necessary to carry this adventure through to a successful issue."

So that very day Tristram departed from Lyonesse for Cornwall, taking with him only Gouvernail as his companion. So, by ship, he reached Cornwall, and the castle of Tintagel, where King Mark was then holding court.

And it was at the sloping of the afternoon when he so came, and at that time King Mark was sitting in hall with many of his knights and lords about him. And the King was brooding in great trouble of spirit. Unto him came an attendant, saying: "Lord, there are two strangers who stand without, and crave to be admitted to your presence. One of them hath great dignity and sobriety of demeanor, and the other, who is a youth, is of so noble and stately an appearance that I do not believe his like is to be found in the entire world."

To this the King said, "Show them in."

So those two were immediately admitted into the hall and came and stood before King Mark; and the one of them was Gouvernail and the other was young Tristram. So Tristram stood forth before Gouver- *Tristram and* nail and Gouvernail bore the harp of Tristram, and the harp *Gouvernail come* was of gold and shone most brightly and beautifully. Then *to Cornwall.* King Mark looked upon Tristram, and marvelled at his size and beauty; for Tristram stood above any man in that place, so that he looked like a hero amongst them. His brow was as white as milk and his lips were red like to coral and his hair was as red as gold and as plentiful as the mane of a young lion, and his neck was thick and sturdy and straight like to a round pillar of white-stone, and he was clad in garments of blue silk embroidered very cunningly with threads of gold and set with a countless multitude of gems of divers colors. So because of all this he glistened with a singular radiance of richness and beauty.

So King Mark marvelled at the haughtiness of Tristram's appearance, and he felt his heart drawn toward Tristram with love and admiration. Then, after a little, he spoke, saying: "Fair youth, who are you, and whence come you, and what is it you would have of me?"

"Lord," said Tristram, "my name is Tristram, and I come from the country of Lyonesse, where your own sister was one time Queen. Touching the purpose of my coming hither, it is this: having heard that *Tristram offers* you are in need of a champion to contend for your rights against *himself as cham-* the champion of Ireland, I come hither to say that if you will *pion for Corn-* make me a knight with your own hand, I will take it upon me *wall.* to stand your champion and to meet Sir Marhaus of Ireland upon your behalf."

Then King Mark was filled with wonder at the courage of Tristram, and he said: "Fair youth, are you not aware that Sir Marhaus of Ireland is a knight well set in years and of such great and accredited deeds of arms that it is supposed that, excepting Sir Launcelot of the Lake, there is not his peer in any court of chivalry in all of the world? How then can you, who are altogether new to the use of arms, hope to stand against so renowned a champion as he?"

"Lord," quoth Tristram, "I am well aware of what sort of knight Sir Marhaus is, and I am very well aware of the great danger of this undertaking. Yet if one who covets knighthood shall fear to face a danger, what virtue would there then be in the chivalry of knighthood? So, Messire, I put my trust in God, His mercies, and I have great hope that He will lend me both courage and strength in my time of need."

Then King Mark began to take great joy, for he said to himself: "Maybe this youth shall indeed bring me forth in safety out of these dangers that menace my honor." So he said: "Tristram, I do believe that you will stand a very excellent chance of success in this undertaking, wherefore it shall be as you desire; I will make you a knight, and besides that I will fit you with armor and accoutrements in all ways becoming to the estate of a knight-royal. Likewise I will provide you a Flemish horse of the best strain, so that you shall be both furnished and horsed as well as any knight in the world hath ever been."

So that night Tristram watched his armor in the chapel of the castle, and the next day he was made knight with all the circumstances apper-
*Tristram is made knight-royal.* taining to a ceremony of such solemnity as that. And upon the afternoon of the day upon which he was thus made knight, King Mark purveyed a ship in all ways befitting the occasion, and in the ship Tristram and Gouvernail set sail for that island where Sir Marhaus was known to be abiding at that time.

Now upon the second day of their voyaging and about the middle of the day they came to a land which they knew must be the place which they were seeking, and there the sailors made a safe harbor. As soon as they were at anchor a gangway was set from the ship to the shore and Sir Tristram and Gouvernail drave their horses across the gangway and so to the dry land.

Thereafter they rode forward for a considerable distance, until about the first slanting of the afternoon they perceived in the distance three very fair ships drawn up close to the shore. And then they were aware of a knight, clad in full armor and seated upon a noble horse under the shadow of those ships, and they wist that that must be he whom Sir Tristram sought.

Then Gouvernail spake to Sir Tristram, saying: "Sir, that knight resting yonder beneath the shelter of the ships must be Sir Marhaus."

"Yea," said Sir Tristram, "that is assuredly he." So he gazed very steadily at the knight for a long while, and by and by he said: "Gouvernail,
*Sir Tristram goes forth to meet Sir Marhaus.* yonder seems to me to be a very great and haughty knight for a knight so young as I am to have to do with in his first battle; yet if God will lend me His strong aid in this affair, I shall assuredly win me great credit at his hands." Then after another short while he said: "Now go, Gouvernail, and leave me alone in this

affair, for I do not choose for anyone to be by when I have to do with yonder knight. For either I shall overcome him in this combat or else I will lay down my life at this place. For the case is thus, Gouvernail; if Sir Marhaus should overcome me and if I should yield me to him as vanquished, then mine uncle must pay truage to the King of Ireland for the land of Cornwall; but if I died without yielding me to mine enemy, then he must yet do battle with another champion at another time, if my uncle the King can find such an one to do battle in his behalf. So I am determined either to win this battle or to die therein."

Now when Gouvernail heard this, he fell a-weeping in great measure; and he cried out: "Sir, let not this battle be of that sort!" To him Sir Tristram said very steadfastly: "Say no more, Gouvernail, but go as I bid thee." Whereupon Gouvernail turned and went away, as he was bidden to do, weeping very bitterly as he went.

Now by this Sir Marhaus had caught sight of Sir Tristram where he stood in that field, and so presently he came riding thitherward to meet Sir Tristram. When he had come nigh, Sir Marhaus said: "Who art thou, Sir Knight?" Unto these Sir Tristram made reply: *Sir Tristram proclaims his degree.* "Sir, I am Sir Tristram of Lyonesse, son of King Meliadus of that land, and nephew of King Mark of Cornwall. I am come to do battle upon behalf of the King of Cornwall, to release him from the demands of truage made by the King of Ireland." Quoth Sir Marhaus: "Messire, are you a knight of approval and of battles?" "Nay," said Sir Tristram, "I have only been created knight these three days."

"Alas!" said Sir Marhaus, "I am very sorry for thee and for thy noble courage that hath brought thee hither to this place. Thou art not fit to have to do with me, for I am one who hath fought in more than twice twenty battles, each one of which was, I believe, greater than this is like to be. Also I have matched me with the very best knights in the world, and have never yet been overcome. So I advise thee, because of thy extreme youth, to return to King Mark and bid him send me another champion in thy stead, who shall be better seasoned than thou art."

"Sir," said Sir Tristram, "I give thee gramercy for thy advice. But I may tell thee that I was made knight for no other purpose than to do battle with thee; so I may not return without having fulfilled mine adventure. Moreover, because of thy great renown and thy courage and prowess, I feel all the more desirous to have to do with thee; for if I should die at thy hand, then there will be no shame to me, but if I should win this battle from thee, then I shall have very great renown in the courts of chivalry."

"Well," said Sir Marhaus, "it is not likely that thou shalt die at my hand. For because of thy youth I will not have it that this battle shall be so desperate as that." "Say not so," said Sir Tristram, "for either I shall die at thy hand, or else I shall overcome thee in this battle, for I make my vow to God that I will not yield myself to thee so long as there is life within my body."

"Alas!" said Sir Marhaus, "that is certes a great pity. But as thou hast foreordained it, so it must needs be." Therewith he saluted Sir Tristram and drew rein and rode aside to a little distance where he straightway made ready for that battle. Nor was Sir Tristram behind him in making preparation, albeit he was filled with doubts as to the outcome of that undertaking.

Then when they were in all ways prepared, each gave shout and drave spurs into his horse and rushed toward the other with such fury that it *Sir Tristram* was terrible to behold. And each smote the other with his *is wounded.* spear in the centre of his shield, and in that encounter Sir Marhaus smote through Sir Tristram's shield and gave Sir Tristram a great wound in his side. Then Sir Tristram felt the blood gush out of that wound in such abundance that it filled his iron boots, so that they were sodden therewith, and he thought he had got his death-wound. But in spite of that grievous bitter stroke, he held his seat and was not overthrown. Then so soon as he had recovered himself he voided his horse and drew his sword and set his shield before him; and when Sir Marhaus saw his preparations, he likewise voided his horse and made ready for battle upon foot. So straightway they came together with terrible fury, lashing at each other with such fearful strength and evil will that it was dreadful to behold. And each gave the other many grievous strokes, so that whole pieces of armor were hewn off from their bodies; and each gave the other many deep wounds, so that that part of the armor that still hung to them became red as though it were painted with red. Likewise the ground was all besprinkled red where they stood, yet neither gave any thought to quitting that battle in which they were engaged.

Now for a while Sir Tristram feared because of the wound which he had at first received that he would die in that battle, but by and by he perceived that he was stouter than Sir Marhaus and better winded; wherefore great hope came to him and uplifted him with redoubled strength. Then presently Sir Marhaus fell back a little and when Sir Tristram perceived that he ran in upon him and smote him several times, such direful strokes that Sir Marhaus could not hold up his shield against that assault. Then Sir Tristram perceived that Sir Marhaus was no longer able to hold up his shield, and therewith he smote him a great blow with his sword upon the helmet, So direful was that blow that the sword of Sir Tristram pierced very deep through the helm of Sir Marhaus and into the brain-pan. And Sir Tristram's sword stuck fast in the helm and the brain-pan of Sir Marhaus so that Sir Tristram could not pull it out again. Then Sir Marhaus, half a-swoon, fell down upon his knees, and therewith a part of the edge of the blade brake off from Sir Tristram's sword, and remained in the wound that he had given to Sir Marhaus. *Sir Tristram gives Sir Marhaus a death-wound.*

Then Sir Marhaus was aware that he had got his death-wound, wherefore a certain strength came to him so that he rose to his feet staggering like a drunken man. And at first he began going about in a circle and crying most dolorously. Then as he wist all that had happed he threw away his sword and his shield, and made away from that place, staggering and stumbling like one who had gone blind; for he was all bewildered with that mortal wound, and wist not very well what he was doing or whither he was going. Then Sir Tristram would have made after him to stop him, but he could not do so because he himself was so sorely wounded and so weak from the loss of blood. Yet he called after Sir Marhaus: "Stay, stay, Sir Knight! Let us finish this battle now we are about it!" But to this Sir Marhaus made no answer, but went on down to his ships, staggering and stumbling like a blind man as aforesaid, for the sore wound which he had received still lent him a false strength of body so that he was able to go his way. Then those who were aboard the ships, beholding him thus coming staggering toward them, came down and met him and lifted him up and bore him away to his own ship. Thereafter, as soon as might be they hoisted sail and lifted anchor and took their way from that place. *Sir Marhaus leaves the field.*

Then by and by came Gouvernail and several others of Sir Tristram's party to where Sir Tristram was; and there they found him leaning upon his sword and groaning very sorely because of the great wound in his side. So presently they perceived that he could not walk, wherefore they lifted him up upon his own shield and bore him thence to that ship that had brought him thither.

And when they had come to the ship they laid him down upon a couch and stripped him of his armor to search his wounds. Then they beheld what a great wound it was that Sir Marhaus had given him in the side, and they lifted up their voices in sorrow, for they all believed that he would die.

So they set sail, and in two days brought him back to King Mark, where he sat at Tintagel in Cornwall. And when King Mark saw how pale and wan and weak Sir Tristram was, he wept and grieved very sorely for sorrow of that sight, for he too thought that Sir Tristram was certainly about to die.

*Sir Tristram returns to Cornwall.*

But Sir Tristram smiled upon King Mark, and he said: "Lord, have I done well for thy sake?" And King Mark said, "Yea," and fell to weeping again.

"Then," quoth Tristram, "it is time for me to tell thee who I am who have saved thy kingdom from the shame of having to pay truage to Ireland, and that I am thine own sister's son. For my father is King Meliadus of Lyonesse, and my mother was the Lady Elizabeth, who was thine own sister till God took her soul to Paradise to dwell there with His angels."

*Sir Tristram proclaims himself to King Mark.*

But when King Mark heard this he went forth from that place and into his own chamber. And when he had come there he fell down upon his knees and cried out aloud: "Alas, alas, that this should be! Rather, God, would I lose my entire kingdom than that my sister's son should come to his death in this wise!"

Now it remaineth to say of Sir Marhaus that those who were with him brought him back to Ireland and that there in a little while he died of the wound that Sir Tristram had given him upon the head. But ere he died, and whilst they were dressing that hurt, the Queen of Ireland, who was sister to Sir Marhaus, discovered the broken piece of the blade still in that grim wound. This she drew forth and set aside, and hid very carefully, saying to herself: "If ever I meet that knight to whose sword this piece of blade fitteth, then it will be an evil day for him."

Thus I have told you all the circumstances of that great battle betwixt Sir Tristram of Lyonesse and Sir Marhaus of Ireland. And now you shall hear how it befell Sir Tristram thereafter; so harken to what followeth.

he Lady Belle Isoult.

# Chapter Third

*How Sir Tristram went to Ireland to be healed of his wound by the King's daughter of Ireland, and of how he came to love the Lady Belle Isoult. Also concerning Sir Palamydes and the Lady Belle Isoult.*

NOW that grievous hurt which Sir Tristram had received at the hands of Sir Marhaus did not heal, but instead grew even more rankled and sore, so that there were many who thought that there had been treachery practised and that the spearhead had been poisoned to cause such a malignant disease as that with which the wounded man suffered. So by and by Sir Tristram grew so grievously sick of his hurt that all those who were near him thought that he must certainly die.

Then King Mark sent everywhere and into all parts for the most wise and learned leeches and chirurgeons to come to Cornwall and search the wounds of Sir Tristram, but of all these no one could bring him any ease.

Now one day there came to the court of King Mark a very wise lady, who had travelled much in the world and had great knowledge of wounds of all sorts. At the bidding of the King, she went to where *How Sir Tristram lieth sick in Cornwall.* Sir Tristram lay, and searched the wound as so many had already done. And when she had done that she came out of Sir Tristram's chamber and unto King Mark, where he was waiting for her. Then King Mark said to her: "Well, how will it be with yonder knight?" "Lord," quoth she, "it is thus; I can do nothing to save his life, nor do I know of any one who may save it unless it be the King's daughter of Ireland, who is known as the Belle Isoult because of her wonderful beauty. She is the most skilful leech in all of the world, and she alone may hope to bring Sir Tristram back to life and health again, for I believe that if she fail no one else can save him."

Then after the aforesaid lady had gone, King Mark went to where Sir Tristram lay, and he told him all that she had said concerning his condition; and King Mark said: "Tristram, wilt thou go to the King's daughter of Ireland and let her search thy wound?"

Then Sir Tristram groaned at the thought of the weariness and pain of moving, and he said: "Lord, this is a great undertaking for one who is so sick.

Moreover, it is a great risk for me, for, if I go to Ireland, and if it be found that I am he who slew Sir Marhaus, then it is hardly likely that I shall ever escape from that country again with my life. Ne'theless, I am so sorely sick of this wound that I would rather die than live as I am living; wherefore I will go to Ireland for the sake of being healed, if such a thing is possible."

Accordingly, a little while after that, King Mark provided a ship to carry Sir Tristram to Ireland. This ship he furnished with sails of silk of divers colors, and he had it hung within with fine embroidered cloth, and fabrics woven with threads of silver and gold, so that in its appearance it was a worthy vessel even for a great king to sail in. Then, when all was ready, King Mark had a number of attendants carry Sir Tristram down to the ship in a litter, and he had them lay Sir Tristram upon a soft couch of crimson satin, which was set upon the deck beneath a canopy of crimson silk, embroidered with threads of silver and garnished with fringe of silver, and Sir Tristram lay there at ease where the breezes of the ocean came pleasantly to him, and breathed upon his face and his temples and his hair and his hands with coolness; and Gouvernail was with Sir Tristram all the while in attendance upon him.

So they set sail for Ireland, the weather being very fair and pleasant, *Sir Tristram sails to Ireland to have his wound searched.* and on the third day, at about the time of sunset, they came to a part of the coast of Ireland where there was a castle built upon the rocks that rose out of the sea.

Now there were several fishermen fishing in boats near that castle, and of these the pilot of Sir Tristram's boat made inquiry what castle that was. To him the fisherman replied: "That castle is the castle of King Angus of Ireland." And the fisherman said: "It so happens that the King and Queen and their daughter, hight the Lady Belle Isoult, and all of their court are there at this very while."

This Sir Tristram heard and said: "This is good news, for indeed I am very sick and am right glad that my voyaging is ended." So he gave orders that the pilot should bring the ship close under the walls of that castle, and that he should there let go anchor; and the pilot did as Sir Tristram had commanded him.

Now, as aforesaid, that ship was of a very wonderful appearance, like to the ship of a king or a high prince, wherefore many people came down *How Sir Tristram came to Ireland.* to the walls of the castle and stood there and gazed at the vessel as it sailed into the harbor. And by that time the sun had set and all the air was illuminated with a marvellous golden light; and in this sky of gold the moon hung like a shield of silver, very bright and steady above the roofs and towers of the castle. And there came from the land a pleasing perfume of blossoms; for it was then in the fulness of the spring-time, and all the fruit-bearing trees were luxuriant with bloom so that the soft air of evening was full of fragrance thereby.

Then there came a great content into the heart of Sir Tristram, wherefore he said to Gouvernail: "Gouvernail, either I shall soon be healed of this wound, or else I shall presently die and enter into Paradise free of pain, for I am become very full of content and of peace toward all men." And then he said: "Bring me hither my harp, that I may play upon it a little, for I have a desire to chant in this pleasant evening-time."

So Gouvernail brought to Sir Tristram his shining harp, and when Sir Tristram had taken it into his hands he tuned it, and when he had tuned it he struck it and sang; and, because of the stillness of the    *Sir Tristram* evening, his voice sounded marvellously clear and sweet across   *sings.* the level water, so that those who stood upon the castle walls and heard it thought that maybe an angel was singing on board of that ship.

That time the Lady Belle Isoult sat at the window of her bower enjoying the pleasantness of the evening. She also heard Sir Tristram singing, and she said to those damsels who were with her, "Ha, what is that I hear?" Therewith she listened for a little while, and then she said: "Meseems that must be the voice of some angel that is singing." They say: "Nay, Lady, it is a wounded knight singing, and he came to this harbor in a wonderful ship some while ago." Then the Lady Belle Isoult said to a page who was in attendance: "Bid the King and Queen come hither, that they may hear this singing also, for never did I think to hear such singing beyond the walls of Paradise."

So the page ran with all speed, and in a little the King and Queen came to the bower of the Lady Belle Isoult; and she and they leaned upon the window-ledge and listened to Sir Tristram whilst he sang in the soft twilight. Then by and by King Angus said: "Now I will have yonder minstrel brought hither to this castle to do us pleasure, for I believe that he must be the greatest minstrel in all the world to sing in that wise." And the Lady Belle Isoult said: "I pray you, sir, do so, for it would be great joy to everybody to have such singing as that in this place."

So King Angus sent a barge to that ship, and besought that he who sang should be brought to the castle. At that Sir Tristram was very glad, for he said: "Now I shall be brought to the Lady the Belle Isoult and maybe she will heal me." So he had them bare him to the barge of the King of Ireland, and so they brought him to the castle of King Angus, where they laid him upon a bed in a fair room of the castle.

Then King Angus came to Sir Tristram where he lay, and he said: "Messire

what can I do for you to put you more at your ease than you are?" "Lord,"
said Sir Tristram, "I pray you to permit the Lady Belle Isoult to search a

*King Angus cometh to Tristram.* great wound in my side that I received in battle. For I hear that she is the most skilful leech in all the world, and so I have come hither from a great distance, being in such pain and dole from
my grievous hurt that I shall die in a little while unless it be healed."

"Messire," said King Angus, "I perceive that you are no ordinary knight,
but somebody of high nobility and estate, so it shall be as you desire." And
then King Angus said: "I pray you, tell me your name and whence you come."

Upon this, Sir Tristram communed within his own mind, saying: "An
I say my name is Tristram, haply there may be someone here will know
me and that I was the cause why the brother of the Queen of this place
hath died." So he said: "Lord, my name is Sir Tramtris, and I am come
from a country called Lyonesse, which is a great distance from this."

Quoth King Angus, "Well, Sir Tramtris, I am glad that you have come
to this place. Now it shall be done to you as you desire, for to-morrow
the Lady Belle Isoult shall search your wound to heal it if possible."

And so it was as King Angus said, for the next day the Lady Belle Isoult
came with her attendants to where Sir Tristram lay, and one of the at-

*My Lady Belle Isoult searches the wound.* tendants bare a silver basin and another bare a silver ewer, and others bare napkins of fine linen. So the Lady Belle Isoult came close to Sir Tristram and kneeled beside the couch whereon he lay and said, "Let me see the wound." Therewith Sir Tristram laid
bare his bosom and his side and she beheld it. Then she felt great pity
for Sir Tristram because of that dolorous wound, and she said: "Alas, that
so young and so fair and so noble a knight should suffer so sore a wound as
this!" Therewith still kneeling beside Sir Tristram she searched the
wound with very gentle, tender touch (for her fingers were like to rose leaves
for softness) and lo! she found a part of the blade of a spear-head embedded
very deep in the wound of Sir Tristram.

This she drew forth very deftly (albeit Sir Tristram groaned with a great
passion of pain) and therewithafter came forth an issue of blood like a
crimson fountain, whereupon Sir Tristram swooned away like one who
had gone dead. But he did not die, for they quickly staunched the flow,
set aromatic spices to his nostrils, so that in a little he revived in spirit to
find himself at great ease and peace in his body (albeit it was for a while
like to the peace of death).

*Sir Tristram is healed.* Thus it was that the Lady Belle Isoult saved the life of Sir Tristram, for in a little while he was able to be about again,
and presently waxed almost entirely hale and strong in limb and body.

And now it is to be told how Sir Tristram loved the Lady Belle Isoult and how she loved Sir Tristram. Also how a famous knight, hight Sir Palamydes the Saracen, loved Belle Isoult and of how she loved not him.

For, as was said, it came about that in a little while Sir Tristram was healed of that grievous wound aforetold of so that he was able to come and go whithersoever he chose. But always he would be with the Lady Belle Isoult, for Sir Tristram loved her with a wonderfully passionate regard. And so likewise the lady *Sir Tristram loves the Lady Belle Isoult.* loved Sir Tristram. For if he loved her because she had saved his life, then she also loved him for the same reason. For she did not ever forget how she had drawn out the head of that spear from the wound at his side, and of how he had groaned when she brought it forth, and of how the blood had gushed out of that wound. Wherefore she loved him very aboundingly for the agony of pain she had one time caused him to suffer.

So they two fair and noble creatures were always together in bower or in hall, and no one in all that while wist that Sir Tramtris was Sir Tristram, and that it was his hand that had slain Sir Marhaus of Ireland.

So Sir Tristram was there in Ireland for a year, and in that time he grew to be altogether well and sturdy again.

Now it was in those days that there came Sir Palamydes the Saracen knight to that place, who was held to be one of the very foremost knights in the world. So great rejoicing was made over him because *Sir Palamydes cometh to Ireland.* he had come thither, and great honor was shown to him by everyone.

But when Sir Palamydes beheld the Lady Belle Isoult and when he saw how fair she was, he came in a short while to love her with almost as passionate a regard as that with which Sir Tristram loved her, so that he also sought ever to be with her whenever the chance offered.

But Belle Isoult felt no regard for Sir Palamydes, but only fear of him, for all of her love was given to Sir Tristram. Nevertheless, because Sir Palamydes was so fierce and powerful a knight, she did not dare to offend him; wherefore she smiled upon him and treated him with all courtesy and kindness although she loved him not, dissembling her regard for him.

All this Sir Tristram beheld from aside and it displeased him a very great deal to see how Sir Palamydes was always beside the lady. But Belle Isoult beheld how Sir Tristram was displeased, wherefore she *Sir Tristram is displeased.* took occasion to say to him: "Tramtris, be not displeased, for what am I to do? You know very well that I do not love this knight, but I am afraid of him because he is so fierce and so strong."

To this Sir Tristram said: "Lady, it would be a great shame to me if I, being by, should suffer any knight to come betwixt you and me and win your regard through fear of him."

She said: "Tramtris, what would you do? Would you give challenge to this knight? Lo, you are not yet entirely healed of your hurt, and Sir Palamydes is in perfect strength of body. For indeed it is for you I am most of all afraid lest you and Sir Palamydes should come to battle and lest he should do you a harm before you are entirely healed."

"Lady," quoth Sir Tristram, "I thank God that I am not at all afraid of this knight, or of any other knight, and I have to thank you that I am now *Sir Tristram de-* entirely recovered and am as strong as ever I was. Wherefore *sires to do battle.* I have now a mind to deal with this knight in your behalf. So if you will provide me with armor I will deal with him so that maybe he will not trouble you again. Now I will devise it in this way:—tell your father, King Angus, to proclaim a great jousting. In that jousting I will seek out Sir Palamydes and will encounter him, and I hope with God's aid that I shall overcome him, so that you shall be free from him."

Belle Isoult said, "Tramtris, are you able for this?" He said, "Yea, I am as ready as ever I shall be in all of my life." Whereat Belle Isoult said, "It shall be as you will have it."

Then Sir Tristram charged Belle Isoult that she should keep secret all this that had been said betwixt them. And more especially she was to keep it secret that he was to take part in such a tournament as that which they had devised. And he said to her: "Lady, I lie here under a great peril to my life, though I cannot tell you what that peril is. But I may tell you that if my enemies should discover me at this place, it would go hard with me to preserve my life from them. Wherefore, if I take part in any such affair as this, it must be altogether a secret betwixt us."

So therewith they parted and Lady Belle Isoult went to her father and besought him to proclaim a great day of jousting in honor of Sir Palamydes, and the King said that he would do so. So the King sent forth proclamation to all the courts of that nation that a great tournament was to be held and that great rewards and great honors were to be given to the best knight thereat. And that tournament was talked about in all the courts of chivalry where there were knights who desired to win glory in such affairs at arms.

And now it shall be told concerning that tournament and how it befell with Sir Tristram thereat, and with Sir Palamydes thereat.

he Queen of Ireland seeks to slay Sir Tristram. ₰₰

# Chapter Fourth

*How Sir Tristram encountered Sir Palamydes at the tournament and of what befell. Also how Sir Tristram was forced to leave the Kingdom of Ireland.*

SO came the time for the tournament that King Angus of Ireland had ordained; and that was a very famous affair at arms indeed. For it hath very rarely happened that so noble a gathering of knights hath ever come together as that company which there presented itself for that occasion at the court of the King of Ireland.

For you may know how excellent was the court of chivalry that foregathered thereat when you shall hear that there came to that tournament, the King of an Hundred Knights and the King of the Scots, and that there came several knights of the Round Table, to wit: Sir Gawaine, Sir Gaheris and Sir Agravaine; and Sir *Of the court of chivalry at Ireland.* Bagdemagus and Sir Kay and Sir Dodinas, and Sir Sagramore le Desirous, and Sir Gumret the Less, and Sir Griflet; and that there came besides these many other knights of great renown.

These and many others gathered at the court of King Angus of Ireland, so that all those meadows and fields coadjacent to the place of battle were gay as beds of flowers with the multitude of tents and pavilions of divers colors that were there emplanted.

And on the day of the tournament there came great crowds of people into the lists, so that all that place was alive with movement. For it was as though a sea of people had arisen to overflow the seats and stalls thereof.

Now that tournament was to last for three days, and upon the third day there was to be a grand mêlée in which all these knights contestant were to take stand upon this side or upon that.

But upon the first two of those three days Sir Tristram sat in the stall of the King and looked down upon the jousting, for, because of the illness from which he had recovered, he was minded to save his body until the right time should come, what time he should be called upon to do his uttermost.

And in those two days, Sir Tristram beheld that Sir Palamydes did more wonderfully in battle than he would have believed it possible for any knight *Sir Palamydes* to do. For Sir Palamydes was aware that the eyes of the Lady *performeth won-* Belle Isoult were gazing upon him, wherefore he felt himself *ders.* uplifted to battle as with the strength of ten. Wherefore he raged about that field like a lion of battle, seeking whom he might overthrow and destroy. And upon the first day he challenged Sir Gawaine to joust with him, and then he challenged Sir Gaheris, and the King of an Hundred Knights, and Sir Griflet, and Sir Sagramore le Desirous and fourteen other knights, and all of these he met and many he overcame, and that without any mishap to himself. And upon the second day he met with great success Sir Agravaine and Sir Griflet and Sir Kay and Sir Dodinas and twelve other knights. Wherefore those who beheld how he did gave great shouts and outcries of applause and acclaim, saying: "Certes, there was never knight in all of the world so great as this knight. Yea; even Sir Launcelot himself could not do more than that knight doeth."

Then Belle Isoult was troubled in her mind, and she said: "Tramtris, yonder in very truth is a most fierce and terrible knight. Now somewhiles I have fear that you may not be able to overcome him."

Thereat Sir Tristram smiled very grimly, and said: "Lady, already I have overcome in battle a bigger knight than ever Sir Palamydes has been or is like to be." But the Lady Belle Isoult wist not that that knight of whom Sir Tristram spake was Sir Marhaus of Ireland.

Now upon the evening of the second day of that tournament, Sir Palamydes came to where the Lady Belle Isoult was, and he said: "Lady, all *Sir Palamydes* these things I have done for your sake. For had it not been *bespeaks the* for my love for you, I would not have been able to do a third *Lady Belle* part of that which I did. Now I think you should have pity *Isoult.* and regard for one who loves you so strongly as that; wherefore I beseech you to bestow some part of your good-will upon me."

"Sir," said the Lady Belle Isoult, "you are not to forget that there is still another day of this battle, and in it you may not happen to have the same fortune that favored you to-day; so I will wait until you have won that battle also before I answer you."

"Well," said Sir Palamydes, "you shall see that I shall do even more worthily to-morrow for your sake than I have done to-day."

But the Lady Belle Isoult was not very well pleased with that saying, for she began again to fear that maybe the will of Sir Palamydes was so strong that Sir Tristram would not have any success against him.

So came the third day of that very famous contest at arms, and when the morning was come there began to gather together in the two parties those who were to contest the one against the other. Of one of these parties, Sir Palamydes was the chiefest knight, and upon that side was also Sir Gawaine and several of the knights who were with him. For these said, "There shall certes be greater credit to be had with Sir Palamydes than against him," and so they joined them with his party. Of the other party the chiefest knights were the King of an Hundred Knights and the King of Scots, and both of these were very famous and well-approved champions, of high courage and remarkable achievements.

Now when the time was nigh ready for that tournament, Sir Tristram went to put on the armor that the Lady Belle Isoult had provided him, and when he was armed he mounted very lightly upon the horse *Belle Isoult* which she had given him. And the armor of Sir Tristram was *arms Sir Tris-* white, shining like to silver, and the horse was altogether white, *tram.* and the furniture and trappings thereof were all white, so that Sir Tristram glistened with extraordinary splendor.

Now when he was armed and prepared in all ways, the Lady Belle Isoult came to where he was and she said, "Tramtris, are you ready?" And he answered "Yea." Therewith she took the horse of Sir Tristram by the bridle and she led him to the postern gate of the castle, and put him out that way into a fair field that lay beyond; and Sir Tristram abided in the fields for some while until the tournament should have begun.

But the Lady Belle Isoult went to the tournament with her father, the King, and her mother, the Queen, and took her station at that place assigned to her whence she might overlook the field.

So in a little while that friendly battle began. And again Sir Palamydes was filled with the vehement fury of contest, wherefore he raged about the field, spreading terror whithersoever he came. For first *How Sir Pala-* he made at the King of an Hundred Knights, and he struck *mydes fought in* that knight so direful a blow that both horse and man fell *the tournament.* to the ground with the force thereof. Then in the same manner he struck the King of Scots with his sword, and smote him straightway out of the saddle also. Then he struck down one after another, seven other knights, all of well-proved strength and prowess, so that all those who looked thereon cried out, "Is he a man or is he a demon?" So, because of the terror of Sir Palamydes, all those in that contest bore away from him as they might do from a lion in anger.

At this time came Sir Tristram, riding at a free pace, shining like to a figure of silver. Then many saw him and observed him and said to one another: "Who is this knight, and what party will he join with to do battle?" These had not long to wait to know what side he would join, for immediately Sir Tristram took stand with that party which was the party of the King of an Hundred Knights and the King of Scots, and at that the one party was very glad, and the other party was sorry; for they deemed that Sir Tristram was certes some great champion.

Then straightway there came against Sir Tristram four knights of the other party, and one of these was Sir Gaheris, and another was Sir Griflet *Sir Tristram* and another was Sir Bagdemagus and another was Sir Kay. *enters the* But Sir Tristram was possessed with a great joy of battle, *tournament.* so that in a very short time he had struck down or overthrown all those knights, beginning with Sir Gaheris, and ending with Sir Kay the seneschal.

This Sir Gawaine beheld, and said to Sir Sagramore: "Yonder is certes a knight of terrible strength; now let us go and see of what mettle he be."

Therewith Sir Gawaine pushed against Sir Tristram from the one side, and Sir Sagramore came against him on the other side, and so they met him both at once. Then first Sir Gawaine struck Sir Tristram such a buffet that the horse of Sir Tristram turned twice about with the force of that stroke; and therewith Sir Sagramore smote him a buffet upon the other side so that Sir Tristram wist not upon which side to defend himself.

Then, at those blows Sir Tristram waxed so exceedingly fierce that it was as though a fire of rage flamed up into his brains and set them into a blaze of rage. So with that he rose up in his stirrups and launched so dreadful a blow upon Sir Gawaine that I believe nothing could have withstood the force of that blow. For it clave through the shield of Sir Gawaine and it descended upon the crown of his helmet and it clave away a part of his helmet and a part of the épaulière of his shoulder; and with the force of that dreadful, terrible blow, Sir Gawaine fell down upon the ground and lay there as though he were dead.

Then Sir Tristram wheeled upon Sir Sagramore (who sat wonder-struck at that blow he had beheld) and thereafter he smote him too, so that he fell down and lay upon the ground in a swoon from which he did not recover for more than two hours.

Now Sir Palamydes also had beheld those two strokes that Sir Tristram had given, wherefore he said: "Hah! Yonder is a very wonderful knight. Now if I do not presently meet him, and that to my credit, he will have more honor in this battle than I."

So therewith Sir Palamydes pushed straight against Sir Tristram, and when Sir Tristram beheld that he was very glad, for he said: "Now it will either be Sir Palamydes his day, or else it will be mine." So he upon his part pushed against Sir Palamydes with good intent to engage him in battle, and then they two met in the midst of the field. *Sir Palamydes rides against Sir Tristram.*

Then immediately Sir Palamydes smote Sir Tristram such a buffet that Sir Tristram thought a bolt of lightning had burst upon him, and for a little while he was altogether bemazed and wist not where he was. But when he came to himself he was so filled with fury that his heart was like to break therewith.

Thereupon he rushed upon Sir Palamydes and smote him again and again and again with such fury and strength that Sir Palamydes was altogether stunned at the blows he received and bare back before them. Then Sir Tristram perceived how that Sir Palamydes bare his shield low because of the fierceness of that assault, and thereupon he rose up in his stirrups and struck Sir Palamydes upon the crown of the helmet so dreadful a buffet that the brains of Sir Palamydes swam like water, and he must needs catch the pommel of his saddle to save himself from falling. Then Sir Tristram smote him another buffet, and therewith darkness came upon the sight of Sir Palamydes and he rolled off from his horse into the dust beneath its feet. *Sir Tristram smites Sir Palamydes.*

Then all who beheld the encounter shouted very loud and with great vehemence, for it was the very best and most notable assault at arms that had been performed in all that battle. But most of those who beheld that assault cried out "The Silver Knight!" For at that time no one but the Lady Belle Isoult wist who that silver knight was. But she wist very well who he was, and was so filled with the glory of his prowess that she wept for joy thereof.

Then the King of Ireland said: "Who is yonder knight who hath so wonderfully overthrown Sir Palamydes? I had not thought there was any knight in the world so great as he; but this must be some great champion whom none of us know." Upon that the Lady Belle Isoult, still weeping for joy, could contain herself *Belle Isoult declares Sir Tristram.* no longer, but cried out: "Sir, that is Tramtris, who came to us so nigh to death and who hath now done us so great honor being of our household! For I knew very well that he was no common knight but some mighty champion when I first beheld him."

At that the King of Ireland was very much astonished and overjoyed, and he said: "If that is indeed so, then it is a very great honor for us all."

Now after that assault Sir Tristram took no more part in that battle but withdrew to one side. But he perceived where the esquires attendant upon Sir Palamydes came to him and lifted him up and took him away. Then by and by he perceived that Sir Palamydes had mounted his horse again with intent to leave that meadow of battle, and in a little he saw Sir Palamydes ride away with his head bowed down like to one whose heart was broken.

All this Sir Tristram beheld and did not try to stay Sir Palamydes in his departure. But some while after Sir Palamydes had quitted that place, Sir Tristram also took his departure, going in that same direction that Sir Palamydes had gone. Then after he had come well away from the meadow of battle, Sir Tristram set spurs to his horse and rode at a hard gallop along that way that Sir Palamydes had taken.

So he rode at such a gait for a considerable pass until, by and by, he perceived Sir Palamydes upon the road before him; and Sir Palamydes was at that time come to the edge of a woods where there were several stone windmills with great sails swinging very slowly around before a strong wind that was blowing.

Now this was a lonely place, and one very fit to do battle in, wherefore Sir Tristram cried out to Sir Palamydes in a loud voice: "Sir Palamydes! Sir Palamydes! Turn you about! For here is the chance for you to recover the honor that you have lost to me." Thereupon Sir Palamydes, hearing that loud voice, turned him about. But when he beheld that the knight who called was he who had just now wrought such shame upon him, he ground his teeth together with rage, and therewith drave his horse at Sir Tristram, drawing his sword so that it flashed like lightning in the bright sunlight. And when he came nigh to Sir Tristram, he stood up in his stirrups and lashed a blow at him with all his might and main; for he said to himself: "Maybe I shall now recover mine honor with one blow which I lost to this knight a while since." But Sir Tristram put aside that blow of Sir Palamydes with his shield with very great skill and dexterity, and thereupon, recovering himself, he lashed at Sir Palamydes upon his part. And at that first stroke Sir Tristram smote down the shield of Sir Palamydes, and gave him such a blow upon the head that Sir Palamydes fell down off his horse upon the earth. Then Sir Tristram voided his own horse very quickly, and running to Sir Palamydes where he lay he plucked off his helmet with great violence. Therewith he cried out very fiercely: "Sir Knight, yield thee to me, or I will slay thee." And therewithal he lifted up his sword as though to strike off the head of Sir Palamydes.

*Sir Tristram overthrows Palamydes again.*

Then when Sir Palamydes saw Sir Tristram standing above him in that wise, he dreaded his buffets so that he said: "Sir Knight, I yield me to thee to do thy commands, if so be thou wilt spare my life."

Thereupon Sir Tristram said, "Arise," and at that Sir Palamydes got him up to his knees with some ado, and so remained kneeling before Sir Tristram.

"Well," said Sir Tristram, "I believe you have saved your life by thus yielding yourself to me. Now this shall be my commandment upon you. First of all, my commandment is that you forsake the Lady Belle Isoult, and that you do not come near her for the space of an entire year. And this is my second commandment; that from this day you do not assume the arms of knighthood for an entire year and a day."

"Alas!" said Sir Palamydes, "why do you not slay me instead of bringing me to such shame as this! Would that I had died instead of yielding myself to you as I did." And therewith he wept for shame and despite.

"Well," said Sir Tristram, "let that pass which was not done. For now you have yielded yourself to me and these are my commands." So with that Sir Tristram set his sword back again into its sheath, and he mounted his horse and rode away, leaving Sir Palamydes where he was.

But after Sir Tristram had gone, Sir Palamydes arose, weeping aloud. And he said: "This is such shame to me that I think there can be no greater shame." Thereupon he drew his misericordia, and *Sir Palamydes* he cut the thongs of his harness and he tore the pieces of armor *disarms himself.* from off his body and flung them away very furiously, upon the right hand and upon the left. And when he had thus stripped himself of all of his armor, he mounted his horse and rode away into the forest, weeping like one altogether brokenhearted.

So Sir Tristram drave Sir Palamydes away from the Lady Belle Isoult as he had promised to do.

Now when Tristram came back to the castle of the King of Ireland once more, he thought to enter privily in by the postern-gate as he had gone out. But lo! instead of that he found a great party waiting for him before the castle and these gave him loud acclaim, crying, "Welcome, Sir Tramtris! Welcome, Sir Tramtris!" And King Angus came forward and took the hand of Sir Tristram, and he also said: "Welcome, Sir Tramtris, for you have brought us great honor this day!"

But Sir Tristram looked at the Lady the Belle Isoult with great reproach and by and by when they were together he said: "Lady, why did you

betray me who I was when you had promised me not to do so?" "Sir,"
she said, "I meant not to betray you, but in the joy of your victory I

*Sir Tristram* know not very well what I said." "Well," said Sir Tris-
*chides Belle* tram, "God grant that no harm come of it." She said,
*Isoult.* "What harm can come of it, Messire?" Sir Tristram said:
"I may not tell you, Lady, but I fear that harm will come of it."

Anon the Queen of Ireland came and said: "Tramtris, one so nigh to death
as you have been should not so soon have done battle as you have done.
Now I will have a bain prepared and you shall bathe therein, for you are not
yet hale and strong."

"Lady," said Tristram, "I do not need any bain, for I believe I am now
strong and well in all wise."

"Nay," said the Queen, "you must have that bain so that no ill may
come to you hereafter from this battle which you have fought."

So she had that bain prepared of tepid water, and it was very strong
and potent with spices and powerful herbs of divers sorts. And when
that bain was prepared, Sir Tristram undressed and entered the bath,
and the Queen and the Lady Belle Isoult were in the adjoining chamber
which was his bed-chamber.

Now whilst Sir Tristram was in that bath, the Queen and Belle Isoult
looked all about his chamber. And they beheld the sword of Sir Tristram

*The Queen of* where it lay, for he had laid it upon the bed when he had
*Ireland beholds* unlatched the belt to make himself ready for that bath. Then
*Sir Tristram's* the Queen said to the Lady Belle Isoult, "See what a great
*sword.* huge sword this is," and thereupon she lifted it and drew the
blade out of its sheath, and she beheld what a fair, bright, glistering sword
it was. Then in a little she saw where, within about a foot and a half from
the point, there was a great piece in the shape of a half-moon broken out
of the edge of the sword; and she looked at that place for a long while. Then
of a sudden she felt a great terror, for she remembered how even such a
piece of sword as that which had been broken off from that blade, she had
found in the wound of Sir Marhaus of which he had died. So she stood
for a while holding that sword of Sir Tristram in her hand and looking as
she had been turned into stone. At this the Lady Belle Isoult was filled
with a sort of fear, wherefore she said, "Lady, what ails you?" The
Queen said, "Nothing that matters," and therewith she laid aside the
sword of Sir Tristram and went very quickly to her own chamber. There
she opened her cabinet and took thence the piece of sword-blade which she
had drawn from the wound of Sir Marhaus, and which she had kept ever
since. With this she hurried back to the chamber of Sir Tristram, and

fitted that piece of the blade to the blade; and lo! it fitted exactly, and without flaw.

Upon that the Queen was seized as with a sudden madness; for she shrieked out in a very loud voice, "Traitor! Traitor! Traitor!" saying that word three times. Therewith she snatched up the sword of Sir Tristram and she ran with great fury into the room where he lay in his bath. And she beheld him where he was there all naked in his bath, and therewith she rushed at him and lashed at him with his sword. But Sir Tristram threw himself to one side and so that blow failed of its purpose. Then the Queen would have lashed at him again or have thrust him through with the weapon; but at that Gouvernail and Sir Helles ran to her and catched her and held her back, struggling and screaming very violently. So they took the sword away from her out of her hands, and all the while she shrieked like one gone entirely distracted. *The Queen assails Sir Tristram.*

Then as soon as Gouvernail and Sir Helles loosed her, she ran very violently out of that room with great outcry of screaming, and so to King Angus and flung herself upon her knees before him, crying out: "Justice! Justice! I have found that man who slew my brother! I beseech of you that you will deal justice upon him."

Then King Angus rose from where he sat, and he said: "Where is that man? Bring me to him." And the Queen said: "It is Tramtris, who hath come hither unknown unto this place."

King Angus said: "Lady, what is this you tell me? I cannot believe that what you say is true." Upon this the Queen cried out: "Go yourself, Lord, and inquire, and find out how true it is."

Then King Angus rose, and went forth from that place, and he went to the chamber of Sir Tristram. And there he found that Sir Tristram had very hastily dressed himself and had armed himself in such wise as he was able. Then King Angus came to Tristram, and he said: "How is this, that I find thee armed? Art thou an enemy to my house?" And Tristram wept, and said: "Nay, Lord, I am not your enemy, but your friend, for I have great love for you and for all that is yours, so that I would be very willing to do battle for you even unto death if so be I were called upon to do so."

Then King Angus said: "If that is so, how is it that I find thee here armed as if for battle, with thy sword in thy hand?" "Lord," said Sir Tristram, "although I be friends with you and yours, yet I know not whether you be friends or enemies unto me; wherefore I have prepared myself so that I may see what is your will with me, for I will not have you slay me without defence upon my part." Then King Angus said: "Thou speakest in a very foolish way, for how could a single knight hope to defend himself against my whole household?   Now I bid thee tell me who thou art, and what is thy name, and why thou camest hither knowing that thou hadst slain my brother?"

Then Sir Tristram said, "Lord, I will tell thee all the truth."   And therewith he confessed everything to King Angus, to wit: who was his

*Sir Tristram*       father and his mother, and how he was born and reared;
*confesses to*       how he fought Sir Marhaus, and for what reason; and of how
*King Angus.*       he came hither to be healed of his wound, from which else he must die in very grievous pain.   And he said: "All this is truth, Lord, and it is truth that I had no ill-will against Sir Marhaus; for I only stood to do battle with him for the sake of mine uncle, King Mark of Cornwall, and to enhance mine own honor; and I took my fortune with him as he took his with me.   Moreover, I fought with Sir Marhaus upon the same day that I was made knight, and that was the first battle which I fought, and in that battle I was wounded so sorely that I was like to die as you very well know.   As for him, he was a knight well-tried and seasoned with many battles, and he suffered by no treachery but only with the fortune of war."

So King Angus listened to all that Sir Tristram said, and when he had ended, quoth he:   "As God sees me, Tristram, I cannot deny that you did with Sir Marhaus as a true knight should.   For it was certes your part to take the cause of your uncle upon you if you had the heart to do so, and it was truly a real knightly thing for you who were so young to seek honor at the hands of so famous a knight as Sir Marhaus.   For I do not believe that until you came his way there was any knight in the world who was greater than he, unless it were Sir Launcelot of the Lake.   Wherefore, from that, and from what I saw you do at the tournament, some time ago, I believe that you are one of the strongest knights in the world, and the peer of Sir Launcelot, or of anybody else.

"But though all this is true, nevertheless it will not be possible for me to maintain you in this country, for if I keep you here I shall greatly displease not only the Queen and her kin, but many of those lords and knights who were kin to Sir Marhaus or who were united to him in pledges of friendship.   So you must even save yourself as you can and leave here straightway, for I may not help or aid you in any way."

Then Sir Tristram said: "Lord, I thank you for your great kindness unto me, and I know not how I shall repay the great goodness that my Lady Belle Isoult hath showed to me.   For I swear to you upon the pommel of my sword which I now hold up before me that I would lay down my life for her sake.   Yea, and my honor too! for she hath the entire love of my heart, so that I would willingly die for her, or give up for her all that

I have in the world. Now as for my knighthood, I do believe that I shall in time become a knight of no small worship, for I feel within my heart that this shall be so. So if my life be spared, it may be that you will gain more having me for your friend and your true servant than you will by taking my life in this outland place. For whithersoever I go I give you my knightly word that I shall be your daughter's servant, and that I shall ever be her true knight in right or in wrong, and that I shall never fail her if I shall be called upon to do her service."

Then King Angus meditated upon this for a while, and he said: "Tristram, what thou sayest is very well said, but how shall I get you away from this place in safety?"

Sir Tristram said: "Lord, there is but one way to get me away with credit unto yourself. Now I beseech you of your grace that I may take leave of my lady your daughter, and that I may then take leave of all your knights and kinsmen as a right knight should. And if there be any among them who chooses to stop me or to challenge my going, then I must face that one at my peril, however great it may be."

"Well," said King Angus, "that is a very knightly way to behave, and so it shall be as you will have it."

So Sir Tristram went down stairs to a certain chamber where Belle Isoult was. And he went straight to her and took her by the hand; and he said: "Lady, I am to go away from this place, if I may do so with credit to my honor; but before I go I must tell you that I shall ever be your own true knight in all ways that a knight may serve a lady. For no other lady shall have my heart but you, so I shall ever be your true knight. Even though I shall haply never see your face again, yet I shall ever carry your face with me in my heart, and the thought of you shall always abide with me withersoever I go."

At this the Lady Belle Isoult fell to weeping in great measure, and thereat the countenance of Sir Tristram also was all writhed with passion, and he said, "Lady, do not weep so!" She said, "Alas I cannot help it!" Then he said: "Lady, you gave me my life when I thought I was to lose it, and you brought me back from pain unto ease, and from sorrow unto joy. Would God I were suffering all those pangs as aforetime, so that there might be no more tears upon your face."

Then, King Angus being by, he took her face into his hands *Sir Tristram* and kissed her upon the forehead, and the eyes, and the lips. *parts from Belle Isoult.* Therewith he turned and went away, all bedazed with his sorrow, and feeling for the latch of the door ere he was able to find it and go out from that place.

After that Sir Tristram went straight unto the hall of the castle, and there he found a great many of the lords of the castle and knights attendant upon the King. For the news of these things had flown fast, and many of them were angry and some were doubtful. But Tristram came in very boldly, clad all in full armor, and when he stood in the midst of them he spoke loud and with great courage, saying: "If there be any man here whom I have offended in any way, let him speak, and I will give him entire satisfaction whoever he may be. But let such speech be now or never, for here is my body to make good my knighthood against the body of any man, whomsoever he may be."

At this all those knights who were there stood still and held their peace, and no man said anything against Sir Tristram (although there were several knights and lords who were kin to the Queen), for the boldness of Tristram overawed them, and no one had the heart to answer him.

So after a little while Sir Tristram left that place, without turning his head to see if any man followed him.

So he left that castle and Gouvernail went with him, and no one stopped *Sir Tristram de-* him in his going. After that, he and Gouvernail came to the *parts from* shore and took a boat and they came to the ship of Sir Tris- *Ireland.* tram, and so they sailed away from Ireland. But the heart of Sir Tristram was so full of sorrow that he wished a great many times that he was dead.

So Sir Tristram, though as to his body he was very whole and sound, was, as to his spirit, very ill at ease; for though he was so well and suffered no pain, yet it appeared to him that all the joy of his life had been left behind him, so that he could nevermore have any more pleasure in this world which lieth outside of the walls of Paradise.

ir Tristram harpeth before
King Mark:

# Chapter Fifth

*How Sir Tristram was sent by command of King Mark to go to
Ireland to bring the Lady the Belle Isoult from Ireland to Cornwall
and how it fared with him.*

SO Sir Tristram came back again to Cornwall, and King Mark and
all the knights and lords of the court of the King gave him great
welcome and made much joy over him because he had returned
safely.

But Sir Tristram took no joy in their joy because he was filled with such
heavy melancholy that it was as though even the blue sky had turned to
sackcloth to his eyes, so that he beheld nothing bright in all the world.

But though he had no great pleasure in life, yet Sir Tristram made many
very good songs about Belle Isoult; about her beauty and her *Sir Tristram*
graciousness; about how he was her sad, loving knight; about *tells of the Lady*
how he was pledged unto her to be true to her all of his life *Belle Isoult.*
even though he might never hope to see her again.

These like words he would sing to the music of his shining, golden harp,
and King Mark loved to listen to him. And sometimes King Mark would
sigh very deeply and maybe say: "Messire, that lady of thine must in
sooth be a very wonderful, beautiful, gracious lady." And Sir Tristram
would say, "Yea, she is all that."

So it was at that time that King Mark had great love for Sir Tristram;
in a little while all that was very different, and his love was turned to bitter
hate, as you shall presently hear tell.

Now in those days the knights of Cornwall were considered to be the least
worthy of all knights in that part of the world, for they had so little skill
and prowess at arms that they were a jest and a laughing-stock to many
courts of chivalry. It was said of them that a knight-champion of Cornwall was maybe a knight, but certes was no champion at all; and this was
great shame to all those of Cornwall, more especially as that saying was
in a great measure true.

One day there came to the court of Cornwall a very noble, haughty knight, hight Sir Bleoberis de Ganys, who was brother to Sir Blamor de

*Sir Bleoberis comes to Cornwall.* Ganys and right cousin to Sir Launcelot of the Lake. This knight was a fellow of King Arthur's Round Table, and so he was received with great honor at Cornwall, and much joy was taken of his being there; for it was not often that knights of such repute as he came to those parts. At that time Sir Tristram was not present at the court, having gone hunting into the forest, but a messenger was sent to him with news that Sir Bleoberis was present at the court of the King and that King Mark wished him to be at court also.

Now whilst Sir Tristram was upon his way to return to the court in obedience to these commands, there was held a feast at the castle of the King in honor of Sir Bleoberis. There was much strong wine drunk at that feast, so that the brains of Sir Bleoberis and of others grew very much heated therewith. Then, what with the heat of the wine and the noise and tumult of the feast, Sir Bleoberis waxed very hot-headed, and boastful. So, being in that condition and not knowing very well how he spake, he made great boast of the prowess of the knights of King Arthur's court above those of Cornwall. And in this boastful humor he said: "It is perfectly true that one single knight of the Round Table is the peer of twenty knights of Cornwall, for so it is said and so I maintain it to be."

Upon that there fell a silence over all that part of the feast, for all the knights and lords who were there heard what Sir Bleoberis said, and yet no one knew how to reply to him. As for King Mark, he looked upon Sir Bleoberis, smiling very sourly, and as though with great distaste of his words, and he said: "Messire, inasmuch as thou art our guest, and sitting here at feast with us, it is not fit that we should take thy words seriously; else what thou sayst might be very easily disproved."

Upon this the blood rushed with great violence into the face and head of Sir Bleoberis, and he laughed very loud. Then he said: "Well, Lord, it need not be that I should be a guest here very long. And as for what I say, you may easily put the truth thereof to the proof."

Therewith Sir Bleoberis arose and looked about him, and he perceived that there was near by where he stood a goblet of gold very beautifully

*Sir Bleoberis challenges the knights of Cornwall.* chased and cunningly carved. This Sir Bleoberis took into his hand, and it was half full of red wine. So he stood up before them all, and he cried in a very loud voice: "Messires, and all you knights of Cornwall, here I drink to your more excellent courage and prowess, and wish that you may have better fortune in arms than you have heretofore proved yourselves to have?" And therewith he drank all the wine that was in the goblet. Then he said: "Now I go away from here and take this goblet with me; and if any knight of Cornwall may take it away from me and bring it back again to the King, then I am very willing to own that there are better knights in this country than I supposed there to be." Therewith he turned and went out from that place very haughtily and scornfully, taking that goblet with him, and not one of all those knights who were there made any move to stay him, or to reprove him for his discourteous speech.

Now after he had come out of the hall and into the cool of the air, the heat of the wine soon left him, and he began to repent him of what he had done; and he said: "Alas! meseems I was not very courteous to King Mark, who was mine host." So for a while he was minded to take that goblet back again and make amends for what he had said; but afterward he could not do this because of his pride. So he went to the chamber that had been allotted to him and clad himself in his armor, and after that he rode away from the court of King Mark carrying the goblet with him.

Now some while after he had gone, Sir Tristram came into the hall where the others were, and there he found them all sitting with ill countenances, and no man daring, for shame, to look at his fellow. So Sir *Sir Tristram* Tristram came to King Mark and said: "Where is Sir Bleo- *is angry.* beris?" And King Mark said, "He is gone away." Sir Tristram said, "Why did he go?" Thereupon King Mark told Sir Tristram of what had befallen, and how Sir Bleoberis had taken away that goblet to the great shame and scorn of all those who were there. Upon this the blood flew very violently into Sir Tristram's face, and he said: "Was there no knight here with spirit enough to call reproof upon Sir Bleoberis, or to stay him in his going?" Therewith he looked all about that hall, and he was like a lion standing among them, and no man dared to look him in the face or to reply to him. Then he said: "Well, if there is no knight in Cornwall who hath the will to defend his King, then is there a knight of Lyonesse who will do so because he received knighthood at the hands of the King of Cornwall." And therewith he turned and went away, and left them very haughtily, and they were all still more abashed than they had been before.

Then Sir Tristram went to his chamber and had himself armed in all wise; and he took his horse and mounted and rode away in the direction that Sir Bleoberis had gone, and Gouvernail went with him.

So Sir Tristram and Gouvernail rode at a good pace for a long *Sir Tristram* time, making inquiry of whomsoever they met if Sir Bleoberis *follows Sir* had passed that way. At last they entered the forest and *Bleoberis.* rode therein a great way, meeting no one till toward the latter part of the afternoon. By and by they saw before them two knights, very large and strong of frame and clad all in bright and shining armor, and each riding a great war-horse of Flemish strain.

"Gouvernail," said Sir Tristram, "ride forward apace and see for me who are yonder knights." So Gouvernail rode forward at a gallop, and so, in a little, came near enough to the two knights to see the devices upon their shields. Upon that he returned to Sir Tristram, and said: "Messire, those are two very famous worthy knights of King Arthur's Court, and of the two you are acquainted with one, but the other is a stranger to you. For the one is Sir Sagramore le Desirous, who was at that tournament in Ireland, and the other is Sir Dodinas le Sauvage."

*Sir Tristram comes to two knights.*

"Well," said Sir Tristram, "those are indeed two very good, worthy knights. Now if you will sit here for a while, I will go forward and have speech with them." "Messire," said Gouvernail, "I would counsel you not to have to do with those knights, for there are hardly any knights more famous at arms than they, so it is not likely that you can have success of them if you should assay them."

But to this Sir Tristram said: "Peace, Gouvernail! Hold thy peace, and bide here while I go forward!"

Now those knights when they became aware that Sir Tristram and Gouvernail were there, had halted at a clear part of the woodland to await what should befall. Unto them Sir Tristram came, riding with great dignity and haughtiness, and when he had come nigh enough he drew rein and spoke with great pride of bearing, saying: "Messires, I require of you to tell me whence you come, and whither you go, and what you do in these marches?"

Unto him Sir Sagramore made reply, speaking very scornfully: "Fair knight, are you a knight of Cornwall?" and Sir Tristram said: "Why do you ask me that?" "Messire," said Sir Sagramore, "I ask you that because it hath seldom been heard tell that a Cornish knight hath courage to call upon two knights to answer such questions as you have asked of us."

"Well," said Sir Tristram, "for the matter of that, I am at this present a knight of Cornwall, and I hereby let you know that you shall not go away from here unless you either answer my question or give me satisfaction at arms."

Then Sir Dodinas spoke very fiercely, saying: "Sir Cornish knight, you shall presently have all the satisfaction at arms that you desire and a great deal more than you desire." Therewith he took a very stout spear in his hand and rode to a little distance, and Sir Tristram, beholding his intent to do battle, also rode to a little distance, and took stand in such a place as seemed to him to be best. Then, when they were in all wise prepared, they rushed together with such astonishing vehemence that the earth shook and trembled beneath them.

Therewith they met in the middle of their course with a great uproar of iron and wood. But in that onset the spear of Sir Dodinas broke into a great many small pieces, but the spear of Sir Tristram held, *Sir Tristram* so that in the encounter he lifted Sir Dodinas entirely out of *does battle with* his saddle, and out behind the crupper of his horse. And he *Sir Dodinas.* flung Sir Dodinas down so violently that his neck was nearly broken, and he lay for a while in a deep swoon like one who has been struck dead.

Then Sir Sagramore said: "Well, Sir Knight, that was certes a very great buffet that you gave my fellow, but now it is my turn to have ado with you."

So therewith he took also his spear in hand and chose his station for an assault as Sir Dodinas had done, and Sir Tristram also took station as he had done before. Then immediately they two ran together *Sir Tristram* with the same terrible force that Sir Tristram and Sir Dodinas *does battle with* had coursed, and in that encounter Sir Tristram struck Sir *Sir Sagramore.* Sagramore so direful a buffet with his spear that he overthrew both horse and man, and the horse, falling upon Sir Sagramore, so bruised his leg that he could not for a while arise from where he lay.

Therewith Sir Tristram, having run his course, came back to where those two knights lay upon the ground, and he said, "Fair Knights, will you have any more fighting?" They said, "No, we have had fighting enough." Then Sir Tristram said: "I pray you, tell me, are there any bigger knights at the court of King Arthur than you? If it is not so, then I should think you would take great shame to yourselves that you have been overthrown the one after the other by a single knight. For this day a knight of Cornwall hath assuredly matched you both to your great despite."

Then Sir Sagramore said: "Sir, I pray you upon your true knighthood to tell us who you are, for you are assuredly one of the greatest knights in the world. Upon this Sir Tristram laughed, "Nay," quoth *Sir Tristram* he, "I am as yet a young knight, who has had but little proof in *acknowledges his* battle. As for my name, since you ask it of me, upon my *degree.* knighthood I am not ashamed to tell you that I am hight Sir Tristram, and that I am King Meliadus' son of Lyonesse."

"Ha!" said Sir Sagramore, "if that be so, then there is little shame in being overthrown by you. For not only do I well remember how at the court of the King of Ireland you overthrew six knights of the Round Table, and how easily you overthrew Sir Palamydes the Saracen, but it is also very well known how you did battle with Sir Marhaus, and of how you overcame him. Now Sir Marhaus and Sir Palamydes were two of the best knights in the world, so it is not astonishing that you should have done as you did with us. But, since you have overthrown us, what is it you would have us do?"

"Messires," said Sir Tristram, "I have only to demand two things of you. One of them is that you give me your word that you will go to Cornwall and confess to King Mark that you have been overthrown by a Cornish knight; and the second thing is that you tell me if you saw Sir Bleoberis de Ganys pass this way?"

They say: "Messire, touching that demand you make upon us to go to King Mark and to confess our fall, that we will do as you desire; and as for Sir Bleoberis, we met him only a short while ago, and he cannot even now be very far from this place."

"Well," said Sir Tristram, "I give you good den, and thank you for your information. I have some words to say to Sir Bleoberis before he leave these marches."

So thereafter he called Gouvernail, and they two rode into the forest and on their way as fast as they were able. As for Sir Dodinas and Sir Sagramore, they betook their course to the court of King Mark, as they had promised to do.

Now, by and by, after Sir Tristram and Gouvernail had gone some considerable distance farther upon that road, they beheld Sir Bleoberis be-
*Sir Tristram comes to Sir Bleoberis.* fore them in a forest path, riding very proudly and at an easy pass upon his way. At that time the sun was setting very low toward the earth, so that all the tops of the forest trees were aflame with a very ruddy light, though all below in the forest was both cool and gray. Now when Sir Tristram and Gouvernail with him had come pretty nigh to Sir Bleoberis, Sir Tristram called to him in a very loud voice, and bade him turn and stand. Therewith Sir Bleoberis turned about and waited for Sir Tristram to come up with him. And when Sir Tristram was come near by, he said to Sir Bleoberis: "Messire, I hear tell that you have with you a very noble goblet which you have taken in a shameful way from the table of King Mark of Cornwall. Now I demand of you that you give me that goblet to take back unto the King again." "Well," said Sir Bleoberis, "you shall freely have that goblet if you can take it from me, and if you will look, you will see where it hangs here from my saddlehorn. But I may tell you that I do not believe that there is any Cornish knight who may take away that goblet against my will."

"As for that," said Sir Tristram, "we shall see in a little while how it may be."

Therewith each knight took his spear in hand and rode a little distance away, and made himself in all wise ready for the assault. Then when they were in all ways prepared, each launched himself against the other, coming together with such violence that sparks of fire flew out from the points of their spears. And in that *Sir Tristram overcometh Sir Bleoberis.* assault the horse of each knight was overthrown, but each knight voided his saddle and leaped very lightly to earth, without either having had a fall. Then each drew his sword and set his shield before him, and therewith came together, foining and lashing with all the power of their might. Each gave the other many sore strokes, so that the armor of each was indented in several places and in other places was stained with red. Then at last Sir Tristram waxed very wode with anger and he rushed at Sir Bleoberis, smiting him so fiercely that Sir Bleoberis bare back and held his shield low before him. This Sir Tristram perceived, and therewith, rushing in upon Sir Bleoberis, he smote that knight such a great buffet upon the head that Sir Bleoberis fell down upon his knees, without having strength to keep his feet. Then Sir Tristram rushed off the helmet of Sir Bleoberis, and he said, "Sir Knight, yield to me or I shall slay you."

"Messire," said Sir Bleoberis, "I yield myself to you, and indeed you are as right a knight as ever I met in all of my life. Then Sir Tristram took Sir Bleoberis by the hand and he lifted him up upon his feet, and he said: "Sir, I am very sorry for to have had to do with you in this fashion, for almost would I rather that you should have overcome me than that I should have overcome you. For I do not at any time forget that you are cousin unto Sir Launcelot of the Lake, and I honor Sir Launcelot above all men else in the world, and would rather have his friendship than that of any man living. So I have had no despite against you in this battle, but have only fought with you because it behooved me to do so for the sake of the King of Cornwall, who is my uncle."

Then Sir Bleoberis said, "Messire, I pray you tell me who you are?" "Lord," said Sir Tristram, "I am a very young knight hight Tristram, and I am the son of King Meliadus of Lyonesse and the Lady Elizabeth, sister unto King Mark of Cornwall."

"Ha," said Sir Bleoberis, "I have heard great report of you, Sir Tristram, and now I know at mine own cost that you are one of the best knights in the world. Yea; I have no doubt that at some time you will be the peer of Sir Launcelot of the Lake himself, *Sir Bleoberis gives the goblet to Sir Tristram.* or of Sir Lamorak of Gales, and they two are, certes, the best knights in the world.   Now I believe that I would have given you this goblet, even without your having to fight for it, had I known who you were; and as it is I herewith give it to you very freely."

So Sir Bleoberis untied the goblet from where it hung at his saddle-bow, and Sir Tristram took the goblet and gave him gramercy for it; and therewith having recovered their horses, each knight mounted, and betook his way whither he was going.

So a little after nightfall Sir Tristram came to the King of Cornwall and his court, and he said to King Mark: "Here is your goblet which I have brought back to you; and I would God that some of your knights who are so much older than I had the courage to do for you what I have had to do." And therewith he went away and left them all sitting ashamed.

Now it chanced some little while after these things happened as aforesaid, that King Mark lay down upon his couch after his midday meal for to sleep a little space during the heat of the day; and it likewise happened that the window near by where he lay was open so that the air might come into the room.   Now at that time three knights of the court sat in the garden beneath where the window was.   These knights talked to one another concerning Sir Tristram, and of how he had brought back that goblet from Sir Bleoberis de Ganys, and of what honor it was to have such a champion in Cornwall for to stand for the honor of that court.   In their talk they said to one another that if only the King of Cornwall were such a knight as Sir Tristram, then there would be plenty of knights of good worth who would come to that court, and Cornwall would no longer have to be ashamed of its chivalry as it was nowadays.   So they said: "Would God our King were such a knight as Sir Tristram!"

All this King Mark overheard, and the words that they said were like a very bitter poison in his heart.   For their words entered into his soul

*King Mark takes hatred to Sir Tristram.*

and abided there, and thereupon at that same hour all his love for Tristram was turned into hate.   Thus it befell that, after that day, King Mark ever pondered and pondered upon that which he had heard, and the longer he pondered it, the more bitter did his life become to him, and the more he hated Sir Tristram.   So it came to pass that whenever he was with Sir Tristram and looked upon him, he would say in his heart: "So they say that you are a better knight than I? Would God you were dead or away from this place, for I believe that some day you will be my undoing!"   Yea; there were times when he would look upon Sir Tristram in that wise and whisper to himself: "Would God would send a blight upon thee, so that thou wouldst wither away!"

But always the King dissembled this hatred for Sir Tristram, so that no one suspected him thereof; least of all did Sir Tristram suspect how changed was the heart of the King toward him.

Now one day Sir Tristram was playing upon his harp and singing before King Mark, and the King sat brooding upon these things as he gazed at Tristram. And Sir Tristram, as he ofttimes did nowadays, sang of the Lady Belle Isoult, and of how her face was like to a rose for fairness, and of how her soul was like to a nightingale in that it uplifted the spirit of whosoever was near her even though the darkness of sorrow as of night might envelop him. And whilst Sir Tristram sang thus, King Mark listened to him, and as he listened a thought entered his heart and therewith he smiled. So when Sir Tristram had ended his song of the Belle Isoult, King Mark said: "Fair nephew, I would that you would undertake a quest for me." Sir Tristram said, "What quest is that, Lord?" "Nay," said King Mark, "I will not tell you what quest it is unless you will promise me upon your knighthood to undertake it upon my behalf." Then Sir Tristram suspected no evil, wherefore he smiled and said: "Dear Lord, if the quest is a thing that it is in my power to undertake, I will undertake it upon your asking, and unto that I pledge my knighthood." King Mark said, "It is a quest that you may undertake." Sir Tristram said, "Then I will undertake it, if you will tell me what it is."

King Mark said: "I have listened to your singing for this long while concerning the Lady Belle Isoult. So the quest I would have you undertake is this: that you go to Ireland, and bring thence the Lady Belle Isoult to be my Queen. For because of your songs and ballads I have come to love her so greatly that I believe that I shall have no happiness in life until I have her for my Queen. *King Mark betrays Sir Tristram to a promise.* So now, since you have pledged me your word upon your knighthood to do my bidding in this case, such is the quest that I would send you upon." And therewith he smiled upon Sir Tristram very strangely.

Then Sir Tristram perceived how he had been betrayed and he put aside his harp and rose from where he sat. And he gazed for a long while at King Mark, and his countenance was wonderfully white like that of a dead man. Then by and by he said: "Sir, I know *How Sir Tristram fell into despair.* not why you have put this upon me, nor do I know why you have betrayed me. For I have ever served you truly as a worthy knight and a kinsman should. Wherefore I know not why you have done this unto me, nor why you seek to compass my death. For you know very well that if I return to Ireland I shall very likely be slain either by the Queen or by some of her kindred, because that for your sake I slew in battle Sir Marhaus, the Queen's brother of Ireland.   Yet, so far as that is concerned, I would rather lose my life than succeed in this quest, for if so be I do not lose my life, then I must do that which I would liever die than do.   Yea; I believe that there was never any knight loved a lady as I love the Lady Belle Isoult. For I love her not only because of her beauty and graciousness, but because she healed mine infirmities and lent ease unto my great sufferings and brought me back from death unto life.   Wherefore that which you bid me fulfil is more bitter to me than death."

"Well," said King Mark, "I know nothing of all this—only I know that you have given me your knightly word to fulfil this quest."

"Very well," said Sir Tristram, "if God will give me His good help in this matter, then I will do that which I have pledged my knighthood to undertake."   Therewith he turned and went out from that place in such great despair that it was as though his heart had been turned into ashes. But King Mark was filled with joy that he should have caused Sir Tristram all that pain, and he said to his heart: "This is some satisfaction for the hate which I feel for this knight; by and by I shall maybe have greater satisfaction than that."

After that Sir Tristram did not come any more where King Mark was, but he went straight away from the King's court and into a small castle that King Mark had given him some while since for his own.   There he abided for several days in great despair of soul, for it seemed to him as though God had deserted him entirely.   There for a while Gouvernail alone was with him and no one else, but after a while several knights came to him and gave him great condolence and offered to join with him as his knights-companion.   And there were eighteen of these knights, and Sir Tristram was very glad of their comradeship.

These said to him: "Sir, you should not lend yourself to such great travail of soul, but should bend yourself as a true knight should to assume that burden that God hath assigned you to bear."

So they spoke, and by and by Sir Tristram aroused himself from his despair and said to himself: "Well, what these gentlemen say is true, and God hath assuredly laid this very heavy burden upon me; as that is so, I must needs assume it for His sake."

So Sir Tristram and the knights who were with him abode in that place
*Sir Tristram departs from Cornwall.*   for a day or two or three, and then one morning Sir Tristram armed himself and they armed themselves, and all took their departure from that castle and went down to the sea.   Then they took ship with intent to depart to Ireland upon that quest Sir Tristram

had promised King Mark he would undertake, and in a little they hoisted sail and departed from Cornwall for Ireland.

But they were not to make their quest upon that pass so speedily as they thought, for, upon the second day of their voyaging, there arose a great storm of wind of such a sort that the sailors of that ship had never seen the like thereof in all of their lives. For the waves rose up like mountains, and anon the waters sank away into deep valleys with hills of water upon either side all crested over with foam as white as snow. And anon that ship would be uplifted as though the huge sea would toss it into the clouds; and anon it would fall down into a gulf so deep that it appeared as though the green waters would swallow it up entirely. The air roared as though it were full of demons and evil spirits out of hell, and the wind was wet and very bitter with brine. So the ship fled away before that tempest, and the hearts of all aboard were melted with fear because of the great storm of wind and the high angry waves.

Then toward evening those who were watching from the lookout beheld a land and a haven, and they saw upon the land overlooking the haven was a noble castle and a fair large town, surrounded by high walls of stone. So they told the others of what they saw, and all gave great rejoicing for that they were so nigh the land. Therewith they sailed the ship toward the haven, and having entered therein in safety, they cast anchor under the walls of the castle and the town, taking great joy that God had brought them safe and sound through that dreadful peril of the tempest.

Then Sir Tristram said to Gouvernail: "Knowest thou, Gouvernail, what place is this to which we have come?" "Messire," said Gouvernail, "I think it is Camelot." And then those knights of Cornwall *Sir Tristram* who stood by said, "Yea, that is true, and it is Camelot." *comes to* And one of them said: "Messire, it is likely that King Arthur *Camelot.* is at that place at this very time, for so it was reported that he was, and so I believe it to be."

"Ha," quoth Tristram, "that is very good news to me, for I believe that it would be the greatest joy to me that the world can now give to behold King Arthur and those noble knights of his court ere I die. More especially do I desire above all things to behold that great, noble champion, Sir Launcelot of the Lake. So let us now go ashore, and mayhap it shall come to pass that I shall see the great King and Sir Launcelot and mayhap shall come to speak with the one or the other." And that saying of Sir Tristram's seemed good to those knights who were with him, for they were weary of the sea, and desired to rest for a while upon the dry land.

So they presently all went ashore and bade their attendants set up their pavilions in a fair level meadow that was somewhat near a league distant away from the castle and the town. In the midst of the other pavilions

*Sir Tristram sets up his pavilion.* upon that plain was set the pavilion of Sir Tristram. It was of fine crimson cloth striped with silver and there was the figure of a gryphon carved upon the summit of the centre pole of the pavilion. The spear of Sir Tristram was emplanted by the point of the truncheon in the ground outside the pavilion, and thereunto his shield was hung so that those who passed that way might clearly behold what was the device thereon.

And now shall be told how Sir Tristram became united in friendship with the brotherhood of good knights at King Arthur's court.

ir Tristram sits with Sir
Launcelot.

# Chapter Sixth

*How Sir Tristram had to do in battle with three knights of the
Round Table. Also how he had speech with King Arthur.*

SO came the next morning, and uprose the sun in all the splendor of
his glory, shedding his beams to every quarter with a rare dazzling
effulgence. For by night the clouds of storm had passed away
and gone, and now all the air was clear and blue, and the level beams of
light fell athwart the meadow-lands so that countless drops of water sparkled
on leaf and blade of grass, like an incredible multitude of shining jewels
scattered all over the earth. Then they who slept were awakened by the
multitudinous voicing of the birds; for at that hour the small fowl sang so
joyous a roundelay that all the early morning was full of the sweet jargon
of their chanting.

At this time, so early in the day, there came two knights riding by where
Sir Tristram and his companions had set up their pavilions. These were
two very famous knights of King Arthur's court and of the Round Table;
for one was Sir Ector de Maris and the other was Sir Morganor of Lisle.

When these two knights perceived the pavilions of Sir Tristram and his
knights-companion, they made halt, and Sir Ector de Maris said, "What
knights are these who have come hither?" Then Sir Morganor *How two knights*
looked and presently he said: "Sir, I perceive by their shields *came to the*
that these are Cornish knights, and he who occupies this central *pavilion of Sir*
*Tristram.*
pavilion must be the champion of this party." "Well," quoth
Sir Ector, "as for that I take no great thought of any Cornish knight, so
do thou strike the shield of that knight and call him forth, and let us see
of what mettle he is made."

"I will do so," said Sir Morganor; and therewith he rode forward to where
the shield of Sir Tristram hung from the spear, and he smote the shield with
the point of his lance, so that it rang with a very loud noise.

Upon this, Sir Tristram immediately came to the door of his pavilion,
and said, "Messires, why did you strike upon my shield?" "Because,"
said Sir Ector, "we are of a mind to try your mettle what sort of a knight

you be." Quoth Sir Tristram: "God forbid that you should not be satisfied. So if you will stay till I put on my armor you shall immediately have your will in this matter."

Thereupon he went back into his tent and armed himself and mounted his horse and took a good stout spear of ash-wood into his hand.

Then all the knights of Cornwall who were with Sir Tristram came forth to behold what their champion would do, and all their esquires, pages, and attendants came forth for the same purpose, and it was a very pleasant time of day for jousting.

Then first of all Sir Morganor essayed Sir Tristram, and in that encounter Sir Tristram smote him so dreadful, terrible a blow that he cast him a full

*Sir Tristram overthrows Sir Morganor.* spear's length over the crupper of his horse, and that so violently that the blood gushed out of the nose and mouth and ears of Sir Morganor, and he groaned very dolorously and could not arise from where he lay.

"Hah," quoth Sir Ector, "that was a very wonderful buffet you struck my fellow. But now it is my turn to have ado with you, and I hope God will send me a better fortune."

So he took stand for battle as did Sir Tristram likewise, and when they were in all wise prepared they rushed very violently to the assault. In

*Sir Tristram overthrows Sir Ector.* that encounter Ector suffered hardly less ill fortune than Sir Morganor had done. For he brake his spear against Sir Tristram into as many as an hundred pieces, whilst Sir Tristram's spear held so that he overthrew both the horse and the knight-rider against whom he drove.

Then all the knights of Cornwall gave loud acclaim that their knight had borne himself so well in those encounters. But Sir Tristram rode back to where those two knights still lay upon the ground, and he said: "Well, Messires, this is no very good hap that you have had with me."

Upon that speech Sir Ector de Maris gathered himself up from the dust and said: "Sir Knight, I pray you of your knighthood to tell us who you be and what is your degree, for I declare to you, I believe you are one of the greatest knights-champion of the world."

"Sir," said Sir Tristram, "I am very willing to tell you my name and my station; I am Sir Tristram, the son of King Meliadus of Lyonesse."

"Ha," quoth Sir Ector, "I would God I had known that before I had ado with you, for your fame hath already reached to these parts, and there hath been such report of your prowess and several songs have been made about you by minstrels and poets. I who speak to you am Sir Ector, surnamed de Maris, and this, my companion, is Sir Morganor of Lisle."

"Alas!" cried out Sir Tristram, "I would that I had known who you were ere I did battle with you. For I have greater love for the knights of the Round Table than all others in the world, and most of all, Sir Ector, do I have reverence for your noble brother Sir Launcelot of the Lake. So I take great shame to myself that any mishap should have befallen you this day through me."

Upon this Sir Ector laughed. "Well," quoth he, "let not that trouble lie with you, for it was we who gave you challenge without inquiry who you were, and you did but defend yourself. We were upon our way to Camelot yonder, when we fell into this mishap, for King Arthur is at this time holding court at that place. So now, if we have your leave to go upon our way, we will betake ourselves to the King and tell him that you are here, for we know that he will be very glad of that news."

Upon this Sir Tristram gave them leave to depart, and they did so with many friendly words of good cheer. And after they had gone Sir Tristram went back into his pavilion again and partook of refreshment that was brought to him.

Now, some while after Sir Ector and Sir Morganor had left that place, and whilst Sir Tristram was still resting in his pavilion, there came a single knight riding that way, and this knight was clad altogether *There comes a* in white armor and his shield was covered over with a cover- *knight in white* ing of white leather, so that one could not see what device *armor.* he bare thereon.

When this white knight came to the place where Sir Tristram and his companions had pitched their pavilions, he also stopped as Sir Ector and Sir Morganor had done, for he desired to know what knights these were. At that time Gouvernail was standing alone in front of Sir Tristram's pavilion, and unto him the white knight said: "Sir, I pray you, tell me who is the knight to whom this pavilion belongs."

Now Gouvernail thought to himself: "Here is another knight who would have ado with my master. Perhaps Sir Tristram may have glory by him also." So he answered the white knight: "Sir, I may not tell you the name of this knight, for he is my master, and if he pleases to tell you his name he must tell it himself."

"Very well," said the white knight, "then I will straightway ask him."

Therewith he rode to where the shield of Sir Tristram hung, and he struck upon the shield so violent a blow that it rang very loud and clear.

Then straightway came forth Sir Tristram and several of his knights-companion from out of the pavilion, and Sir Tristram said, "Sir Knight, wherefore did you strike upon my shield?"

"Messire," quoth the white knight, "I struck upon your shield so that I might summon you hither for to tell me your name, for I have asked it of your esquire and he will not tell me."

"Fair Knight," quoth Sir Tristram, "neither will I tell you my name until I have wiped out that affront which you have set upon my shield by that stroke you gave it. For no man may touch my shield without my having to do with him because of the affront he gives me thereby."

"Well," said the white knight, "I am satisfied to have it as you please."

So therewith Sir Tristram went back into his pavilion and several went with him. These put his helmet upon his head and they armed him for *Sir Tristram* battle in all ways. After that Sir Tristram came forth and *does battle with* mounted his horse and took his spear in hand and made himself *the white knight.* in all ways ready for battle, and all that while the white knight awaited his coming very calmly and steadfastly. Then Sir Tristram took ground for battle, and the white knight did so likewise. So being in all ways prepared, each launched forth against the other with such amazing and terrible violence that those who beheld that encounter stood as though terrified with the thunder of the onset.

Therewith the two knights met in the midst of the course, and each knight smote the other directly in the centre of the shield. In that encounter the spear of each knight broke all to small pieces, even to the truncheon which he held in his fist. And so terrible was the blow that each struck the other that the horse of each fell back upon his haunches, and it was only because of the great address of the knight-rider that the steed was able to recover his footing. As for Sir Tristram, that was the most terrible buffet he ever had struck him in all his life before that time.

Then straightway Sir Tristram voided his saddle and drew his sword and dressed his shield. And he cried out: "Ha, Sir Knight! I demand of you that you descend from your horse and do me battle afoot."

"Very well," said the white knight, "thou shalt have thy will." And thereupon he likewise voided his horse and drew his sword and dressed his shield and made himself in all ways ready for battle as Sir Tristram had done.

Therewith they two came together and presently fell to fighting with such ardor that sparks of fire flew from every stroke. And if Sir Tristram struck hard and often, the white knight struck as hard and as often as he, so that all the knights of Cornwall who stood about marvelled at the strength and fierceness of the knights-combatant. Each knight gave the other many sore buffets so that the armor was here and there dinted and here and there was broken through by the edge of the sword so that the red blood flowed out therefrom and down over the armor. turning its brightness in places into an ensanguined red. Thus they fought for above an hour and in all that time neither knight gave ground or gained any vantage over the other.

Then after a while Sir Tristram grew more weary of fighting than ever he had been in all of his life before, and he was aware that this was the greatest knight whom he had ever met. But still he would *Sir Tristram* not give ground, but fought from this side and from that *falls in the* side with great skill and address until of a sudden, he slipped *battle.* upon some of that blood that he himself had shed, and because of his great weariness, fell down upon his knees, and could not for the instant rise again.

Then that white knight might easily have struck him down if he had been minded to do so. But, instead, he withheld the blow and gave Sir Tristram his hand and said: "Sir Knight, rise up and stand upon thy feet and let us go at this battle again if it is thy pleasure to do so; for I do not choose to take advantage of thy fall."

Then Sir Tristram was as greatly astonished at the extraordinary courtesy of his enemy as he had been at his prowess. And because of that courtesy he would not fight again, but stood leaning upon his sword panting. Then he said: "Sir Knight, I pray thee of thy knighthood to tell me what is thy name and who thou art."

"Messire," said the white knight, "since you ask me that upon my knighthood, I cannot refuse to tell you my name. And so I will do, provided you, upon your part, will do me a like courtesy and will first tell me your name and degree."

Quoth Sir Tristram: "I will tell you that. My name is Sir Tristram of Lyonesse, and I am the son of King Meliadus of that land whereby I have my surname."

"Ha, Sir Tristram," said the white knight, "often have I heard of thee and of thy skill at arms, and well have I proved thy fame this day and that all that is said of thee is true. I must tell thee that I *Sir Launcelot* have never yet met my match until I met thee this day. For *confesses him-* I know not how this battle might have ended hadst thou not *self.* slipped and fallen by chance as thou didst. My name is Sir Launcelot, surnamed of the Lake, and I am King Ban's son of Benwick."

At this Sir Tristram cried out in a loud voice: "Sir Launcelot! Sir Launcelot! Is it thou against whom I have been doing battle! Rather I would that anything should have happened to me than that, for of all men in the world I most desire thy love and friendship."

Then, having so spoken, Sir Tristram immediately kneeled down upon his knees and said: "Messire, I yield myself unto thee, being overcome not more by thy prowess than by thy courtesy. For I freely confess that thou art the greatest knight in the world, against whom no other knight can hope to stand; for I could fight no more and thou mightest easily have slain me when I fell down a while since."

*Sir Tristram yields to Sir Launcelot.*

"Nay, Sir Tristram," said Sir Launcelot, "arise, and kneel not to me, for I am not willing to accept thy submission, for indeed it is yet to be proved which of us is the better knight, thou or I. Wherefore let neither of us yield to the other, but let us henceforth be as dear as brothers-in-arms the one toward the other."

Then Sir Tristram rose up to his feet again. "Well, Sir Launcelot," he said, "whatsoever thou shalt ordain shall be as thou wouldst have it. But there is one thing I must do because of this battle."

Then he looked upon his sword which he held naked and ensanguined in his hand and he said: "Good sword; thou hast stood my friend and hast served me well in several battles, but this day thou hast served me for the last time." Therewith he suddenly took the blade of the sword in both hands—the one at the point and the other nigh the haft—and he brake the blade across his knee and flung the pieces away.

*Sir Tristram breaks his sword.*

Upon this Sir Launcelot cried out in a loud voice: "Ha, Messire! why didst thou do such a thing as that? To break thine own fair sword?"

"Sir," quoth Sir Tristram, "this sword hath this day received the greatest honor that is possible for any blade to receive; for it hath been baptized in thy blood. So, because aught else that might happen to it would diminish that honor, I have broken it so that its honor might never be made less than it is at this present time."

Upon this Sir Launcelot ran to Sir Tristram and catched him in his arms, and he cried out: "Tristram, I believe that thou art the noblest knight whom ever I beheld!" And Sir Tristram replied: "And thou, Launcelot, I love better than father or kindred." Therewith each kissed the other upon the face, and all they who stood by were so moved at that sight that several of them wept for pure joy.

Thereafter they two went into Sir Tristram's pavilion and disarmed themselves. Then there came sundry attendants who were excellent leeches and these searched their hurts and bathed them and dressed them. And several other attendants came and fetched soft robes and clothed the knights therein so that they were very comfortable in their bodies. Then still other attendants brought them good strong wine and manchets of bread and they sat together at table and ate very cheerfully and were greatly refreshed.

*Sir Tristram and Sir Launcelot feast together.*

So I have told you of that famous affair-at-arms betwixt Sir Launcelot and Sir Tristram, and I pray God that you may have the same pleasure in reading of it that I had in writing of it.

Now, as Sir Launcelot and Sir Tristram sat in the pavilion of Sir Tristram making pleasant converse together, there suddenly entered an esquire to where they were sitting. This esquire proclaimed: "Messires, hither cometh King Arthur, and he is very near at hand." Thereupon, even as that esquire spoke, there came from without the pavilion a great noise of trampling horses and the pleasant sound of ringing armor, and then immediately a loud noise of many voices uplifted in acclamation.

Therewith Sir Launcelot and Sir Tristram arose from where they sat, and as they did so the curtains at the doorway of the pavilion were parted and there entered King Arthur himself enveloped, as it were, with all the glory of his royal estate. *King Arthur comes to Sir Tristram's pavilion.*

Unto him Sir Tristram ran, and would have fallen upon his knees, but King Arthur stayed him from so doing. For the great king held him by the hand and lifted him up, and he said, "Sir, are you Sir Tristram of Lyonesse?" "Yea," said Sir Tristram, "I am he." "Ha," said King Arthur, "I am gladder to see you than almost any man I know of in the world," and therewith he kissed Sir Tristram upon the face, and he said: "Welcome, Messire, to these parts! Welcome! And thrice welcome!"

Then Sir Tristram besought King Arthur that he would refresh himself, and the King said he would do so. So Sir Tristram brought him to the chiefest place, and there King Arthur sat him down. And Sir Tristram would have served him with wine and with manchets of bread with his own hand, but King Arthur would not have it so, but bade Sir Tristram to sit beside him on his right hand, and Sir Tristram did so. After that, King Arthur spake to Sir Tristram about many things, and chiefly about King Meliadus, the father of Sir Tristram, and about the court of Lyonesse.

Then, after a while King Arthur said: "Messire, I hear tell that you are a wonderful harper." And Sir Tristram said, "Lord, so men say of me." King Arthur said, "I would fain hear your minstrelsy." To which Sir Tristram made reply: "Lord, I will gladly do anything at all that will give you pleasure."

So therewith Sir Tristram gave orders to Gouvernail, and Gouvernail brought him his shining golden harp, and the harp glistered with great splendor in the dim light of the pavilion.

Sir Tristram took the harp in his hands and tuned it and struck upon it.

And he played upon the harp, and he sang to the music thereof so wonderfully that they who sat there listened in silence as though they were *Sir Tristram* without breath.  For not one of them had ever heard such *sings before* singing as that music which Sir Tristram sang; for it was as *King Arthur.* though some angel were singing to those who sat there harkening to his chanting.

So after Sir Tristram had ended, all who were there gave loud acclaim and much praise to his singing.  "Ha, Messire!" quoth King Arthur, "many times in my life have I heard excellent singing, but never before in my life have I heard such singing as that.  Now I wish that we might always have you at this court and that you would never leave us."  And Sir Tristram said: "Lord, I too would wish that I might always be with you and with these noble knights of your court, for I have never met any whom I love as I love them."

So they sat there in great joy and friendliness of spirit, and, for the while, Sir Tristram forgot the mission he was upon and was happy in heart and glad of that terrible storm that had driven him thitherward.

And now I shall tell you the conclusion of all these adventures, and of how it fared with Sir Tristram.

Belle Isoult and Sir Tristram drink the love draught

# Chapter Seventh

*How Sir Tristram had speech with King Angus of Ireland; how he
undertook to champion the cause of King Angus and of what happened thereafter.*

NOW, as Sir Tristram and King Arthur and Sir Launcelot sat
together in the pavilion of Sir Tristram in pleasant, friendly
discourse, as aforetold, there came Gouvernail of a sudden into
that place. He, coming to Sir Tristram, leaned over his shoulder and he
whispered into his ear: "Sir, I have just been told that King Angus of Ireland is at this very time at Camelot at the court of the King."

Upon this Sir Tristram turned to King Arthur and said: "Lord, my esquire
telleth me that King Angus of Ireland is here at Camelot; now I pray you
tell me, is that saying true?" "Yea," said King Arthur, *Sir Tristram*
"that is true; but what of it?" "Well," said Sir Tristram, *hears news of*
"I had set forth to seek King Angus in Ireland, when I and *King Angus.*
my companions were driven hither by a great storm of wind. Yet when I
find him, I know not whether King Angus may look upon me as a friend or
as an unfriend."

"Ha," said King Arthur, "you need not take trouble concerning the
regard in which King Angus shall hold you. For he is at this time in such
anxiety of spirit that he needs to have every man his friend *How Sir Ber-*
who will be his friend, and no man his enemy whom he can *trand was killed*
reconcile to him. He is not just now in very good grace, *in Ireland.*
either with me or with my court, for the case with him is thus: Some while
ago, after you left the court of Ireland, there came to that place Sir Blamor
de Ganys (who is right cousin to Sir Launcelot of the Lake) and with Sir
Blamor a knight-companion hight Sir Bertrand de la Riviere Rouge. These
two knights went to Ireland with intent to win themselves honor at the court
of Ireland. Whilst they were in that kingdom there were held many jousts
and tourneys, and in all of them Sir Blamor and Sir Bertrand were victorious,
and all the knights of Ireland who came against them were put to shame
at their hands. Many of the Irish knights were exceedingly angry at this,
and so likewise was the King of Ireland. Now it happened one day that Sir

Bertrand was found dead and murdered at a certain pass in the King's forest, and when the news thereof was brought to Sir Blamor, he was very wroth that his knight-companion should have been thus treacherously slain. So he immediately quitted Ireland and returned hither straightway, and when he had come before me he accused King Angus of treason because of that murder. Now at this time King Angus is here upon my summons for to answer that charge and to defend himself therefrom; for Sir Blamor offers his body to defend the truth of his accusation, and as for the King of Ireland, he can find no knight to take his part in that contention. For not only is Sir Blamor, as you very well know, one of the best knights in the world, but also nearly everybody here hath doubt of the innocence of King Angus in this affair. Now from this you may see that King Angus is very much more in need of a friend at this time than he is of an enemy."

"Lord," said Sir Tristram, "what you tell me is very excellent good news, for now I know that I may have talk with King Angus with safety to myself, and that he will no doubt receive me as a friend."

So after King Arthur and his court had taken their departure—it being then in the early sloping of the afternoon—Sir Tristram called Gouvernail to him and bade him make ready their horses, and when Gouvernail had done so, they two mounted and rode away by themselves toward that place where King Angus had taken up his lodging. When they had come there, Sir Tristram made demand to have speech with the King, and therewith they in attendance ushered him in to where the King Angus was.

But when King Angus saw Sir Tristram who he was, and when he beheld a face that was both familiar and kind, he gave a great cry of joy, and
*King Angus welcomes Sir Tristram.* ran to Sir Tristram and flung his arms about him, and kissed him upon the cheek; for he was rejoiced beyond measure to find a friend in that unfriendly place.

Then Sir Tristram said, "Lord, what cheer have you?" Unto that King Angus replied: "Tristram, I have very poor cheer; for I am alone amongst enemies with no one to befriend me, and unless I find some knight who will stand my champion to-morrow or the next day I am like to lose my life for the murder of Sir Bertrand de la Riviere Rouge. And where am I to find any one to act as my champion in defence of my innocence in this place, where I behold an enemy in every man whom I meet? Alas, Tristram! There is no one in all the world who will aid me unless it be you, for you alone of all the knights in the world beyond the circle of the knights of the Round Table may hope to stand against so excellent and so strong a hero!"

"Lord," quoth Sir Tristram, "I know very well what great trouble overclouds you at this time, and it is because of that that I am come hither for to visit you. For I have not at any time forgotten how that I told you when you spared my life in Ireland that mayhap the time might come when I might serve as your friend in your day of need. So if you will satisfy me upon two points, then I myself will stand for your champion upon this occasion."

"Ah, Tristram," quoth King Angus, "what you say is very good news to me indeed. For I believe there is no other knight in all the world (unless it be Sir Launcelot of the Lake) who is so strong and worthy a knight as you. So tell me what are those two matters concerning which you would seek satisfaction, and, if it is possible for me to do so, I will give you such an answer as may please you."

"Lord," said Sir Tristram, "the first matter is this: that you shall satisfy me that you are altogether innocent of the death of Sir Bertrand. And the second matter is this: that you shall grant me whatsoever favor it is that I shall have to ask of you."

Then King Angus arose and drew his sword and he said: "Tristram, behold; here is my sword—and the guard thereof and the blade thereof and the handle thereof make that holy sign of the cross unto *King Angus* which all Christian men bow down to worship. Look! See! *swears innocence* Here I kiss that holy sign and herewith I swear an oath upon *to Sir Tristram.* that sacred symbol, and I furthermore swear upon the honor of my knighthood, that I am altogether guiltless of the death of that noble, honorable knight aforesaid. Nor do I at all know how it was he met his death, for I am innocent of all evil knowledge thereof. Now, Messire, art thou satisfied upon that point?" And Sir Tristram said, "I am satisfied."

Then King Angus said: "As to the matter of granting you a favor, that I would do in any case for the love I bear you. So let me hear what it is that you have to ask of me."

"Lord," cried out Sir Tristram, "the favor is one I had liever die than ask. It is this: that you give me your daughter, the Lady *Sir Tristram* Belle Isoult, for wife unto mine uncle, King Mark of Cornwall." *asks his boon.*

Upon these words, King Angus sat in silence for a long while, gazing very strangely upon Sir Tristram. Then by and by he said: "Messire, this is a very singular thing you ask of me; for from what you said to me aforetime and from what you said to my daughter I had thought that you desired the Lady Belle Isoult for yourself. Now I can in no wise understand why you do not ask for her in your name instead of asking for her in the name of King Mark."

Then Sir Tristram cried out as in great despair: "Messire, I love that dear lady a great deal more than I love my life; but in this affair I am fulfilling a pledge made upon the honor of my knighthood and unto the King of Cornwall, who himself made me knight. For I pledged him unaware, and now I am paying for my hastiness. Yet I would God that you might take the sword which you hold in your hand and thrust it through my heart; for I had liefer die than fulfil this obligation to which I am pledged."

"Well," said King Angus, "you know very well that I will not slay you, but that I will fulfil your boon as I have promised. As for what you do in this affair, you must answer for it to God and to the honor of your own knighthood whether it is better to keep that promise which you made to the King of Cornwall or to break it."

Then Sir Tristram cried out again in great travail of soul: "Lord, you know not what you say, nor what torments I am at this present moment enduring." And therewith he arose and went forth from that place, for he was ashamed that anyone should behold the passion that moved him.

And now is to be told of that famous battle betwixt Sir Tristram and Sir Blamor de Ganys of which so much hath been written in all the several histories of chivalry that deal with these matters.

Now when the next morning had come—clear and fair and with the sun shining wonderfully bright—a great concourse of people began to betake themselves to that place where the lists had been set up in preparation for that ordeal of battle. That place was on a level meadow of grass very fair bedight with flowers and not far from the walls of the town nor from the high road that led to the gate of the same.

And, indeed, that was a very beautiful place for battle, for upon the one hand was the open countryside, all gay with spring blossoms and flowers; and *Of the meadow of battle.* upon the other hand were the walls of the town. Over above the top of those walls was to be seen a great many tall towers— some built of stone and some of brick—that rose high up into the clear, shining sky all full of slow-drifting clouds, that floated, as it were, like full-breasted swans in a sea of blue. And beyond the walls of the town you might behold a great many fair houses with bright windows of glass all shining against the sky. So you may see how fair was all that place, where that fierce battle was presently to be fought.

Meanwhile, great multitudes of people had gathered all about the meadow of battle, and others stood like flies upon the walls of the town and looked down into that fair, pleasant meadow-land, spread with its carpet of flowers. All along one side of the ground of battle was a scaffolding of seats fair bedraped with fabrics of various colors and textures. In the midst of all the other seats were two seats hung with cloth of scarlet, and these seats were the one for King Arthur and the other for King Angus of Ireland.

In the centre of the meadow-land Sir Blamor rode up and down very proudly. He was clad in red armor, and the trappings and the furniture of his horse were all of red, so that he paraded the field like a crimson flame of fire.

"Sir," quoth King Arthur to King Angus, "yon is a very strong, powerful, noble knight; now where mayst thou find one who can hope to stand against him in this coming battle?"

"Lord," said King Angus, "I do believe that God hath raised up a defender for me in this extremity. For Sir Tristram of Lyonesse came to me yesterday, and offered for to take this *King Angus* quarrel of mine upon him. Now I do not believe that there *presents Sir* is any better knight in all of Christendom than he, wherefore *Tristram for his champion.* I am to-day uplifted with great hopes that mine innocence shall be proved against mine accuser."

"Ha!" quoth King Arthur, "if Sir Tristram is to stand thy champion in this affair, then I do believe that thou hast indeed found for thyself a very excellent, worthy defender."

So anon there came Sir Tristram riding to that place, attended only by Gouvernail. And he was clad all in bright, polished armor so that he shone like a star of great splendor as he entered the field of battle. He came straight to where King Arthur sat and saluted before him. King Arthur said, "Sir, what knight art thou?" "Lord," answered he, "I am Sir Tristram of Lyonesse, and I am come to champion King Angus who sits beside you. For I believe him to be innocent of that matter of which he is accused, and I will emperil my body in that belief for to prove the truth of the same."

"Well," quoth King Arthur, "this King accused hath, certes, a very noble champion in thee. So go and do thy devoirs, and may God defend the right."

Thereupon each knight took a good stout spear into his hand and chose his place for the encounter, and each set his shield before him and feutered his lance in rest. Then, when each was ready, the marshal *Sir Tristram* blew a great blast upon his trumpet, and thereupon, in an *does battle with* instant, each knight launched against the other like a bolt of *Sir Blamor.* thunder. So they met in the very middle of the course with such violence that the spear of each knight was shattered all into pieces unto the very truncheon thereof. Each horse fell back upon his haunches, and each would no doubt, have fallen entirely, had not the knight-rider recovered his steed with the greatest skill and address.

Then each knight voided his saddle and each drew his sword and set his shield before him. Therewith they came to battle on foot like two wild boars—so fiercely and felly that it was terrible to behold. For they traced this way and that and foined and struck at one another so that whole pieces of armor were hewn from the bodies of each.

But in all this battle Sir Tristram had so much the better that, by and by after they had fought for above an hour, Sir Blamor de Ganys began *Sir Tristram* to bare back before him, and to give ground, holding his *overcomes Sir* shield low for weariness. This Sir Tristram perceived, and, *Blamor.* running in suddenly upon Sir Blamor, he struck him so terrible a blow upon the right shoulder that Sir Blamor's arm was altogether benumbed thereby, and he could no longer hold his sword in his hand.

So the sword of Sir Blamor fell down into the grass, and Sir Tristram, perceiving this, ran and set his foot upon it. Then Sir Blamor could not stand any longer, but fell down upon his knees because of a great weariness and faintness that lay upon him like the weariness and faintness of approaching death.

Then Sir Tristram said: "Sir Knight, thou canst fight no longer. Now I bid thee for to yield thyself to me as overcome in this battle."

Thereunto Sir Blamor made reply, speaking very deep and hollow from out of his helmet: "Sir Knight, thou hast overcome me by thy strength and prowess, but I will not yield myself to thee now nor at any time. For that would be so great shame that I would rather die than endure it. I am a knight of the Round Table, and have never yet been overcome in this wise by any man. So thou mayst slay me, but I will not yield myself to thee."

Then Sir Tristram cried out: "Sir Knight, I beseech thee to yield thyself, for thou art not fit to fight any more this day."

Sir Blamor said, "I will not yield, so strike and have done with it."

So Sir Tristram wist not what to do, but stood there in doubt looking down upon Sir Blamor. Then Sir Blamor said, again: "Strike, Sir Knight, and have done with it."

Upon this Sir Tristram said: "I may not strike thee, Sir Blamor de Ganys, to slay thee, for thou art very nigh of blood to Sir Launcelot of the Lake, and unto him I have sworn brotherhood in arms; wherefore I pray thee now to yield thyself to me."

Sir Blamor said, "Nay, I will not yield me to thee."

"Well," said Sir Tristram, "then I must fain act this day in a manner like as I acted yesterday."

Therewith speaking, he took his sword into both his hands and he swung it several times around his head and when he had done that he flung it to a great distance away, so that he was now entirely unarmed saving only for his misericordia. After that he gave Sir Blamor his hand and *Sir Tristram* lifted him up upon his feet. And he stooped and picked up Sir *gives Sir Blamor* Blamor's sword out of the grass and gave it back to Sir Blamor *back his sword.* into his hands, and he said: "Sir Knight, now thou art armed and I am entirely unarmed, and so thou hast me at thy mercy. Now thou shalt either yield thyself to me or slay me as I stand here without any weapon; for I cannot now strike thee, and though I have overcome thee fairly yet thou hast it now in thy power to slay me. So now do thy will with me in this matter."

Then Sir Blamor was greatly astonished at the magnanimity of Sir Tristram, and he said, "Sir Knight, what is thy name?" Sir Tristram said, "It is Tristram, surnamed of Lyonesse."

Upon this Sir Blamor came to Sir Tristram and put his arms about his shoulders, and he said: "Tristram, I yield myself to thee, but in love and not in hate. For I yield myself not because of thy strength of arms (and yet I believe there is no knight in the world, unless it be my cousin Sir Launcelot of the Lake, who is thy peer), but I yield me because of thy exceeding nobility. Yet I would that I might only be satisfied that this King of Ireland is no traitor."

"Messire," said Sir Tristram, "of that I have assured myself very strongly ere I entered into this contest, wherefore I may now freely avouch upon mine own knightly word that he is innocent."

"Then," said Sir Blamor, "I also am satisfied, and I herewith withdraw all my impeachment against him."

Then those two noble, excellent knights took one another by the hand and went forward together to where King Arthur sat in high estate, and all those who looked on and beheld that reconciliation gave loud *Sir Tristram* acclaim. And when King Arthur beheld them coming thus, *and Sir Blamor* he arose from where he sat and met them and embraced them *are reconciled.* both, and he said: "I do not believe that any king can have greater glory in his life than this, to have such knights about him as ye be."

So ended this famous battle with great glory to Sir Tristram and yet with no disregard to that famous knight against whom he did battle.

After that, they and King Arthur and King Angus of Ireland and all the court went up unto the castle of Camelot, and there the two knights-combatant were bathed in tepid water and their wounds were searched and dressed and they were put at their ease in all ways that it was possible.

Now that very day, as they all sat at feast in the castle of Camelot, there came one with news that the name of Sir Tristram had suddenly appeared upon one of the seats of the Round Table. So after they had ended their feast they all immediately went to see how that might be. When they came to the pavilion of the Round Table, there, behold! was his name indeed upon that seat that had once been the seat of King Pellinore. For this was the name that now was upon that seat:

# SIR TRISTRAM
## OF
## LYONESSE

So the next day Sir Tristram was duly installed as a knight-companion

*Sir Tristram becomes knight of the Round Table.*

of the Round Table with a great pomp and estate of circumstance, and a day or two after that he set sail for Ireland with King Angus, taking with him Gouvernail and those Cornish knights who were his companions.

So they all reached Ireland in safety, and, because Sir Tristram had aided the King of Ireland in the day of his extremity, the Queen forgave him all the despite she held against him, so that he was received at the court of the King and Queen with great friendship and high honor.

For a while Sir Tristram dwelt in Ireland and said nothing concerning

*How Sir Tristram dwelt in Ireland.*

that purpose for which he had come. Then one day he said to King Angus: "Lord, thou art not to forget to fulfil that promise which thou madst to me concerning the Lady Belle Isoult."

To this King Angus made reply: "I had hoped that now we were come to Ireland you had changed your purpose in that matter. Are you yet of the same mind as when you first spake to me?"

"Yea," said Sir Tristram, "for it cannot be otherwise."

"Well, then," said King Angus, "I shall go to prepare my daughter for this ill-hap that is to befall her, though indeed it doth go against my heart to do such a thing. After I have first spoken to her, you are to take the matter into your own hands, for, to tell you the truth, I have not the heart to contrive it further."

So King Angus went away from where Sir Tristram was, and he was gone a long while. When he returned he said: "Sir, go you that way and the Lady Belle Isoult will see you."

So Sir Tristram went in the direction King Angus had said, and a page showed him the way. So by and by he came to where the Lady Belle

Isoult was, and it was a great chamber in a certain tower of the castle and high up under the eaves of the roof.

The Lady Belle Isoult stood upon the farther side of this chamber so that the light from the windows shone full upon her face, and Sir Tristram perceived that she was extraordinarily beautiful, and rather *How Lady Belle* like to a shining spirit than to a lady of flesh and blood. For *Isoult appeared* she was clad altogether in white and her face was like to wax *to Sir Tristram.* for whiteness and clearness, and she wore ornaments of gold set with shining stones of divers colors about her neck and about her arms so that they glistered with a wonderful lustre. Her eyes shone very bright and clear like one with a fever, and Sir Tristram beheld that there were channels of tears upon her face and several tears stood upon her white cheeks like to shining jewels hanging suspended there.

So, for a while, Sir Tristram stood still without speaking and regarded her from afar. Then after a while she spake and said, "Sir, what is this you have done?" "Lady," he said, "I have done what God set me to do, though I would rather die than do it."

She said, "Tristram, you have betrayed me." Upon the which he cried out in a very loud and piercing voice, "Lady, say not so!"

She said: "Tristram, tell me, is it better to fulfil this pledge you have made, knowing that in so doing you sacrifice both my happiness and your happiness to satisfy your pride of honor; or is it better that you sacrifice your pride and break this promise so that we may both be happy? Tristram, I beseech you to break this promise you have made and let us be happy together."

At this Sir Tristram cried out in a very loud voice: "Lady, did you put your hand into my bosom and tear my naked heart, you could not cause me so much pain as that which I this moment endure. It cannot be as you would have it, for it is thus with me: were it but myself whom I might consider, I would freely sacrifice both my life and my honor for your sake. But it may not be so, lady; for I am held to be one of the chiefest of that order of knighthood to which I belong, wherefore I may not consider myself, but must ever consider that order. For if I should violate a pledge given upon my knighthood, then would I dishonor not myself, but that entire order to which I belong. For, did I so, all the world would say, what virtue is there in the order of knighthood when one of the chiefest of that order may violate his pledge when it pleases him to do so? So, lady, having assumed that great honor of knighthood I must perform its obligations even to the uttermost; yea, though in fulfilling my pledge I sacrifice both thee and myself."

Then Belle Isoult looked upon Sir Tristram for some little while, and by and by she smiled very pitifully and said: "Ah, Tristram, I believe I am more sorry for thee than I am for myself."

"Lady," said Tristram, "I would God that I lay here dead before you. But I am not able to die, but am altogether strong and hale—only very sorrowful at heart." And therewith he turned and left that place. Only when he had come to a place where he was entirely by himself with no one but God to see him, he hid his face in his hands and wept as though his heart were altogether broken. So it was that Sir Tristram fulfilled his pledge.

After that, King Angus furnished a very noble and beautiful ship with sails of satin embroidered with figures of divers sorts, and he *Belle Isoult and* fitted the ship in all ways such as became the daughter of a *Sir Tristram* *depart for Corn-* king and the wife of a king to embark upon. And that ship *wall.* was intended for the Lady Belle Isoult and Sir Tristram in which to sail to the court of Cornwall.

And it was ordained that a certain very excellent lady of the court of the Queen, who had been attendant upon the Lady Belle Isoult when she was a little child and who had been with her in attendance ever since that time, should accompany her to the Court of Cornwall. And the name of this lady was the Lady Bragwaine.

Now the day before the Lady Belle Isoult was to take her departure from Ireland, the Queen of Ireland came to the Lady Bragwaine and she bare *The Queen of* with her a flagon of gold very curiously wrought. And the *Ireland provides* Queen said: "Bragwaine, here is a flask of a very singular and *a love potion for* *King Mark and* precious sort of an elixir; for that liquor it is of such a sort that *Belle Isoult.* when a man and a woman drink of it together, they two shall thereafter never cease to love one another as long as they shall have life. Take this flask, and when you have come to Cornwall, and when the Lady Belle Isoult and King Mark have been wedded, then give them both to drink of this elixir; for after they have drunk they shall forget all else in the world and cleave only to one another. This I give you to the intent that the Lady Isoult may forget Sir Tristram, and may become happy in the love of King Mark whom she shall marry."

Soon thereafter the Lady Belle Isoult took leave of the King and the Queen and entered into that ship that had been prepared for her. Thus, with Sir Tristram and with Dame Bragwaine and with their attendants, she set sail for Cornwall.

Now it happened that, whilst they were upon that voyage, the Lady Bragwaine came of a sudden into the cabin of that ship and there she beheld the Lady Belle Isoult lying upon a couch weeping. Dame Bragwaine said, "Lady, why do you weep?" Whereunto the Lady Belle Isoult made reply: "Alas, Bragwaine, how can I help but weep seeing that I am to be parted from the man I love and am to be married unto another whom I do not love?"

Dame Bragwaine laughed and said: "Do you then weep for that? See! Here is a wonderful flask as it were of precious wine. When you are married to the King of Cornwall, then you are to quaff of it and he is to quaff of it and after that you will forget all others in the world and cleave only to one another. For it is a wonderful love potion and it hath been given to me to use in that very way. Wherefore dry your eyes, for happiness may still lay before you."

When the Lady Belle Isoult heard these words she wept no more but smiled very strangely. Then by and by she arose and went away to where Sir Tristram was.

When she came to him she said, "Tristram, will you drink of a draught with me?" He said, "Yea, lady, though it were death in the draught." She said, "There is not death in it, but something very different," and thereupon she went away into the cabin where that chalice aforesaid was hidden. And at that time Dame Bragwaine was not there.

Then the Lady Belle Isoult took the flagon from where it was hidden, and poured the elixir out into a chalice of gold and crystal and she brought it to where Sir Tristram was. When she had come there, she said, "Tristram, I drink to thee," and therewith she drank the half of the elixir there was in the chalice. Then she said, "Now drink thou the rest to me."

Upon that Sir Tristram took the chalice and lifted it to his lips, and drank all the rest of that liquor that was therein.

Now immediately Sir Tristram had drunk that elixir he felt it run like fire through every vein in his body. Thereupon he cried out, "Lady, what is this you have given me to drink?" She said: "Tristram, that was a powerful love potion intended for King Mark and me. But now thou and I have drunk of it and *potion.* never henceforth can either of us love anybody in all of the world but the other."

*Sir Tristram and Belle Isoult drink the love potion.*

Then Sir Tristram catched her into his arms and he cried out: "Isoult! Isoult! what hast thou done to us both? Was it not enough that I should have been unhappy but that thou shouldst have chosen to be unhappy also?"

Thereat the Lady Belle Isoult both wept and smiled, looking up into Sir Tristram's face, and she said: "Nay, Tristram; I would rather be sorry with thee than happy with another." He said, "Isoult, there is much woe in this for us both." She said, "I care not, so I may share it with thee."

Thereupon Sir Tristram kissed her thrice upon the face, and then immediately put her away from him and he left her and went away by himself in much agony of spirit.

Thereafter they reached the kingdom of Cornwall in safety, and the Lady Belle Isoult and King Mark were wedded with much pomp and ceremony and after that there was much feasting and every appearance of rejoicing.

# PART II

# The Story of Sir Tristram and Sir Lamorack

*A*ND *now shall be told the story of Sir Tristram and Sir Lamorack of Gales, how they became brothers-in-arms; how Sir Lamorack took offence at Sir Tristram, and how they became reconciled again.*

*But first of all you must know that Sir Lamorack of Gales was deemed to be one of the greatest knights alive. For it was said that there were three knights that were the greatest in all of the world, and those three were Sir Launcelot of the Lake, Sir Tristram of Lyonesse, and Sir Lamorack of Gales.*

*Sir Lamorack was the son of King Pellinore, of whom it hath already been told in the Book of King Arthur that he was the greatest knight during that time; and he was the brother of Sir Percival, of whom it is to be told hereinafter that he was the peer even of Sir Launcelot of the Lake. So because that house produced three such great and famous knights, the house of King Pellinore hath always been singularly renowned in all histories of chivalry. For indeed there was not any house so famous as it saving only the house of King Ban of Benwick, which brought forth those two peerless knights beyond all compare:—to wit, Sir Launcelot of the Lake and Sir Galahad, who achieved the quest of the San Grail.*

*So I hope that you may find pleasure in the story of how Sir Tristram and Sir Lamorack became acquainted, and of how they became brothers-in-arms.*

Sir Lamorack of Gales

# Chapter First

*How Sir Lamorack of Gales came to Tintagel and how he and Sir Tristram sware friendship together in the forest.*

AFTER these happenings, Sir Tristram abode for awhile at the Court of Cornwall, for so King Mark commanded him to do. And he sought in every way to distract his mind from his sorrows by deeds of prowess. So during this time he performed several adventures of which there is not now space to tell you. But these adventures won such credit to his knighthood that all the world talked of his greatness.

And ever as he grew more and more famous, King Mark hated him more and more. For he could not bear to see Sir Tristram so noble and so sorrowful with love of the Lady Belle Isoult.

Also Sir Tristram spent a great deal of time at chase with hawk and hound; for he hoped by means also of such sports to drive away, in some measure, his grief for the loss of Belle Isoult.

Now the season whereof this chapter speaketh was in the autumn of the year, what time all the earth is glorious with the brown and gold of the woodlands. For anon, when the wind would blow, then the leaves would fall down from the trees like showers of gold so that everywhere they lay heaped like flakes of gold upon the russet sward, rustling dry and warm beneath the feet, and carpeting all the world with splendor. And the deep blue sky overhead was heaped full of white, slow-moving clouds, and everywhere the warm air was fragrant with the perfume of the forest, and at every strong breeze the nuts would fall pattering down upon the ground like hailstones.

And because the world was so beautiful and so lusty, Sir Tristram took great pleasure in life in spite of that trouble that lay upon him. So he *Sir Tristram* and his court rode very joyfully amid the trees and thickets, *rides ahunting.* making the woodlands merry with the music of winding horns and loud-calling voices and with the baying of hounds sounding like sweet tolling bells in the remoter aisles of the forest spaces.

Thus Sir Tristram made sport all one morning, in such an autumn season, and when noon had come he found himself to be anhungered. So he gave orders to those who were in attendance upon him that food should be spread at a certain open space in the forest; and therewith, in accordance with those orders, they in attendance immediately opened sundry hampers of wicker, and therefrom brought forth a noble pasty of venison, and manchets of bread and nuts and apples and several flasks and flagons of noble wine of France and the Rhine countries. This abundance of good things they set upon a cloth as white as snow which they had laid out upon the ground.

Now just as Sir Tristram was about to seat himself at this goodly feast he beheld amid the thin yellow foliage that there rode through a forest path not far away a very noble-seeming knight clad all in shining armor and with vestments and trappings of scarlet so that he shone like a flame of fire in the woodlands.

Then Sir Tristram said to those who stood near him, "Know ye who is yonder knight who rides alone?" They say, "No, Lord, we know him not." Sir Tristram said, "Go and bid that knight of his courtesy that he come hither and eat with me."

So three or four esquires ran to where that knight was riding, and in a little they came attending him to where Sir Tristram was, and Sir Tristram went to meet him.

Then Sir Tristram said: "Sir Knight, I pray you for to tell me your name and degree, for it seems to me that you are someone very high in order of knighthood."

"Messire," quoth the other, "I shall be very glad to tell you my name if so be you will do the like courtesy unto me. I am Sir Lamorack of Gales, *Sir Lamorack* and I am son of the late King Pellinore, who was in his days *meets Sir* held to be the foremost knight in this realm. I come to *Tristram.* these parts seeking Sir Tristram of Lyonesse, of whose fame I hear told in every court of chivalry whither I go. For I have never beheld Sir Tristram, and I have a great desire to do so."

"Well," quoth Sir Tristram, "meseems I should be greatly honored that you should take so much trouble for nothing else than that; for lo! I am that very Sir Tristram of Lyonesse whom you seek."

Then Sir Lamorack immediately leaped down from his war-horse and putting up the umbril of his helmet, he came to Sir Tristram and took him by the hand and kissed him upon the cheek. And Sir Tristram kissed Sir Lamorack again, and each made great joy of the other.

After that, Sir Lamorack, with the aid of these esquires attendant upon Sir Tristram, put aside his armor, and bathed his face and neck and hands in a cold forest brook, as clear as crystal, that came brawling down out of the woodlands. Therewith, being greatly refreshed he and Sir Tristram sat down to that bountiful feast together, and ate and drank with great joy and content of spirit. And whiles they ate each made inquiry of the other what he did, and each told the other many things concerning the goodly adventures that had befallen him.

And after they were through eating and drinking, Sir Tristram took his harp in hand and sang several excellent ballads and rondels *Sir Tristram* which he had made in honor of Belle Isoult, and Sir Lamorack *sings to Sir* listened and made great applause at each song that Sir Tris- *Lamorack.* tram sang. And so each knight loved the other more and more the longer they sat together.

Then, after a while, Sir Tristram said: "Dear friend, let us swear brotherhood to one another, for I find that my heart goeth out to thee with a wonderful strength."

"Ha, Tristram," said Sir Lamorack, "I would rather live in brotherhood with thee than with any man whom I know, for I find that the longer I am with thee, the greater and the stronger my love groweth for thee."

Then Sir Tristram drew from his finger a very splendid ring (for the ring held an emerald carved into the likeness of the head of a beautiful woman, and that emerald was set into the gold of the ring) and Sir Tristram said: "Give me that ring upon thy finger, O Lamorack! and take thou this ring in its stead; so we shall have confirmed our brotherhood to one another."

Then Sir Lamorack did very joyfully as Sir Tristram bade him, and he took the ring that Sir Tristram gave him and kissed it and put it upon his finger; and Sir Tristram kissed the ring that Sir Lamorack gave him and put it upon his finger.

Thus they confirmed brotherhood with one another that day as they sat together in the forest at feast, with the golden leaves falling about them. And so they sat together all that afternoon and until the sun began to hang low in the west; after that, they arose and took horse, and rode away together toward Tintagel in great pleasure of companionship.

Now all the court at Tintagel was greatly rejoiced at the presence of so famous a knight as Sir Lamorack of Gales; so there was great celebration *Sir Lamorack* upon that account, and everybody did the most that he was *is honored at* able to give pleasure to Sir Lamorack. And during the time *Tintagel.* that Sir Lamorack was at Tintagel there were several joustings held in his honor, and in all these assays at arms Sir Lamorack himself took part and overthrew everyone who came against him, so that he approved himself to be so wonderful a champion that all men who beheld his performance exclaimed with astonishment at his prowess.

But from all these affairs at arms Sir Tristram held himself aloof, and would not take part in them. For he took such pleasure in Sir Lamorack's glory that he would not do anything that might imperil the credit that his friend thus gained by his prowess. For though Sir Tristram dearly loved such affairs, he would ever say to himself: "Perhaps if I should enter the lists against my friend it might be my mishap to overthrow him and then his glory would be forfeited unto me."

Now upon a certain time there was held a great day of jousting in honor of Sir Lamorack, and in that affair at arms twenty of the best knights, *Sir Lamorack* both of Cornwall and the countries circumadjacent, took the *does famous* the field to hold it against all comers. Of these knights, *battle.* several were well-known champions, so that they maintained the field for a long while, to the great credit both of themselves and of Cornwall. But some while after the prime of day, there came Sir Lamorack into that field, and, the day being cool and fresh, he was filled with a wonderful strength and spirit of battle. So he challenged first one of those Cornish champions and then another, and in all such challenges he was successful, so that he overthrew of those knights, the one after the other, fifteen men, some of whom were sorely hurt in the encounter. Upon this, the other five of those champions, beholding the prowess and strength and skill of Sir Lamorack said to one another: "Why should we venture against this man? Of a verity, this knight is no mere man, but a demon of strength and skill. Wherefore no man may hope to stand against him in an assault of arms; for lo! if he doth but touch a man with his lance that man straightway falleth from his saddle." So they withdrew themselves from that encounter and would not have to do with Sir Lamorack.

Now at that time Sir Tristram was sitting with the court of the King, and not far from the Lady Belle Isoult, overlooking the meadow of battle.

To him King Mark said: "Messire, why do you take no part against this knight? Is it that you fear him?"

To this Sir Tristram replied with great calmness: "Nay, I fear not him nor any man alive, and that you know, Lord, better than anyone in all of the world."

"I am glad to hear of your courage and fearlessness," quoth King Mark, "for meseems it is a great shame to all of us that this gentleman, who is a stranger amongst us, should win so much credit to the disadvantage of all the knights of Cornwall. Now, as you say you have no fear of him, I pray you go down into the field and do battle with him in our behalf." So said King Mark, for he thought to himself: "Perhaps Sir Lamorack may overthrow Sir Tristram, and so bring him into disrepute with those who praise him so greatly."

But Sir Tristram said: "No; I will not go down to battle against Sir Lamorack this day whatever I may do another day. For I have sworn brotherhood to that noble and gentle champion, and it would ill beseem me to assault him now, when he is weary and short of breath from this great battle which he hath done to-day against such odds. For if I should overthrow him now, it would bring great shame upon him. Some other day and in some other place I may assay him in friendliness, with honor and credit both to myself and him."

"Well," said King Mark, "as for that, I do not choose to wait. Nor am I pleased that you should sit by and suffer this knight to carry away all the credit of arms from Cornwall in despite of the knights of Cornwall. For not only would this be a great shame to the knights of Cornwall (of whom you are the acknowledged champion), but it would be equally a shame unto this lady whom you have fetched hither from Ireland to be Queen of Cornwall. *King Mark commands Sir Tristram to do battle.* So I lay this command upon you—not only because I am your King, but because I am he who made you knight—that you straightway go down into yonder meadow and do battle with this knight who beareth himself so proudly in our midst."

Then Sir Tristram looked upon King Mark with great anger and bitterness, and he said: "This is great shame and despite which you seek to put upon me by giving such commands unto me. Verily, it would seem that in all ways you seek to put shame and sorrow upon me. And yet I have ever been your true knight, and have saved your kingdom from truage to Ireland and have served you very faithfully in all ways. Would to God I had been made knight by any man in the world rather than by you."

At this King Mark smiled very bitterly upon Tristram. "Sirrah," quoth he, "meseems you speak very outrageously to me who am your King. Now I herewith command you to go straightway down into that field without any further words and to do my bidding against yonder knight."

Then Sir Tristram groaned in spirit, and then he said, "I go."

So Sir Tristram arose and went away from that place very full of bitterness and anger against the King and his court. For whiles there were some of that court who were sorry for the affront that King Mark had put upon him in public before the eyes of the entire court, yet there were others who smiled and were glad of his humiliation. For even so true and noble a gentleman as Sir Tristram, when he groweth great and famous, is like to have as many enemies as friends. For there are ever those who envy truth and nobility in a man, as well as others who hate meanness and falsity, and so Sir Tristram ever had many enemies whithersoever he went. And that also was the case with Sir Launcelot and Sir Lamorack, and with other noble knights at that time.

But though Sir Tristram was so filled with indignation he said nothing *Sir Tristram* to any man, but went to his lodging and summoned Gouvernail, *arms himself.* and bade Gouvernail to help him to his armor and his horse.

Gouvernail said: "Lord, what would you do for to arm and horse yourself at this hour?" Sir Tristram made reply: "The King hath commanded me to do battle with Sir Lamorack, and yet Sir Lamorack is my very dear friend and sworn brother-in-arms. He is already weary with battle, and of a surety I shall be very likely to overthrow him in an assault at arms at this time." Gouvernail said, "Lord, that would be great shame to you as well as to him." And Sir Tristram said, "Yea, it is great shame." Then Gouvernail beheld Sir Tristram's face, how it was all filled with a passion of shame and indignation, and so he guessed what had passed, and held his peace.

So when Sir Tristram was armed and mounted, he rode down into the meadow of battle, where was Sir Lamorack parading with great glory before the applause of all who looked down upon that field.

But when Sir Lamorack beheld that it was Sir Tristram who came against *Sir Lamorack* him, he was greatly astonished, and cried out: "Ha, Tristram, *speaks to Sir* how is this? Is it you who come against me? Have you *Tristram.* then forgot that I am your brother-in-arms and a fellow of the Round Table?"

To this Sir Tristram said: "Messire, I come not of my own free will, but only because I must needs come, being so commanded by the King of Cornwall."

"Very well," said Sir Lamorack, "so be it as you will, though I am very much surprised that you should do battle against me, after all that hath passed betwixt us. More especially at this season when, as you very well know, I am weary and winded with battle."

Thereupon and without further parley, each knight took stand for the encounter at the position assigned to him. Then when they were in all ways prepared, the marshal of the field blew upon his trumpet a call for the assault.

So rushed those two together like two stones, flung each out of a catapult; and therewith they two smote together in the midst of their course like to a clap of thunder.

In that encounter the spear of Sir Lamorack brake into as many as twenty or thirty pieces; but the spear of Sir Tristram held, *Sir Tristram* so that the horse of Sir Lamorack, which was weary with the *overthrows Sir* several charges he had made, was overthrown into a great *Lamorack.* cloud of dust.

But Sir Lamorack did not fall with his steed; for he voided his saddle with a very wonderful agility and dexterity, so that he himself kept his feet, although his horse fell as aforesaid. Then he was filled with great rage and shame that he had been so overthrown before all those who looked upon him; wherefore he immediately drew his sword and cried out aloud: "Come down, Sir Knight, and do battle with me afoot, for though my horse hath failed me because of his weariness, yet you shall find that my body shall not so fail me."

But that while Sir Tristram sat very sorrowful, and he said: "Nay, I will not have to do with thee again this day, for it was against my will that I came hither to do battle with thee, and it is to my shame that I did so. Wherefore I will not now do further battle with thee. But wait until to-morrow and until thou art fresh, and then I will give thee the chance of battle again."

To this Sir Lamorack made answer very bitterly: "Sir, I think you talk to amuse me; for first you put shame upon me in this encounter, and then you bid me wait until to-morrow ere I purge me of that shame. Now I demand of you to do battle with me upon this moment and not to-morrow."

Sir Tristram said: "I will not do battle with thee, Lamorack, for I have done wrong already, and I will not do more wrong."

Upon this, Sir Lamorack was so filled with anger that he scarce knew what to say or to do. Wherefore he turned him to several who had come down into the meadow of battle, and he said: "Hear ye all, *Sir Lamorack* and listen to my words: This knight came against me in this *reproves Sir* field after I had had to do with fifteen other knights. In that *Tristram.* encounter he overthrew me, because of the weariness of my horse. Having done that unknightly deed, he now refuseth me any further test of battle, but allows me to lie beneath that shame which he put upon me.   Now I bid you who stand here to take this word to Sir Launcelot of the Lake; I bid ye tell Sir Launcelot that Sir Tristram of Lyonesse, having sworn brotherhood-in-arms to me, and being a fellow-knight of the Round Table, hath come against me when I was weary with battle and he was fresh.   Tell Sir Launcelot that so Sir Tristram overthrew me with shame to himself and with discredit to me, and that he then refused me all satisfaction such as one true knight should afford another.''

Then Sir Tristram cried out in a loud voice, "I pray you, hear me speak, Messire!"   But Sir Lamorack replied, "I will not hear thee!" and therewith turned and went away, leaving Sir Tristram where he was. And Sir Tristram sat there without movement, like to a statue of stone.

After that Sir Lamorack did not tarry longer at Tintagel, but immediately left the King's court without making speech with anyone.   And there-
*Sir Lamorack* after he went down to the seashore and embarked in a boat *leaves Tintagel.* with intent to sail to Camelot where King Arthur was then holding court.   For his heart was still so bitter against Sir Tristram that he intended to lay complaint against him before the court of chivalry at Camelot.

But Sir Lamorack did not reach Camelot upon that voyage; for, whilst he was in passage, there suddenly arose a great tempest of wind, and in spite of all that the mariners could do, that small ship wherein he sailed was driven upon a cruel headland of rocks and cliffs where it was dashed to pieces.

But Sir Lamorack had foreseen that that small boat was to be wrecked, wherefore, before the end came, he stripped himself entirely naked and leaped into the waters and swam for his life.

So he swam for a long time until he was wellnigh exhausted and upon the point of drowning in the waters.   But at that moment he came by
*Sir Lamorack* good hap to where was a little bay of quiet water, whereinto he *is shipwrecked* swam and so made shift to come safe to land—but faint and *upon a strange* weak, and so sick that he feared that he was nigh to death. *land.*
Then Sir Lamorack perceived that there was heather at that place growing upon the rocks of the hillside, so he crawled into the heather and lay him down therein in a dry spot and immediately fell into such a deep sleep of weariness that it was more like to the swoon of death than to slumber.

Now the lord of that country whereunto Sir Lamorack had come was

a very wicked knight, huge of frame and very cruel and hard of heart.
The name of this knight was Sir Nabon, surnamed le Noir; for he was
very swarth of hue, and he always wore armor entirely of *Of Sir Nabon*
black.   This knight had several years before slain the lord *le Noir.*
of that land, and had seized upon all of the island as his own possession,
and no one dared to come against him for to recover these possessions,
for his prowess was so remarkable and his body so huge that all the
world was afraid of him.   So he dwelt there unmolested in a strong castle
of stone built up upon a rock near to the seashore, whence he might behold
all the ships that passed him by.   Then, whenever he would see such a ship
pass by, he would issue forth in his own ships and seize upon that other
vessel, and either levy toll upon it or sink it with all upon board.   And
if he found any folk of high quality aboard such a ship, that one he would
seize and hold for ransom.   So Sir Nabon made himself the terror of all
that part of the world, and all men avoided the coasts of so inhospitable a
country.   Such was the land upon which Sir Lamorack had been cast by
the tempest.

Now whilst Sir Lamorack lay sleeping in the heather in that wise as
aforetold, there came by that way several fisher-folk; these, *The fisher-folk*
when they saw him lying there, thought at first that he was *disarm Sir*
dead.   But as they stood talking concerning him, Sir Lamorack *Lamorack.*
was aware of their voices and woke and sat up and beheld them.

Then the chiefest of those fisher-folk spake and said, "Who are you, and
how came you here?"   Him Sir Lamorack answered: "Alas! friend!   I
am a poor soul who was cast ashore from a shipwreck, naked as you see
me.   Now I pray you, give me some clothes to cover my nakedness, and
give me some food to eat, and lend me such succor as man may give to
man in distress."

Then the chief fisherman perceived the ring upon Sir Lamorack's finger
that Sir Tristram had given him, and he said, "How got you that ring upon
your finger?"   Sir Lamorack said, "He who was my friend gave it to me."
"Well," quoth the fisherman, "I will give you clothes to wear and food to
eat, but if I do so you must give me that ring that I see upon your hand.
As for lending you aid, I must tell you that the lord of this island hath
ordained upon peril of our lives that all who come hither must straightway
be brought before him to be dealt with as he may deem fitting.   Wherefore,
after I have fed you and clothed you I must immediately take you to
him."

"Alas!" quoth Sir Lamorack, "this is certes an inhospitable land into
which I have come!   Ne'ertheless, as I am naked and starving, I see that

I have no choice other than that which ye put upon me." So therewith he gave the chief of the fisher-folk the ring that Sir Tristram had

*The fisher-folk give Sir Lamorack clothes and food.*

given him, and in return the fishermen gave him such garments as they could spare to cover his nakedness; and they gave him black bread and cheese to eat, and bitter ale to drink from a skin that they carried with them. After that they tied Sir Lamorack's hands behind his back, and so, having made him prisoner, they brought him to the castle of Sir Nabon, and before Sir Nabon who was there at that time.

Now it chanced that the swineherd of Sir Nabon's castle had been slain in a quarrel with one of his fellows, so that when Sir Nabon beheld Sir Lamorack, that he was big and sturdy of frame, he said: " I will spare this fellow his life, but I will make him my swineherd. So take ye him away and let him herd my swine."

So they led Sir Lamorack away, and he became swineherd to Sir Nabon

*Sir Lamorack turns swineherd.*

surnamed le Noir, and presently in a little while he grew so rough and shaggy that his own mother would hardly have known him had she beheld him.

So endeth this adventure of Sir Lamorack. And now it shall be told how it befel with Sir Tristram after Sir Lamorack had left Tintagel as aforetold.

ir Tristram cometh to y̆ castle
of Sir Nabon

# Chapter Second

*How Sir Tristram started to go to Camelot, and how he stayed by the
way to do battle with Sir Nabon le Noir.*

NOW after Sir Lamorack had quit the court of King Mark of Cornwall as aforetold, Sir Tristram was very sad at heart for a long
while. Nevertheless, he tried to comfort himself by saying:
"Well, it was not by my will that I did battle with my friend and brother-
in-arms, for I had no choice as to that which I was compelled to do." So
he spake to himself, and took what comfort he was able from such con-
siderations, and that comfort was not very great.

Then one day there came from Sir Launcelot of the Lake a letter in
which Sir Launcelot said that he had heard that Sir Tristram had assailed
Sir Lamorack when that knight was weary and spent with *Sir Launcelot*
battle. And in that letter Sir Launcelot further said: "It is *sends a letter*
very strange to me, Messire, that such things should be said *to Sir Tristram.*
of you, and that by several mouths. Now, I pray you, set this matter
at right, for I do not choose to have such a thing said of you; that you
would wait until a knight was weary with fighting before you would do
battle with him. Moreover, Sir Lamorack is your sworn brother-at-arms,
and a fellow-knight of the Round Table, and is, besides, one of the noblest
and gentlest knights in Christendom. Wherefore I beseech you to set
this matter right, so that those who accuse you of unknightliness may be
brought to confusion."

So wrote Sir Launcelot, and at those words Sir Tristram was cast into a
great deal of pain and trouble of spirit; for he wist not how to answer that
letter of Sir Launcelot's so as to make the matter clear to that knight.
Wherefore he said: "I will straightway go to Camelot and to Sir Launcelot
and will speak to him by word of mouth, and so will make him understand
why I did that which I had to do."

So when the next day had come Sir Tristram arose and took horse and
rode away from Tintagel with intent to betake himself to *Sir Tristram*
Camelot where King Arthur was then holding court, and *rides to Camelot.*
where he might hope to find Sir Launcelot abiding. And Sir Tristram
took no companion with him, not even Gouvernail.

And now I shall tell you how Sir Tristram rode: the way that he took led him down by the seashore, and by and by to a deep forest, which was then nearly altogether devoid of leaves, so that the branches above him were in some places like to the meshes of a net spread against the sky. Here that young knight rode upon a deep carpet of leaves, so that the steps of his war-horse were silenced save only for the loud and continued rustling of his footfalls in the dry and yellow foliage. And as Sir Tristram rode he sang several songs in praise of the Lady Belle Isoult, chanting in a voice that was both clear and loud and very sweet, and that sounded to a great distance through the deep, silent aisles of the forest.

Thus he travelled, anon singing as aforetold of, and anon sank in meditation, so travelling until the day declined and the early gray of the evening began to fall. Then he began to bethink him how he should spend the night, and he thought he would have to sleep abroad in the forest. But just as the gray of the evening was fading away into darkness he came to a certain place of open land, where, before him, he perceived a tall castle, partly of stone and partly of red bricks, built up upon a steep hill of rocks. And upon one side of this castle was the forest, and upon the other side was the wide and open stretch of sea.

And Sir Tristram perceived that there were lights shining from several windows of that castle, and that all within was aglow with red as of a great fire in the hall of the castle; and at these signs of good cheer, his heart was greatly expanded with joy that he should not after all have to spend that night in the darkness and in the chill of the autumn wilds.

So Sir Tristram set spurs to his good horse and rode up to the castle and *Sir Tristram* made request for rest and refreshment for the night. Then, *comes to a* after a little parley, the drawbridge was lowered, and the *friendly castle.* portcullis was raised, and he rode with a great noise into the stone-paved courtyard of the castle.

Thereupon there came several attendants of the castle, and took his horse and aided him to descend from the saddle; and then other attendants came and led him away into the castle and so to an apartment where there was a warm bath of tepid water, and where were soft towels and napkins of linen for to dry himself upon after he was bathed. And when he had bathed and refreshed himself, there came still other attendants bearing soft warm robes for him in which to clothe himself after his journey; and Sir Tristram clothed himself and felt greatly at his ease, and was glad that he had come to that place.

For thus it was that worthy knights like Sir Tristram travelled the world in those days so long ago; and so they were received in castle and hall with great pleasure and hospitality. For all folk knew the worth of these noble gentlemen and were glad to make them welcome whithersoever they went. And so I have told to you how Sir Tristram travelled, that you might, perchance, find pleasure in the thought thereof.

Now after Sir Tristram had refreshed himself and clothed himself as aforesaid, there came the steward of the castle and besought him that he would come to where the lady of the castle was awaiting him *Sir Tristram* for to welcome him. And Sir Tristram went with the steward, *meets the lady* and the steward brought him where the lady sat at a table *of the castle.* prepared for supper. And Sir Tristram perceived that the lady was very beautiful, but that she was clad in the deep weeds of a widow.

When the lady perceived Sir Tristram, she arose and went to meet him, and gave him welcome, speaking in a voice both soft and very sweet. "Messire," quoth she, "I am grieved that there is no man here to welcome you in such a manner as is fitting. But, alas! as you may see by the weeds in which I am clad, I am alone in the world and without any lord of the castle to do the courtesies thereof as is fitting. Yet such as I am, I give you welcome with my entire heart."

"Lady," quoth Sir Tristram, "I give you gramercy for your courtesy. And indeed I am grieved to see you in such sorrow as your dress foretells. Now if there is any service I may render to you, I beseech you to call upon me for whatever aid I may give you."

"Nay," quoth she, "there is nothing you can do to help me." And therewith the lady, who was hight Loise, took Sir Tristram by the hand and led him to the table and sat him down beside her. Then *Sir Tristram* straightway there came sundry attendants, and set a noble *feasts with the* feast before them, with good excellent wines, both white and *chatelaine.* red; and they two ate and drank together with great appetite and enjoyment.

Now after that feast was over and done, Sir Tristram said: "Lady, will you not of your courtesy tell me why you wear the weeds of sorrow in which you are clad? This I ask, not from idle humor, but because, as I said before, I may haply be able to aid you in whatever trouble it is under which you lie."

"Alas, Sir Knight!" quoth she, "my trouble lieth beyond your power to aid or to amend. For can you conquer death, or can you *The Lady telleth* bring the dead back to life again? Nevertheless, I will tell you *Sir Tristram of* what my sorrow is, and how it came unto me. You must *Sir Nabon le* know that some distance away across the sea, which you may *Noir.* behold from yonder window, there lieth an island. The present lord of that island is a very wicked and cruel knight, huge of frame and big of limb, hight Sir Nabon surnamed le Noir. One time the noble and gentle knight who was my husband was the lord of that island and the castle thereon, and of several other castles and manors and estates upon this mainland as well. But one evil day when I and my lord were together upon that island, this Sir Nabon came thither by night, and with certain evil-disposed folk of the island he overcame my lord and slew him very treacherously. Me also he would have slain, or else have taken into shameful captivity, but, hearing the noise of that assault in which my lord was slain, I happily escaped, and so, when night had come, I got away from that island with several attendants who were faithful to me, and thus came to this castle where we are. Since that time Sir Nabon has held that castle as his own, ruling it in a very evil fashion. For you are to know that the castle sits very high upon the crags overlooking the sea, and whenever a vessel passeth by that way, Sir Nabon goeth forth to meet it; and upon some of these crafts he levies toll, and other ships he sinks after slaying the mariners and sailor-folk who may by evil hap be aboard thereof. And if anyone is by chance cast ashore upon that island, that one he either slays or holds for ransom, or makes thereof a slave for to serve him. Because of this, very few ships now go by that way, for all people shun the coasts of so evil a country as that. So Sir Nabon took that land away from me; nor have I any kin who will take up this quarrel for me, and so I must endure my losses as best I may."

"Ha!" quoth Sir Tristram, "and is there then no good knight-champion in this country who will rid the world of such an evil being as that Sir Nabon of whom you speak?"

"Nay," said the lady, "there is no one who cares to offer challenge to that knight, for he is as strong and as doughty as he is huge of frame, and he is as fierce and cruel as he is strong and masterful, wherefore all men hold him in terror and avoid him."

"Well," said Sir Tristram, "meseems it is the business of any knight to rid the world of such a monster as that, whatever may be the danger to himself. Now as there is no knight hereabouts who hath heart to undertake such an adventure, I myself shall undertake it so soon as to-morrow shall have come."

"Sir," said the lady, "I beseech you to think twice before you enter into such an affair as that. Or rather be ruled by me and do not undertake this quest at all; for I misdoubt that anyone could conquer this huge and powerful champion, even if that knight were such as Sir Launcelot of the Lake or Sir Tristram of Lyonesse."

At this Sir Tristram laughed with great good-will, and he said, "Lady, do you not then know who I am?" "Nay," said she, "I know you not." "Well," said Sir Tristram, "then I may tell you *Sir Tristram* that I am that Sir Tristram of Lyonesse of whom you *confesses his de-* spoke just now. And I also tell you that I shall undertake *gree to the* this adventure to-morrow morning." *chatelaine.*

Now when the lady found that the stranger she had taken in was Sir Tristram of Lyonesse, she made great exclamation of surprise and pleasure at having him at that place, for at that time all the world was talking of Sir Tristram's performances. So she took great pleasure and pride that her castle should have given him shelter. She made many inquiries concerning his adventures, and Sir Tristram told her all she asked of him.

Then the lady said: "Messire, I hear tell that you sing very sweetly, and that you are a wonderful harper upon the harp. Now will you not chaunt for me a song or two or three?" And Sir Tristram said: "Lady, I will do whatsoever you ask me that may give you pleasure."

So the lady bade them bring a harp and they did so. And Sir Tristram took the harp and set it before him and tuned it and played upon it, and sang so sweetly that they of the castle said: "Certes, this is no *Sir Tristram* knight-errant who sings, but an angel from Paradise who hath *sings to the lady.* come among us. For surely no one save an angel from Paradise could sing so enchantingly."

So passed that evening very pleasantly until the hours waxed late. Then Sir Tristram retired to a very noble apartment where a soft couch spread with flame-colored linen had been prepared for him, and where he slept a soft sleep without disturbance of any kind.

Now when the next morning had come, Sir Tristram armed himself and mounted upon his war-horse, and rode him to a certain place on the shore. There he found some mariners in haven with a large *Sir Tristram* boat, and to these he paid ten pieces of silver money to bear *departs for the* him across the sea to that island where Sir Nabon le Noir *island of Sir* abided. At first these mariners said they would not sail to *Nabon.* such a coast of danger and death; but afterward they said they would, and they did do so. But still they would not bring Sir Tristram to land nigh to the castle, but only at a place that was a great way off, and where they deemed themselves to be more safe from the cruel lord of that land.

As for Sir Tristram he made merry with their fear, saying: "It is well that we who are knights-errant have more courage than you who are sailor-men, else it would not be possible that monsters such as this Sir Nabon should ever be made an end of."

Upon this the captain of these sailors replied: "Well, Messire, for the matter of that, it is true that mariners such as we have not much courage, for we are the first of our order who have dared to come hither. But it is also true that you are the first errant-knight who hath ever had courage to come hither. So what say you for the courage of your own order?" And at that Sir Tristram laughed with great good will and rode his way.

Thereafter he rode forward along the coast of that land for several leagues, with the noise of the sea ever beating in his ears, and the shrill clamor of the sea-fowl ever sounding in the air about him. By and by he came to a place of certain high fells, and therefrom perceived before him in the distance a tall and forbidding castle standing upon a high headland of the coast. And the castle was built of stone, that was like the rocks upon which it stood, so that at first one could not tell whether what one beheld was a part of the cliffs or whether it was the habitation of man. But when Sir Tristram had come somewhat nearer, he perceived the windows of the castle shining against the sky, and he saw the gateway thereof, and the roofs and the chimneys thereof, so that he knew that it was a castle of great size and strength and no wall of rock as he had at first supposed it to be; and he wist that this must be the castle of that wicked and malignant knight, Sir Nabon, whom he sought.

*Sir Tristram arrives at the castle of Sir Nabon.*

Now as Sir Tristram wended his way toward that castle by a crooked path meditating how he should come at Sir Nabon for to challenge him to battle, he was by and by aware of a fellow clad in pied black and white, who walked along the way in the direction that he himself was taking. At the first that fellow was not aware of Sir Tristram; then presently he was aware of him and turned him about, and beheld that a strange knight was riding rapidly down toward him upon a horse.

Then at first that fellow stood like one struck with amazement; but in a moment he cried out aloud as with a great fear, and instantly turned again and ran away, yelling like one who had gone mad.

But Sir Tristram thundered after him at speed, and, in a little, came up with him, and catched him by the collar of his jerkin and held him fast. And Sir Tristram said: "Fellow, who are you?"

"Lord," quoth the fellow, "I am an attendant upon the knight of yonder castle, which same is hight Sir Nabon surnamed le Noir."

*Sir Tristram talks with a knave of the earth.*

Then Sir Tristram said: "Sirrah, why did you run from me when you first beheld me?" And the fellow replied: "Messire, you are the first stranger who hath dared to come hither to this country; wherefore, seeing you, and seeing that you rode upon horseback, and not knowing how you came to this land, I wist not whether you were a man of flesh and blood, or whether you were a spirit come hither for to punish us for our sins; so I ran away from you."

"Well," said Sir Tristram, "as you see, I am no spirit, but a man of flesh and blood. Yet I have great hope that I have indeed been sent hither for to punish those who have done evil, for I come hither seeking the knight of yonder castle for to do battle with him in behalf of that lady whose lord he slew so treacherously as I have heard tell. And I hope to take away from him this island and return it to the Lady Loise, to whom it belongeth."

"Alas, Messire," quoth the fellow, "this is for you a very sorry quest upon which you have come. For this Sir Nabon whom you seek is accounted to be the most potent knight in all of the world. Yea; he is held to be a bigger knight than even Sir Launcelot of the Lake or Sir Tristram of Lyonesse or Sir Lamorack of Gales. Wherefore I beseech you to turn about and go away whither you have come whilst there is still the chance for you to escape."

"Gramercy for your pity, good fellow," quoth Sir Tristram, "and may God grant that it may not be deserved. Nevertheless, in spite of the danger in this quest, I am still of the same mind as I was when I came hither. So do you presently go to your lord and tell *Sir Tristram sends challenge to Sir Nabon.* him from me that a knight hath come to do battle with him upon the behalf of the lady to whom this island by rights belongeth."

Therewith Sir Tristram let the fellow go, and he ran off with great speed and so away to the postern of the castle and entered in and shut the door behind him.

Now at that time Sir Nabon le Noir was walking along the wall of the castle, and his son, who was a lad of seventeen years, was with him. There the messenger from Sir Tristram found him and delivered his message. Thereupon Sir Nabon looked over the battlements and down below and he beheld that there was indeed a tall and noble knight seated upon horseback in a level meadow that reached away, descending inland from the foot of the crags whereon the castle stood.

But when Sir Nabon perceived that a stranger knight had dared to come thus into his country, he was filled with amazement at the boldness of that knight that he wist not what to think. Then, presently a great rage got hold upon him, and he ground his teeth together, and the cords on his neck stood out like knots on the trunk of a tree. For a while he stood as though bereft of speech; then anon he roared out in a voice like that of a bull, crying to those who were near him: "Go! Haste ye! Fetch me straightway my horse and armor and I will go immediately forth and so deal with yonder champion of ladies that he shall never take trouble upon their account again."

Then those who were in attendance upon Sir Nabon were terrified at his words and ran with all speed to do his bidding, and presently fetched his armor and clad him in it; and they fetched his horse into the courtyard of the castle and helped him to mount upon it. And lo! the armor of Sir Nabon was as black as ink; and the great horse upon which he sat was black; and all the trappings and furniture of the armor and of the horse were black, so that from top to toe he was altogether as black and as forbidding as Death himself.

So when Sir Nabon was thus in all wise prepared for battle, the portcullis of the castle was lifted up, and he rode forth to meet Sir Tristram; and his *Sir Nabon rides* young son rode with him as his esquire. Then all the people of *forth to meet* the castle gathered together upon the walls to see that battle *Sir Tristram.* that was to be, and not one of those several score of folk thought otherwise than that Sir Tristram would certainly be overcome in that encounter.

Sir Nabon rode straight up to Sir Tristram and he said very fiercely, "Sirrah, what is it brings you hither to this land?"

"As to that," said Sir Tristram, "the messenger whom I have sent to you hath, I believe, told you what I come for, and that it is to redeem this island from your possession, and to restore it to the Lady Loise, to whom it belongeth. Likewise that I come to punish you for all the evil you have done."

"And what business is all this of yours?" quoth Sir Nabon, speaking with great fury of voice.

"Messire," quoth Sir Tristram, "know ye not that it is the business of every true knight to rid the world of all such evil monsters as you be?"

"Ha!" quoth Sir Nabon, "that was very well said, for whatever mercy I should have been willing before this to show you hath now been forfeited unto you. For now I shall have no mercy upon you but shall slay you."

"Well," quoth Sir Tristram, "as for that, meseems it will be time enough to offer me mercy after you have overcome me in battle."

So thereupon each knight took his place for assault, and when they were in *Sir Tristram* all ways prepared, each set spurs to his horse and dashed the *does battle with* one against the other, with a dreadful, terrible fury of onset. *Sir Nabon.* Each smote the other in the very midst of his shield, and at that blow the lance of each was altogether shivered into pieces to the very truncheon thereof. But each knight recovered his horse from the fall and each leaped to earth and drew his sword, and each rushed against the other with such fury that it was as though sparks of pure fire flew out from the oculariums of the helmets. Therewith they met together, and each lashed and smote at the other such fell strokes that the noise thereof might easily have been heard several furlongs away. Now in the beginning of that battle Sir Tristram was at first sore bestead and wist that he had met the biggest knight that ever he had encountered in all of his life, unless it was Sir Launcelot of the Lake, whom he had encountered as aforetold of in this history. So at first he bore back somewhat from the might of the blows of Sir Nabon. For Sir Nabon was so huge of frame and the blows he struck were so heavy that they drove Sir Tristram back as it were in spite of himself.

Then Sir Tristram began to say to himself: "Tristram, if you indeed lose this battle, then there will be no one to defend your honor before Sir Launcelot who hath impeached it." Therewith it was as *Sir Tristram* though new strength and life came back to him, and of a sudden *slays Sir Nabon.* he rushed that battle, and struck with threefold fury, and gave stroke upon stroke with such fierceness of strength that Sir Nabon was astonished and fell back before his assault. Then Sir Tristram perceived how Sir Nabon held his shield passing low, and therewith he rushed in upon him and smote him again and again and yet again. And so he smote Sir Nabon down upon his knees. Then he rushed in upon him and catched his helmet and plucked it off from his head. And he catched Sir Nabon by the hair of his head and drew his head forward. And Sir Tristram lifted his sword on high and he smote Sir Nabon's head from off his body so that it rolled down into the dust upon the ground.

Now when the son of Sir Nabon perceived how that his father was slain, he shrieked like a woman. And he fell down upon his knees and crawled upon his knees to Sir Tristram and catched him about the thighs, crying out to him, "Spare me, and slay me not!"

But Sir Tristram thrust him away and said, "Who art thou?"

"Messire," said the youth, "I am the son of him whom thou hast just slain."

Then Sir Tristram looked closely into his face, and he perceived that it was wicked and treacherous and malevolent like to the face of Sir Nabon. Thereupon Sir Tristram said: "If a man shall slay the wolf and *Sir Tristram* spare the whelp of the wolf, what shall the world be the better *slays the son of* therefor?" Therewith he catched the son of Sir Nabon by the *Sir Nabon.* hair and dragged him down and smote off his head likewise as he had smitten off the head of his father, so that it fell upon the ground beside the head of Sir Nabon.

And now it shall be told how Sir Tristram discovered Sir Lamorack upon the island and how he made amends to him, so that they became friends and brethren-in-arms once more as they had been before.

Sir Lamorack herds the swine of Sir Nabon

# Chapter Third

*How Sir Tristram did justice in the island, and thereby released
Sir Lamorack from captivity. Also how Sir Tristram and Sir
Lamorack renewed their great tenderness toward one another.*

NOW after Sir Tristram had overcome Sir Nabon le Noir, and had
slain the son of Sir Nabon as has been just told, he went straightway to the castle that had been Sir Nabon's, and commanded
that they should bring forth the seneschal and the officers thereof unto him.
Meantime, being a little wounded in that battle, he sat himself down upon
a bench of wood that stood in the hall of the castle, and there he held his
court.

So, in a little while, there came the seneschal and several of the officers
of the household to where Sir Tristram was, and when the seneschal came
before Sir Tristram, he fell down upon his knees and besought pardon and
mercy.

Then Sir Tristram said: "I will consider thy case anon, and if I may
assure myself that thou and these others are truly repentant, and if I may
have assurity that ye will henceforth be faithful in your duty *Sir Tristram*
toward that lady who is now again the mistress of this castle *talks with the*
and land, then I shall have mercy. But if ye show yourselves *castle help.*
recreant and treacherous, according to the manners of this Sir Nabon who
is dead, then I shall of a surety return hither and shall punish you even
as ye beheld me punish that wicked knight and his young son."

Then Sir Tristram said, "Who is the porter of this castle?" And the
porter lifted his hand and said, "Lord, I am he." Sir Tristram said, "What
captives have ye in this place?" The porter said: "Lord, there be four
knights and three ladies who are held captive here for ransom." Then
Sir Tristram said, "Bring them forth hither to me."

So the porter and several other of the castle folk departed with all speed
and presently returned bringing with them those miserable captives whom
they had liberated from the dungeons of the castle. These *Sir Tristram*
they led to where Sir Tristram still sat in justice upon the *comforts the*
bench of wood. And Sir Tristram looked upon them with *captives.*
pity and beheld that they were in a very sad and forlorn condition and so

sorrowful from their captivity that some of them wept from pure weakness of heart. Then Sir Tristram said: "Comfort ye, and take no more sorrow to yourselves, for now your troubles are past and gone, and happiness lieth before you. Sir Nabon is dead, and so is his son, and there is no one now to torment you. Moreover, I dare say that there is much treasure gathered at this place by Sir Nabon, and all that treasure shall be divided amongst you, for to comfort ye, wherefore when ye leave this place, ye shall go away a great deal richer than ye were when ye came."

So spake Sir Tristram, promising them much for to comfort them a little.

As to that treasure he spake of, ye shall immediately be told how it was. For when Sir Tristram had summoned the treasurer of that place, be brought Sir Tristram down into the vaults of the castle and there he beheld seven strong chests bolted and locked. Then Sir Tristram summoned the locksmith of that castle; and the smith came and burst open the chests; and lo! the eyes of all were astonished and bedazzled with the treasure which they therewith beheld; for in those chests was heaped an incalculable treasure of gold and silver and precious gems of many divers sorts.

And besides this treasure, you are to know that they found in that vault many bales of cloths—some of silk and velvet, and some of tissues of cloth of gold and silver; and they found many precious ornaments, and many fine suits of armor, and many other valuable things. For in several years Sir Nabon had gathered all that treasure in toll from those ships that had sailed past that land.

All this treasure Sir Tristram had them bring forth into the light of day, and he divided it into seven equal parcels. Then he said to those sad, sorrowful captives: "Look! See! all this shall be yours for to comfort ye! Take each of you one parcel and depart hence in joy!" Then all they were greatly astonished at Sir Tristram's generosity, and they said: "Lord, how is this? Do you not then take any of this treasure for yourself?"

*Sir Tristram divides the treasure amongst the captives.*

To them Sir Tristram made reply: "Nay, why should I take it? I am not sad, nor sick, nor troubled at heart as you poor captives are. All this I have taken for to comfort you, and not for to satisfy my own covetousness. So let each take his share of it and see that ye all use it in comfort and peace and for the advantage of other men and women who are in trouble as ye have been. For, as hitherto this treasure hath been used for evil purpose, so shall it be henceforth that it shall be used to good purpose."

So there was great rejoicing amongst all those poor people who had been so sad and sorrowful before.

Now, after all this had been settled, Sir Tristram cast about how he might put that land under good government upon behalf of the Lady Loise.  To this intent he chose from amongst those captives *Sir Tristram* whom he had liberated a certain very worthy honorable knight *appoints Sir* of Cornwall hight Sir Segwarides.  Him Sir Tristram ap- *ernor of the* pointed to be governor of that island, giving him liberty to rule *castle.* it as he chose saving only that he should do homage to the Lady Loise as lady paramount.  And Sir Tristram ordained that Sir Segwarides should pay tribute to that lady every year such an amount as should be justly determined upon betwixt them.  For Sir Tristram wist that some strong worthy knight should rule that island, or else, from its position, it might again some time fall from the Lady Loise's possession into the hands of such an evil and malignant overlord as Sir Nabon had been.

So it was done as Sir Tristram had ordained.  And it may here be said that Sir Segwarides ruled that land very justly and that he and the Lady Loise became dear friends, so that at the end of three years from that time he and she were made husband and wife.

Now Sir Tristram remained in that island several days, with intent to see to it that the power of Sir Segwarides should be established.  And he made all the people of that land come before Sir Segwarides for to pledge obedience to him.

Amongst these came Sir Lamorack in the guise of a swineherd, and Sir Tristram knew him not, because that he was clad in rags and in the skins of animals and because that his beard and his hair were uncut and unkempt, and hung down very shaggy upon his breast.  But Sir Lamorack knew Sir Tristram yet would not acknowledge him, being ashamed that Sir Tristram should discover him in such a guise and so ragged and forlorn as he then was.  So he kept his eyes from Sir Tristram, and Sir Tristram passed him by and knew him not.

But amongst other of the people of the castle that passed before Sir Tristram, there came a woman, very fair to look upon, and she had been a house-slave to Sir Nabon.  As this woman passed before Sir *Sir Tristram* Tristram, he beheld that she wore upon her thumb a very fair *beholds Sir* and shining ring, that bare a green stone set in wrought gold. *Lamorack's ring.* And when he looked again he saw it was that ring of carven emerald that he had given to Sir Lamorack as aforetold.

At this Sir Tristram was astonished beyond measure, and he ordered that woman to come before him, and she came and stood before him trembling.  Then Sir Tristram said: " Fear not, but tell me where got ye that ring that I behold upon your hand?"  And the woman said: "Lord, I will tell you the very truth. My husband is the chief fisherman of this place, and one day, some while ago, he gave me this ring when I had favor in his sight."

Sir Tristram said, "Where is your husband?" The slave-woman said, "Yonder he stands." Then Sir Tristram said: "Come hither, Sirrah!" And therewith the fisherman came and stood before Sir Tristram as his wife had done, and he also trembled with fear as she had done.

To him Sir Tristram said, "Why do you tremble so?" And the fisherman said, "Lord, I am afeard!" Sir Tristram said: "Have no fear, unless you have done wrong, but tell me the truth. Where got ye *Sir Tristram questions the fisherman.* that ring that yonder woman weareth?" "Lord," said the fisherman, "I will tell you the perfect truth. One day I and several of my fellows found a man lying naked in a bed of heather near the seaside. At first we thought he was dead, but he awoke and arose when he heard our voices. He was naked and hungry, and he besought us for clothes to cover his nakedness and for food to eat. So we gave him what we could, demanding that ring in payment. So he gave the ring to me, who am the chief of the fishermen, and I gave it to that woman who is my wife; and that, lord, is the very truth."

Then Sir Tristram was very much disturbed in mind, for he feared that it might have gone ill with Sir Lamorack. And he said, "Where now is that man of whom ye speak?" The fisherman replied: "Lord, he was set to keep the swine, and he is the swineherd of the castle to this day."

At this Sir Tristram was very glad that no more ill had befallen Sir Lamorack, and that he was yet alive.

Then, after the fisherman had departed from that place, Sir Tristram sat for a while sunk into deep thought. And he said to himself: "Alas, that so noble a knight should be brought to such a pass as that! How greatly must my friend be abased when he would not acknowledge himself to me nor claim my assistance because of the shame of his appearance! Meseems it is not fitting for me to send for him to come to me in the guise which he now wears, for it would be discourteous a thing for me to do, to make him so declare himself. So first I shall see to it that he is clothed in such a manner as shall be fitting to his high estate, and then haply he will be willing to make himself manifest to me. After that, perhaps his love will return to me again, and remain with me as it was at first."

So Sir Tristram called to him several of the people of that castle, and he bade them do certain things according to his command, and straightway they departed to do as he ordained.

Now turn we to Sir Lamorack: whilst he sat keeping watch over his swine there came to him four men from the castle. These say to him, "You must come straightway with us." Sir Lamorack said, "Whither would you take me?" They say: "That we are not permitted to tell you, only that you are to go with us as we bid you."

So Sir Lamorack arose and went with those four, much wondering what it was that was to befall him, and whether that which was to happen was good or evil.

The four men brought him to the castle and they entered in thereat, and they escorted Sir Lamorack, still greatly wondering, up the stairway of the castle, and so into a noble and stately apartment, hung *Sir Lamorack* with tapestries and embroidered hangings. And there Sir *is brought to* Lamorack beheld a great bath of tepid water, hung within *the castle.* and without with linen. There were at this place several attendants; these took Sir Lamorack and unclothed him and brought him to the bath, and bathed him and dried him with soft linen and with fine towels. Then there came the barber and he shaved Sir Lamorack and clipped his hair, and when he was thus bathed and trimmed, his nobility shone forth again as the sun shines forth from a thick cloud that hides its effulgence for a while, only to withdraw so that the glorious day-star may shine forth again with redoubled splendor.

Then there came divers other attendants and clothed Sir Lamorack in rich and handsome garments such as were altogether fitting for a knight-royal to wear. And after that there came several esquires *Sir Lamorack is* and brought a very splendid suit of armor; and they clad Sir *armed in armor.* Lamorack in that armor; and the armor gleamed as bright as daylight, being polished to a wonderful clearness, and inlaid with figures of arabesqued silver.

Then Sir Lamorack said, "What means all this that ye do to me?" And they said, "Wait, Messire, and you shall see."

So after all these things were done, five other esquires appeared to conduct Sir Lamorack away from that place. These led him through several passages and hallways until at last they came to a great space of hall wherein stood a single man; and that man was Sir Tristram.

And Sir Tristram gazed upon Sir Lamorack and his heart yearned over him with great loving-kindness. But he would not betray his love to those who had come with Sir Lamorack, so he contained himself for a little, and he said to those in attendance, "Get ye gone," and straightway they departed.

Then Sir Lamorack lifted up his eyes and he came to where Sir Tristram was standing and he said: "Is it thou, Tristram, who hath bestowed all these benefits upon me?" And he said: "From thy nobility of soul such things may be expected."

Then Sir Tristram wept for joy, and he said: "Lamorack, it is little that I have done to pleasure thee, and much that I have done to affront thee."

*Sir Tristram and Sir Lamorack are reconciled.* Then Sir Lamorack said: "Nay; it is much that thou hast done to comfort me, and little to cause me discomfort. For lo! thou hast uplifted me from misery into happiness, and thou hast brought me from nakedness and want into prosperity and ease, and what more may one man do for another man than that?"

"Lamorack," said Sir Tristram, "there is much more than one man may do for another man than that. For if one man hath given offence to another man, he may be reconciled to that one so offended, and so the soul of that other shall be clothed with peace and joy, even as thy body hath been clothed with garments of silk and fine linen." Then Sir Tristram took Sir Lamorack by the hand, and he said, "Dear friend, art thou now strong and fresh of body?" And Sir Lamorack, greatly wondering, said, "Ay."

"Then," said Sir Tristram, "I may now offer thee reparation for that offence which I one time unwillingly committed against thee. For lo! I have had thee clad in the best armor that it is possible to provide, and now that thou art fresh and hale and strong, I am ready to do battle with thee at any time thou mayst assign. For if, before, thou wert overcome because thou wert weary with battle, now thou mayst prove thy prowess upon me being both strong and sound in wind and limb."

But upon this Sir Lamorack ran to Sir Tristram and catched him in his arms and kissed him upon the cheek. And he said: "Tristram, thou art indeed a very noble soul. I will do no battle with thee, but instead I will take thee into my heart and cherish thee there forever."

Sir Tristram said, "Art thou altogether satisfied?" And Sir Lamorack said, "Yea." And therewith Sir Tristram wept for pure joy.

Then Sir Tristram said: "Let us go to Sir Launcelot of the Lake, so that I may make my peace with him also. For he hath writ me a letter chiding *Sir Tristram and Sir Lamorack depart from the island.* me for having done battle with thee when thou wert weary and winded with fighting. And I was upon my way to see Sir Launcelot and to plead my cause with him when I came hither by good hap, and was able to uplift thee out of thy distress." To this Sir Lamorack said: "I will go with thee to Sir Launcelot whenever it shall please thee; and I will bear full testimony to thy knightliness and to thy courtesy."

So when the next morning had come they took boat and sailed away from that island. And the night of that day they abided at the castle of the Lady Loise, who gave thanks without measure to Sir Tristram for ridding the world of so wicked and malign a being as Sir Nabon, and for restoring her inheritance of that land unto her again. And upon the morning of the next day those two good knights betook their way to Camelot, where they found Sir Launcelot. There Sir Lamorack exculpated Sir Tristram, and Sir Launcelot immediately withdrew his rebuke for that battle which Sir Tristram had aforetime done against Sir Lamorack.

After that Sir Tristram and Sir Lamorack abode at the court of King Arthur for nigh a year, and during that time they went upon many quests and adventures of various sorts—sometimes alone, sometimes together. All these have been set down in ancient histories that tell of the adventures of Sir Tristram and Sir Lamorack. Some of them I would like right well to tell you of, but should I undertake to do so, the story of those happenings would fill several volumes such as this. Nevertheless, I may tell you that they did together many knightly deeds, the fame whereof hath been handed down to us in several histories of chivalry. Therein you may read of those things if you should care to do so.

All this I leave to tell you how Sir Tristram returned into Cornwall, and likewise to tell you of one more famous adventure that he did at this time.

Sir Tristram had been at the court of King Arthur for about a year when one day there came a messenger unto the court at Camelot with news that Sir Palamydes, the Saracen knight aforetold of in this history, *Sir Tristram* had through a cunning trick seized the Lady Belle Isoult and *hears from Corn-* had carried her away to a lonely tower in the forest of Cornwall. *wall of Sir* The messenger bore a letter from King Mark beseeching Sir *Palamydes.* Tristram to return as immediately as possible unto Cornwall and to rescue that lady from her captivity. And the letter further said that two knights of Cornwall had already essayed to rescue the Lady Belle Isoult, but that they had failed, having been overcome and sorely wounded in battle by Sir Palamydes. And the letter said that it was acknowledged by all men that Sir Tristram was the only knight of Cornwall who could achieve the rescue of Belle Isoult from so wonderful and puissant a knight as Sir Palamydes.

So in answer to that letter, Sir Tristram immediately left the court of King Arthur and returned in all haste to Cornwall, and there he found them all in great perturbation that the Lady Belle Isoult had thus been stolen away.

But Sir Tristram did not remain at court very long for, after he had obtained such information as he desired, he immediately left Tintagel and plunged into the forest with Gouvernail as his companion in quest of that lonely tower where Belle Isoult was said to be held prisoner.

After several adventures of no great note he came at last very, very deep into the forest and into an open space thereof; and in the midst of that open space he beheld a lonely tower surrounded by a moat. And he wist that that must be the place where the Lady Belle Isoult was held prisoner.

But when Sir Tristram drew nigh to this tower he perceived a single knight sitting at the base of the tower with head hanging down upon his breast as though he were broken-hearted with sorrow. And when he came still more nigh, Sir Tristram was astonished to perceive that that mournful knight was Sir Palamydes the Saracen, and he wondered why Sir Palamydes should be so broken-hearted.

*Sir Tristram finds Sir Palamydes in the forest.*

And now it must be told why it was that Sir Palamydes came to be in such a sorry case as that; for the truth was that he was locked and shut outside of the tower, whilst the Lady Belle Isoult was shut and locked inside thereof.

Now it hath already been told how the letter of King Mark had said to Sir Tristram that two knights of Cornwall went both against Sir Palamydes for to challenge him and to rescue the Lady Belle Isoult.

The second of these knights was Sir Adthorp, and he had followed Sir Palamydes so closely through the forest that he had come to the forest tower not more than an hour after Sir Palamydes had brought the Lady Belle Isoult thither.

Therewith Sir Anthorp gave loud challenge to Sir Palamydes to come forth and do him battle, and therewith Sir Palamydes came immediately out against him, full of anger that Sir Adthorp should have meddled in that affair.

But immediately Sir Palamydes had thus issued forth to do battle with Sir Adthorp, the Lady Belle Isoult ran down the tower stairs and immediately shut the door through which he had passed, and she locked it and set a great bar of oak across the door.

So when Sir Palamydes had overthrown the Cornish knight, and when he would have returned to the tower, he could not, for lo! it was fastened against him. So now for three days he had set there at the foot of the tower and beside the moat, sunk in sorrow like to one who had gone out of his mind.

*How Sir Pala-mydes came with-out the tower.*

So Sir Tristram found him, and perceiving that it was Sir Palamydes who was sitting there, he said to Gouvernail: "Go thou and bid that knight to come and do battle with me."

So Gouvernail went to Sir Palamydes and he said: "Sir, arise, for here is a knight would speak with you!" But Sir Palamydes would not move. Then Gouvernail touched him with his lance, and said: "Sir Palamydes, arise and bestir yourself, for here is Sir Tristram come to do battle with you." With that, Sir Palamydes awoke from his stupor and arose very slowly and stiffly. And he gathered up his helmet which was lying beside him and put it upon his head. Then he took down his shield from where it hung against the wall and he mounted upon his horse, doing all as though he were moving in a dream.

But as soon as he was upon horseback he suddenly aroused himself, for his fierce spirit had come back to him once more. Then he gnashed his teeth, crying out in a loud voice, "Tristram, this time either thou or I shall perish."

Therewith he rushed upon Sir Tristram and smote him so violently that Sir Tristram had much ado to defend himself. And Sir Palamydes smote him again and again; and with that Sir Tristram smote in *Sir Tristram* return. And if the blows of Sir Palamydes were terrible, the *overcomes Sir* blows of Sir Tristram were terrible likewise. Then by and *Palamydes.* by Sir Tristram smote Sir Palamydes so sore a buffet that the Saracen knight fell down from his horse and was unable immediately to arise. Then Sir Tristram ran to him and rushed off his helmet and catched him by the hair with intent to cut his head from off his body.

But with that the Lady Belle Isoult came running from out the tower and cried out: "Tristram, is it thou? Spare that mistaken knight and have mercy upon him as thou hopest for mercy."

"Lady," said Sir Tristram, "for thy sake and at thy bidding I will spare him." Then he said to Sir Palamydes, "Arise." And Sir Palamydes arose very painfully, and Sir Tristram said: "Get thee hence, and go to the court of King Arthur and make thy confession to the King and ask him to forgive thee, and if he forgive thee, then also I will forgive thee."

Therewith Sir Palamydes mounted upon his horse and rode away without speaking another word, his head bowed with sorrow upon his breast for shame and despair.

Then Sir Tristram took the Lady Belle Isoult up behind him on his horse, and he and she and Gouvernail departed from that place.

So Sir Tristram brought the Lady Isoult back to Cornwall, *Sir Tristram* and there he was received with loud praise and great rejoicing, *brings Belle* for everybody was glad that Belle Isoult had been brought *Isoult back to* safely back again. *Cornwall.*

And now it shall be told what reward Sir Tristram received for this deed of arms.

For, though at first King Mark was greatly beholden to Sir Tristram, that he had thus rescued the Lady Belle Isoult, yet, by little and little, he grew to hate that noble knight more bitterly than ever. For he heard men say to one another: "Lo, Sir Tristram is, certes, the very champion of Cornwall, for who is there in this country is his equal?" So King Mark, hearing these things said to himself: "The more noble Tristram is, the more ignoble will men deem me to be who am under obligations to such an enemy." So he would say in his heart, "Yea, Tristram; I hate thee more than death."

# PART III

# The Madness of Sir Tristram

*H*ERE *followeth the story of how Sir Tristram was driven out of Cornwall and of how he went mad because of his troubles. Likewise it shall
be told how he performed several very wonderful adventures whilst he was in
that state, and of how he was brought back into his senses again.*

**S**ir Tristram assaults King
Mark

# Chapter First

*How Sir Tristram was discovered with the Lady Belle Isoult; how he assaulted King Mark, and how he escaped from Tintagel into the forest.*

AFTER Sir Tristram had thus rescued the Lady Belle Isoult from the hand of Sir Palamydes, he dwelt very peacefully at the court of Cornwall for all of that winter and until the spring that followed, and during that time he was given every meed of praise and honor. But although King Mark and his court gave praise to Sir Tristram with the lips, yet he and many of his people hated Sir Tristram at heart, and there were many mischief-makers about the court who were ever ready to blow the embers of the King's wrath into a flame.

Now the chiefest of all these mischief-makers was Sir Andred, who was nephew unto King Mark, and cousin-germaine unto Sir Tristram. Sir Andred was a fierce strong knight, and one very dextrous at arms; but he was as mean and as treacherous as Sir Tristram was generous and noble, wherefore he hated Sir Tristram with great bitterness (though he dissembled that hatred) and sought for every opportunity to do Sir Tristram a harm by bringing him and the King into conflict.

So Sir Andred set spies upon Sir Tristram, and he himself spied upon his cousin, yet neither he nor they were able to find anything with which to accuse Sir Tristram. Then one day Sir Andred came to Sir Tristram and said: "Sir, the Lady Belle Isoult wishes to see you to talk with you." Sir Tristram said, "Where is she?" *Sir Andred of Cornwall sets spies upon Sir Tristram.*

And Sir Andred said, " She is in her bower."    Then Sir Tristram said, " Very well, I will go to her."

So Sir Tristram arose and departed from where he was with intent to find the lady; and therewith Sir Andred hurried to where King Mark was, and said: " Lord, arise, for Sir Tristram and the Lady Isoult are holding converse together."

King Mark said, " Where are they?"    And Sir Andred said, " They are in the bower of the Queen."    At that King Mark's rage and jealousy blazed up into a flame, so that he was like one seized with a sudden frensy.    So, in that madness of rage, he looked about for some weapon with which to destroy Sir Tristram, and he perceived a great sword where it hung against the wall.    Thereupon he ran to the sword and took it down from where it was, and ran with all speed to that place where Sir Tristram and the Lady Isoult were, and Sir Andred guided him thither.

And when King Mark reached the bower of the Lady Isoult he flung open the door and found Sir Tristram and the Lady Isoult sitting together *King Mark assaults Sir Tristram.* in the seat of a deep window.    And he perceived that the Lady Isoult wept and that Sir Tristram's face was very sorrowful because of her sorrow.    Then King Mark twisted him about and bent double as with a great pain, and then he cried out thrice in a voice very hoarse and loud: " Traitor! Traitor! Traitor!"    Saying those words three times.    Therewith he ran at Sir Tristram and struck furiously at him with that sword he held, with intent to slay him.

Now Sir Tristram was at that time altogether without armor and was clad in clothes of scarlet silk.    Accordingly, he was able to be very quick and alert in his movements.    So perceiving King Mark rushing upon him with intent to slay him he leaped aside and so avoided the blow.    Then immediately he rushed in upon King Mark and catched him by the wrist and wrenched the sword out of his hand.

Then Sir Tristram was blinded with his rage and might have slain his uncle, but the Lady Isoult, beholding the fury in his face, shrieked in a very piercing voice, " Forbear! Forbear!"    And therewith he remembered him how that King Mark was his mother's brother and that it was his hand that had made him a knight.

So he turned the sword in his hand and he smote King Mark with the flat thereof again and again, and at those blows King Mark was filled with *Sir Tristram beats King Mark.* terror so that he howled like a wild beast.    And King Mark fled away from that place, striving to escape, but Sir Tristram ever pursued him, grinding his teeth like a wild boar in rage, and smiting the King as he ran, over and over again, with the flat of the sword so that the whole castle was filled with the tumult and uproar of that assault.

Then many of the knights of Cornwall came running with intent to defend the King, and with them came Sir Andred. But when Sir Tristram saw them, his rage suddenly left the King and went out toward them; so therewith, naked of armor as he was, he rushed at them, and he struck at them so fiercely that they were filled with the terror of his fury, and fled away from before his face. And Sir Tristram chased them through the courts of the castle, striking right and left until he was weary with striking, and many he struck down with the fierceness of his blows, and amongst them was Sir Andred who was sorely wounded. So after a while Sir Tristram grew weary of that battle, and he cried out, "Certes, these are not knights, but swine!" And therewith he ceased striking, and allowed those who could do so to escape.

Thereafter he went to his chamber and armed himself without summoning Gouvernail, and after that he took horse and rode away altogether from that place. And not even Gouvernail went with him, but only *Sir Tristram* his favorite hound, hight Houdaine, which same followed him *departs from* into the forest as he rode thitherward. And in his going Sir *Tintagel.* Tristram looked neither to the right nor to the left but straight before him very proudly and haughtily, and no one dared to stay him in his going.

Yet, though he appeared so steadfast, he was like one who was broken-hearted, for he wist that in going away from that place he was leaving behind him all that he held dear in the world, wherefore he was like one who rode forth from a pleasant garden into an empty wilderness of sorrow and repining.

Then, some little while after Sir Tristram had gone, Gouvernail also took horse and rode into the forest, and he searched for a long while in the forest without finding his master. But after a while he *Gouvernail finds* came upon Sir Tristram seated under a tree with his head *Sir Tristram in* hanging down upon his breast. And Houdaine lay beside Sir *the forest.* Tristram and licked his hand, but Sir Tristram paid no heed to him, being so deeply sunk in his sorrow that he was unaware that Houdaine licked his hand in that wise.

Then Gouvernail dismounted from his horse and came to where Sir Tristram was, and Gouvernail wept at beholding the sorrow of Sir Tristram. And Gouvernail said: "Messire, look up and take cheer, for there must yet be joy for thee in the world."

Then Sir Tristram raised his eyes very slowly (for they were heavy and dull like lead) and he looked at Gouvernail for some while as though not seeing him. Then by and by he said: "Gouvernail, what evil have I done that I should have so heavy a curse laid upon me?" Gouvernail said, still weeping: "Lord, thou hast done no ill, but art in all wise a very noble, honorable gentleman." "Alas!" quoth Sir Tristram, "I must unwittingly have done some great evil in God's sight, for certes the hand of God lieth grievously heavy upon me." Gouvernail said: "Lord, take heart, and tell me whither shall we go now?" And Sir Tristram said, "I know not."

Then Gouvernail said: "Lord, let us go hence, I care not where, for I reckon nothing of storm or rain or snow or hail if it so be that I am with you."

Then Sir Tristram looked upon Gouvernail and smiled, and he said: "Gouvernail, it is great joy to me that you should love me so greatly as you do. But this time you may not go with me whither I go, *Sir Tristram* for the Lady Belle Isoult hath few friends at the court of *bids Gouvernail* Cornwall, and many enemies, wherefore I would have you *return to Tin-* *tagel.* return unto her for my sake, so that you may befriend her and cherish her when that I am no longer by her for to stand her friend in her hour of need. And take this dog Houdaine with you and bid the Lady Belle Isoult for to keep him by her to remind her of my faithfulness unto her. For even as this creature is faithful unto me under all circumstances, so am I faithful unto her whether she be glad or sorry, or in good or evil case. So return to Tintagel as I bid thee, and see that thou pay thy duty unto that lady even as thou payst it unto me. For she is so singularly dear unto me that, even as a man's heart is the life of his body, so is her happiness the life of my life."

Then Gouvernail wept again in very great measure, and he said, "Lord, I obey." Therewith he mounted his horse, still weeping with a great passion of sorrow, and rode away from that place, and Houdaine followed after him and Sir Tristram was left sitting alone in the deep forest.

After that Sir Tristram wandered for several days in the forest, he knew not whither for he was bewildered with that which had happened; so that *Sir Tristram* he ate no food and took no rest of any sort for all that time. *wanders in the* Wherefore, because of the hardship he then endured, he by and *forest mad.* by became distraught in his mind. So, after a while, he forgot who he himself was, and what was his condition, or whence he came or whither he wended. And because his armor weighed heavily upon him, he took it off and cast it away from him, and thereafter roamed half naked through the woodlands.

Now upon the sixth day of this wandering he came to the outskirts of the forest and nigh to the coast of the sea at a spot that was not very far away was the castle of the Lady Loise, where he had once stayed at the time that he undertook the adventure against Sir Nabon as aforetold. There, being exhausted with hunger and weariness, he laid himself down in the sunlight out beyond the borders of the forest and presently fell into a deep sleep that was like to a swoon.

Now it chanced at that time that there came that way a certain damsel attendant upon the Lady Loise. She perceiving that a man lay there on the grass at the edge of the forest was at first of a mind to quit that place. Then, seeing that the man lay very strangely still as though he were dead, she went forward very softly and looked into his face.

Now that damsel had beheld Sir Tristram a great many times when he was at the castle of the Lady Loise; wherefore now, in spite of his being so starved and shrunken, and so unkempt and unshaved, she remembered his face and she knew that this was Sir Tristram.

Therewith the damsel hurried away to the Lady Loise (and the lady was not a very great distance away) and she said: "Lady, yonder way there lieth a man by the forest side and I believe that it is Sir Tristram of Lyonesse. Yet he is but half-clad and in great distress of body so that I know not of a surety whether it is really Sir Tristram or not. Now I pray you come with me and look upon his face and see if you may know him."

So the Lady Loise went with the damsel to where Sir Tristram lay and looked into his face, and she knew Sir Tristram in spite of his ill condition.

Then the Lady Loise touched Sir Tristram upon the shoulder and shook him, and thereupon Sir Tristram awoke and sat up. Then the Lady Loise said, "Sir Tristram, is it thou who liest here?" And Sir *The Lady Loise* Tristram said, "I know not who I am." The Lady Loise said, *finds Sir Tris-* "Messire, how came you here in this sad case?" And Sir *tram.* Tristram said: "I know not whence I came, nor how I came hither, nor who I am, nor what it is that ails me, for I cannot hold my mind with enough steadiness to remember those things." Then the lady sighed for sorrow of Sir Tristram, and she said: "Alas, Sir Tristram, that I should find you thus! Now I pray you, lord, for to come with me to my castle which is hard by. There we may care for you and may perhaps bring you back to health again."

To this Sir Tristram said: "Lady, I may not go with you. For though I cannot remember whence I came, nor who I am, this much I know—I know that I am mad, and that the forest is the only fit place for such as I am come to be."

The lady said: "Alas, Sir Tristram, thou wilt die if thou art left alone here in the forest." And Sir Tristram said: "Lady, I know not what you

mean when you say I am to die.  What is it to die?"  So at these words
the Lady Loise saw how it was with Sir Tristram; that his brains were
altogether turned; and she wist that some sore trouble must have befallen
to bring him to such a pass.  Then she bethought her of how dearly he
loved the music of the harp, and she said to herself: "Mayhap by means
of music I may bring him back into his senses again."  So she said to that
damsel who had brought her thither: "Go thou and bring hither my little
harp of gold, and let us see if music may charm him to remembrance."

So the damsel ran to the castle and brought the harp thence, and the
Lady Loise took the harp and tuned it and struck it and played upon
it.  And the lady sang very sweetly a ballad that she knew Sir Tristram
loved.

Then when Sir Tristram heard the sound of the music and singing he
*The Lady Loise* aroused himself.  For first he listened with great pleasure, and
*harps to Sir* then he said, "Give it to me!  Give it to me!" and he reached
*Tristram.* out his hands and would have taken the harp from the lady.

But the Lady Loise laughed and shook her head, and she walked away
from Sir Tristram and toward the castle, still playing upon the little harp
*Sir Tristram* and singing; and Sir Tristram followed close after, saying
*comes to the* ever, "Give it to me!  Give it to me!" and reaching out his
*Lady's castle.* hands for the harp.  So the Lady Loise led him away from
that place across the meadows; and she led him to the castle and into the
castle; and ever Sir Tristram followed after her, beseeching her for to give
the harp unto him.  And the lady led Sir Tristram that way until she had
brought him to a fair room, and there she gave him the harp, and Sir Tristram took it very eagerly into his hands and struck upon it and played
and sang most sweetly and with great joy and pleasure.

Afterward, being so much comforted, he ate and drank with appetite,
and then fell into a fair sound sleep.

Yet, though he so slept, still Sir Tristram's wits in no wise recovered
themselves; for when he awoke from that slumber he still could not remember who he was or whence he came, neither could he remember the
faces of any of those who were around about him.  But, though he was
thus mad, he was still gentle and kind in his madness and courteous and
civil to all those who came nigh him.

So Sir Tristram remained a gentle captive in the castle of the Lady Loise
for nigh upon a month, and somewhiles she would sing and harp to him,
and otherwhiles he himself would harp and sing.  But ever and anon,
when he found the chance for to do so, he would escape from the captivity
of the castle and seek the forest; for he was aware of his madness and he

ever sought to hide that madness in the deep and shady woodland where only the wild creatures of the forest might see him.

Yet always when he so escaped the Lady Loise would take her little golden harp and go forth to the skirts of the forest and play upon it, and when the music thereof would reach Sir Tristram's ears he would return to the castle, being led thither by the music.

But one day he wandered so far astray that the music of the harp could not reach his ears, and then he wandered on farther and farther until he was altogether lost. At that Lady Loise took much sorrow *Sir Tristram* for she had much love for Sir Tristram. So she sent many *quits the Lady's* of her people to search the forest for him, but none of these *castle.* were able to find him and thereafter he came no more to the castle.

Thus Sir Tristram escaped from that castle and after that he wandered in the forest as he had done at the first. And in that time he took no food and but little rest. And the brambles tore his clothes, so that in a short time he was wellnigh altogether naked.

And somewhiles during this time of wandering he would be seized as with a fury of battle, and in such case he would shout aloud as though in challenge to an enemy. And then he would rend and tear great branches from the trees in the fury of his imaginings. But otherwhiles he would wander through the leafy aisles of the forest in gentler mood, singing so sweetly that had you heard him you would have thought that it was some fairy spirit of the forest chanting in those solitudes.

So he wandered until he failed with faintness, and sank down into the leaves; and I believe that he would then have died, had it not been that there chanced to come that way certain swineherds of the *Sir Tristram* forest who fed their swine upon acorns that were to be therein *dwells with the* found. These found Sir Tristram lying there as though dead, *swineherds.* and they gave him to eat and to drink so that he revived once more. After that they took him with them, and he dwelt with them in those woodlands. There these forest folk played with him and made merry with him, and he made them great sport. For he was ever gentle and mild like a little child for innocence so that he did no harm to anyone, but only talked in such a way that the swineherds found great sport in him.

Now Sir Andred of Cornwall very greatly coveted the possessions of Sir Tristram, so that when several months had passed by and Sir Tristram did not return to Tintagel, he said to himself: "Of a surety, Tristram must now be dead in the forest, and, as there is no one nigher of kin to him than I, it is altogether fitting that I should inherit his possessions."

But as Sir Andred could not inherit without proof of the death of Sir Tristram, he suborned a certain very beautiful but wicked lady who dwelt in the forest, persuading her that she should give false evidence of Sir Tristram's death. Accordingly, he one day brought that lady before King Mark, and she gave it as her evidence that Sir Tristram had died in the forest and that she had been with him when he died. And she showed them a new-made grave in the forest, and she said: "That is the grave of Sir Tristram, for I saw him die and I saw him buried there with mine own eyes."

*Sir Andred seizes Sir Tristram's possessions.*

So everybody believed this evidence, and thought that Sir Tristram was really dead, and so Sir Andred seized upon all the possessions of Sir Tristram. And there were many who were very sorry that Sir Tristram was dead and there were others who were glad thereof in the same measure. But when the news was brought to Belle Isoult that Sir Tristram was dead, she shrieked aloud and swooned away. And she lay in that swoon so long that they thought for a while she would never recover from it. But by and by she awoke therefrom, crying, "Would to God that I were dead with Tristram and had never awakened!"

And thereafter she mourned continually for Sir Tristram and would not be comforted; for she was like to a woman who hath been widowed from a lover of her youth.

And now it shall be told of how it fared with Sir Tristram in the forest where he dwelt with the swineherds, and of how he achieved a very notable adventure therein.

# ir Kay and the Forest Mad: man

# Chapter Second

*How Sir Tristram got him a sword from Sir Kay and how he slew therewith a huge knight in the forest and rescued a lady in very great distress. Also how Sir Launcelot found Sir Tristram in the forest and brought him thence to Tintagel again.*

NOW it chanced one day that Sir Kay the Seneschal came riding through those parts of the forest where Sir Tristram abided with the swineherds, and with Sir Kay there came a considerable court of esquires. And with him besides there travelled Sir Dagonet, King Arthur's Fool.

Now, you are to know that though Sir Dagonet was the King's jester, and though he was slack of wit, yet he was also a knight of no mean prowess. For he had performed several deeds of good repute and was well held in all courts of chivalry. So Sir Dagonet always went armed; though he bore upon his shield the device of a cockerel's head as a symbol of his calling.

The time that Sir Kay and his court travelled as aforesaid was in the summer season and the day was very warm, so that Sir Kay was minded to take rest during the midday and until the coolness of the afternoon should come. So they all dismounted from their horses and sat them down under the shade of the trees where it was cool and pleasant and where the breezes reached them to breathe upon their faces.

But whilst Sir Kay and his court thus rested themselves, Sir Dagonet must needs be gadding, for he was of a very restless, meddlesome disposition. So, being at that time clad only in half armor, he wandered hither and thither through the forest as his fancy led him. For somewhiles he would whistle and somewhiles he would gape, and otherwhiles he would cut a caper or two. So, as chance would have it, he came by and by to that open glade of the forest where the swineherds were gathered; and at that time they were eating their midday meal of black bread and cheese, and were drinking beer; some talking and laughing and others silent as they ate their food. Unto these Sir Dagonet appeared, coming out of the forest in very gay attire, and shining in the half armor he wore, so that he appeared like a bright bird of the woodland.

Then Sir Dagonet, seeing where those rude boors were eating their meal of food, came to them and stood amongst them. And he said, "Who are ye fellows?" Whereunto they replied, "We are swineherds, Messire; who be ye?"

Quoth Sir Dagonet: "I am King Arthur's Fool. And whilst there are haply many in the world with no more wits than I possess, yet there are few so honest as I to confess that they are fools."

At these words those swineherds laughed very loudly. "Well," quoth one, "if King Arthur hath his fool, so have we, and yonder he is," and therewith he pointed to where Sir Tristram lay in the shade of the trees some distance away and beside a deep well of the forest.

Upon that Sir Dagonet must needs go to where Sir Tristram lay, nearly naked, upon the ground. And when he had come there he said, "Arise, fool." Whereunto Sir Tristram replied: "Why should I arise? Lo! I am weary."

Then Sir Dagonet said: "It is not fitting that thou, who art the fool of swineherds shouldst lie upon the grass, whilst I who am the fool of a king stand upright upon my shanks. So, fool, I bid thee bestir thyself and arise."

But Sir Tristram said, "I will not arise." And therewith Sir Dagonet took his sword and pricked the thigh of Sir Tristram with the point thereof with intent to make him bestir himself.

Now when Sir Tristram felt the prick of Sir Dagonet's sword, a certain *Sir Tristram* part of his memory of knighthood came back to him and he *souses Sir* was seized with a sudden fury against Sir Dagonet. So he arose *Dagonet in the* and ran at Sir Dagonet and catched him in his arms, and lifted *well.* Sir Dagonet off his feet and he soused him in the well four or five times so that he was like to have drowned him.

As for those swineherds, when they saw what their fool did to that other fool, they roared with laughter so that some of them rolled down upon the ground and lay grovelling there for pure mirth. But others of them called out to Sir Tristram, "Let be, or thou wilt drown that man"; and therewith Sir Tristram let Sir Dagonet go, and Sir Dagonet ran away.

Nor did Sir Dagonet cease to run until he came to his party under the shade of the trees. But when Sir Kay perceived what a sorry plight it was in which Sir Dagonet appeared, he said, "What hath befallen thee?"

To this Sir Dagonet replied as follows: "Messire, I, who am a fool, went into the forest and met another fool. I fool would have a jest with he fool, but he fool catched I fool and soused I fool in a well of cold water. So it came about that while I fool had the jest, he fool had the sport of the jest."

Then Sir Kay understood in some manner what had befallen, and he was very angry that Sir Dagonet should have been so served. Wherefore he said, "Where did this befall thee?" And Sir Dagonet *Sir Kay seeks* said, "Over yonder ways." Then Sir Kay said: "I will avenge *to avenge Sir* thee for the affront that hath been put upon thee. For no *Dagonet.* boor shall serve a knight of King Arthur's court in such a fashion!" So therewith Sir Kay arose and put on his armor and mounted his horse and rode away; and after a while he came to that place where the swineherds were.

Then Sir Kay said very sternly: "Which of ye is that boor who put so grievous an affront upon a gentleman of my party?" The swineherds say: "Yonder he is lying by the well; but he is slack of wit, wherefore we beseech you to do him no harm."

Then Sir Kay rode to where Sir Tristram was, and he said: "Sirrah, why did you souse Sir Dagonet into the water?" To this Sir Tristram did not reply, but only looked at Sir Kay and laughed, for it pleased him wonderfully to behold that knight all in shining armor. But when Sir Kay beheld Sir Tristram laugh in that wise, he waxed exceedingly wroth. Wherefore he drew his sword straightway, and rode at Sir Tristram with intent to strike him with the blade thereof. But when Sir Tristram saw the sword of Sir Kay shining like lightning in the sunlight, somewhat of his knightly spirit arose within him and took wing like to a bird springing up out of the marish grass into the clear air. For beholding that bright flashing sword he cried out aloud and arose and came very steadily toward Sir Kay, and Sir Kay rode toward Sir Tristram. Then when Sir Kay had come near enough to strike, he arose in his stirrups and lifted the blade on high with intent to strike Sir Tristram with it. But therewith Sir Tristram ran very quickly in beneath the blow, so that the stroke of Sir Kay failed of its mark. Then Sir Tristram leaped up and catched Sir Kay around the body and dragged him down from off his horse very violently upon the ground, and with that the sword of Sir Kay fell down out of his hands and lay in the grass. Then Sir Tristram lifted up Sir Kay very easily and ran with him to the well of water and soused him *Sir Tristram* therein several times until Sir Kay cried out, "Fellow, spare *souses Sir Kay* me or I strangle!" Upon that Sir Tristram let go Sir Kay, *in the water.* and Sir Kay ran to his horse and mounted thereon and rode away from that place with might and main, all streaming with water like to a fountain.

And all that while those swineherds roared with great laughter, ten times louder than they had laughed when Sir Tristram had soused Sir Dagonet into the well.

Then Sir Tristram beheld the sword of Sir Kay where it lay in the grass and forthwith he ran to it and picked it up. And when he held it in his hands he loved it with a great passion of love, wherefore he hugged it to his bosom and kissed the pommel thereof.

But when the swineherds beheld the sword in Sir Tristram's hands, they said, "That is no fit plaything for a madman to have," and they would have taken it from him, but Sir Tristram would not permit them, for he would not give them the sword, and no one dared to try to take it from him.

So thereafter he kept that sword ever by him both by night and by *Sir Tristram* day, and ever he loved it and kissed it and fondled it; for, as *keeps the sword* aforesaid, it aroused his knightly spirit to life within him, *of Sir Kay.* wherefore it was he loved it.

So it hath been told how Sir Tristram got him a sword, and now it shall be told how well he used it.

Now there was at that time in the woodlands of that part of Cornwall a gigantic knight hight Sir Tauleas, and he was the terror of all that district. For not only was he a head and shoulders taller than the tallest of Cornish men, but his strength and fierceness were great in the same degree that he was big of frame. Many knights had undertaken to rid the world of this Sir Tauleas, but no knight had ever yet encountered him without meeting some mishap at his hands.

(Yet it is to be said that heretofore no such knight as Sir Launcelot or Sir Lamorack had come against Sir Tauleas, but only the knights of Cornwall and Wales, whose borders marched upon that district where Sir Tauleas ranged afield.)

Now one day there came riding through the forest a very noble, gallant young knight, hight Sir Daynant, and with him rode his lady, a beautiful *Sir Daynant and* dame to whom he had lately been wedded with a great deal of *his lady come to* love. These wayfarers in their travelling came to that part *the forest.* of the forest where the swineherds abode, and where were the open glade of grass and the fair well of water aforespoken of.

Hereunto coming, and the day being very warm, these two travellers dismounted and besought refreshment of the swineherds who were there, and those rude good fellows gladly gave them to eat and to drink of the best they had.

Whilst they ate, Sir Tristram came and sat nigh to Sir Daynant and his lady and smiled upon them, for he loved them very greatly because of their nobility and beauty. Then Sir Daynant looked upon Sir *Sir Daynant* Tristram and beheld how strong and beautiful of body and *regards Sir* how noble of countenance he was, and he saw that beautiful *Tristram.* shining sword that Sir Tristram carried ever with him. And Sir Daynant said, " Fair friend, who are you, and where gat ye that sword?"

"I know not who I am," said Sir Tristram, "nor know I whence I came nor whither I go. As for this sword, I had it from a gentleman who came hither to us no great while ago."

Then the chiefest of the swineherds said: " Lord, this is a poor madman whom we found naked and starving in the forest. As for that sword, I may tell you that he took it away from a knight who came hither to threaten his life, and he soused that knight into the well so that he was wellnigh drowned."

Sir Daynant said: " That is a very strange story, that a naked madman should take the sword out of the hands of an armed knight and treat that knight as ye tell me. Now maybe this is some famous hero or knight who hath lost his wits through sorrow or because of some other reason, and who hath so come to this sorry pass."

(So said Sir Daynant, and it may here be said that from that time those rude swineherds began to look upon Sir Tristram with different eyes than before, saying amongst themselves: " Maybe what that knight said is true, and this is indeed no common madman.")

Now whilst Sir Daynant sat there with his lady, holding converse with the swineherds concerning Sir Tristram in that wise, there came a great noise in the forest, and out therefrom there came riding with great speed that huge savage knight Sir Tauleas aforetold of. Then Sir Daynant cried out, " Alas, here is misfortune!" And therewith he made all haste to put his helmet upon his head.

But ere he could arm himself in any sufficient wise, Sir Tauleas drave down very fiercely upon him. And Sir Tauleas rose up in his *Sir Tauleas* stirrups and lashed so terrible a blow at Sir Daynant that it *strikes down Sir* struck through Sir Daynant's helmet and into his brain-pan, *Daynant.* wherefore Sir Daynant immediately fell down to the ground as though he had been struck dead.

Then Sir Tauleas rode straightway to where the lady of Sir Daynant was, and he said: " Lady, thou art a prize that it is very well *Sir Tauleas* worth while fighting for! And lo! I have won thee." There- *bears away the* with he catched her and lifted her up, shrieking and scream- *lady.* ing and struggling, and sat her upon the saddle before him and held her there maugre all her struggles. Then straightway he rode away into the forest, carrying her with him; and all that while Sir Tristram stood as though in a maze, gazing with a sort of terror upon what befell and not rightly knowing what it all meant. For there lay Sir Daynant as though dead upon the ground, and he could yet hear the shrieks of the lady sounding out from the forest whither Sir Tauleas had carried her.

Then the chief of the swineherds came to Sir Tristram, and said: "Fellow, as thou hast a sword, let us see if thou canst use it. If thou art a hero as that knight said of thee a while since, and not a pure madman, then follow after that knight and bring that lady back hither again."

Then Sir Tristram awoke from that maze and said, "I will do so." And therewith he ran away very rapidly into the forest, pursuing the direction

*Sir Tristram follows Sir Tauleas.* that Sir Tauleas had taken. And he ran for a great distance, and by and by, after a while, he beheld Sir Tauleas before him where he rode. And by that time the lady was in a deep swoon and lay as though dead across the saddle of Sir Tauleas. Then Sir Tristram cried out in a great voice: "Stay, Sir Knight, and turn this way, for I come to take that lady away from thee and to bring her back unto her friend again!"

Then Sir Tauleas turned him and beheld a naked man running after him with a sword in his hand, whereupon he was seized with a great

*Sir Tristram slays Sir Tauleas.* rage of anger, so that he put that lady he carried down to the ground. And he drew his sword and rushed at Sir Tristram very violently with intent to slay him. And when he came nigh to Sir Tristram he arose up on his stirrups and lashed so terrible a blow at him that, had it met its mark, it would have cloven Sir Tristram in twain. But Sir Tristram leaped aside and turned the blow very skilfully; and therewith a memory of his knightly prowess came upon him and he, upon his part, lashed a blow at Sir Tauleas that Sir Tauleas received very unexpectedly. And that blow struck Sir Tauleas so terrible a buffet upon the head that the brain of Sir Tauleas swam, and he swayed about and then fell down from off his horse. Therewith Sir Tristram ran to him and rushed his helmet from off his head. And when he beheld the naked head of Sir Tauleas he catched it by the hair and drew the neck of Sir Tauleas forward. Then Sir Tauleas cried out, "Spare me, fellow!" But Sir Tristram said, "I will not spare thee for thou art a wicked man!" And therewith he lifted his sword on high and smote off the head of Sir Tauleas so that it rolled down upon the ground.

After that, Sir Tristram went to the Lady and he chafed her hands and her face so that she revived from her swoon. And when she was revived, he said: "Lady, take cheer; for look yonder and thou wilt see thy enemy is dead, and so now I may take thee back again unto thy friend." And therewith the lady smiled upon Sir Tristram and catched his hand in hers and kissed it.

Then Sir Tristram lifted the lady upon the horse of Sir Tauleas, and after that he went back again to where he had left Sir Daynant and the swineherds; and he led the horse of Sir Tauleas by the bridle with the lady upon the back thereof and he bore the head of Sir Tauleas in his hand by the hair.

But when those swineherds saw Sir Tristram come forth thus out of the forest bringing that lady and bearing the head of Sir Tauleas, they were amazed beyond measure, and they said to one another: "Of a certainty what this young knight hath just said is sooth and this madman is indeed some great champion in distress. But who he is no one may know, since he himself doth not know."

And when Sir Daynant had recovered from that blow that Sir Tauleas had given him, he also gave Sir Tristram great praise for what he had done. And Sir Tristram was abashed at all the praise that was bestowed upon him.

Then Sir Daynant and his lady besought Sir Tristram that he would go with them to their castle so that they might care for him, but Sir Tristram would not, for he said: "I wist very well that I am mad, and so this forest is a fit place for me to dwell and these kind rude fellows are fit companions for me at this time whilst my wits are wandering."

Thus it was with this adventure. And now you shall hear how Sir Launcelot found Sir Tristram in the forest and how he brought him out thence and likewise what befell thereafter.

For only the next day after all these things had happened, Sir Launcelot came riding through the forest that way, seeking for Sir Tauleas with intent to do battle with him because of his many evil deeds. For *Sir Launcelot* Sir Launcelot purposed either to slay him or else to bring him *enters the forest.* captive to King Arthur.

So it came to pass that Sir Launcelot came to that place where Sir Tristram and the swineherds abode.

There Sir Launcelot made pause for to rest and to refresh himself, and whilst he sat with his helmet lying beside him so that the breezes might cool his face, all those rude swineherds gathered about and stared at him. And Sir Launcelot smiled upon them, and he said: "Good fellows, I pray you tell me; do you know where, hereabouts, I shall find a knight whom men call Sir Tauleas?"

Unto this the chief swineherd made reply, saying: "Lord, if you come hither seeking Sir Tauleas, you shall seek him in vain. For yesterday he was slain, and if you look yonder way you may see his head hanging from a branch of a tree at the edge of the glade."

Upon this Sir Launcelot cried out in great amazement, "How hath that come to pass?" and therewith he immediately arose from where he sat and went to that tree where the head hung. And he looked into the face of the head, and therewith he saw that it was indeed the head of Sir Tauleas that hung there. Then Sir Launcelot said: "This is very wonderful. Now I pray you, tell me what knight was it who slew this wicked wretch, and how his head came to be left hanging here?"

To this the chief of the swineherds made reply: "Messire, he who slew Sir Tauleas was no knight, but a poor madman whom we found in the forest and who has dwelt with us now for a year past. Yonder you may see him, lying half naked, sleeping beside that well of water."

Sir Launcelot said, "Was it he who did indeed slay Sir Tauleas?" And the swineherd said, "Yea, lord, it was he."

Sir Launcelot said, "Do ye not then know who he is?" The swineherd replied: "No, lord, we only know that one day we found him lying in the forest naked and nigh to death from hunger and that we fed him and clothed him, and that since then he hath dwelt ever with us, showing great love for us all."

Then Sir Launcelot went to where Sir Tristram lay, and he looked upon him as he slept and he knew him not; for the beard and the hair of Sir

*Sir Launcelot regards Sir Tristram.* Tristram had grown down all over his breast and shoulders and he was very ragged and beaten by the weather. But though Sir Launcelot knew him not, yet he beheld that the body of Sir Tristram was very beautiful and strong, for he saw how all the muscles and thews thereof were cut very smooth and clean as you might cut them out of wax, wherefore Sir Launcelot gazed for a long while and felt great admiration for his appearance.

Then Sir Launcelot beheld how the sleeping man held a naked sword in his arms very caressingly, as though he loved it, and thereat he was very much surprised to find such a sword as that in the hands of this forest madman. Wherefore he said to those swineherds, "Where got this man that sword?"

"Messire," said the swineherd who had afore spoken, "some while since there came a knight hitherward who ill-treated him. Thereupon this poor man ran at the knight and overthrew him and took the sword away from him and soused him several times in the well. After that he hath ever held fast to this sword and would not give it up to any of us."

"Ha!" said Sir Launcelot, "that is a very wonderful story, that a naked man should overthrow an armed knight and take his sword away from him. Now I deem that this is no mere madman, but some noble knight in misfortune."

Therewith he reached forward and touched Sir Tristram very gently on the shoulder, and at that Sir Tristram awoke and opened his eyes and sat up. And Sir Tristram looked upon Sir Launcelot, but knew him not, albeit some small memory moved very deeply within him. Nevertheless, though he knew not Sir Launcelot, yet he felt great tenderness for that noble knight in arms, and he smiled very lovingly upon him. And Sir Launcelot felt in return a very great deal of regard for Sir Tristram, but wist not why that was; yet it seemed to Sir Launcelot that he should know the face of Sir Tristram, and that it was not altogether strange to him. *Sir Launcelot awakens Sir Tristram.*

Then Sir Launcelot said, "Fair friend, was it thou who slew Sir Tauleas?" And Sir Tristram said, "Ay." Sir Launcelot said, "Who art thou?" Whereunto Sir Tristram made reply: "I know not who I am, nor whence I come, nor how I came hither."

Then Sir Launcelot felt great pity and tenderness for Sir Tristram, and he said: "Friend, wilt thou go with me away from this place and into the habitations of men? There I believe thy mind may be made whole again, and that it may be with thee as it was beforetime. And verily, I believe that when that shall come to pass, the world shall find in thee some great knight it hath lost."

Sir Tristram said: "Sir Knight, though I know not who I am, yet I know that I am not sound in my mind; wherefore I am ashamed to go out in the world and amongst mankind, but would fain hide myself away in this forest. Yet I love thee so much that, if thou wert to bid me go with thee to the ends of the world, I believe I would go with thee."

Then Sir Launcelot smiled upon Sir Tristram very kindly and said, "I do bid thee come with me away from here," and Sir Tristram said, "I will go."

So Sir Launcelot bade the swineherds clothe Sir Tristram in such a wise that his nakedness might be covered, and he bade them give Sir Tristram hosen and shoon, and when Sir Tristram was thus decently clad, Sir Launcelot made ready to take his departure from that place. *Sir Tristram quits the forest with Sir Launcelot.*

But ere the two left, all those good fellows crowded around Sir Tristram, and embraced him and kissed him upon the cheek; for they had come to love him a very great deal.

Then the two went away through the forest, Sir Launcelot proudly riding upon his great horse and Sir Tristram running very lightly beside him.

But Sir Launcelot had other business at that time than to seek out Sir Tauleas as aforetold. For at that time there were three knights of very ill-repute who harried the west coast of that land that overlooked the sea toward the Kingdom of Ireland, and Sir Launcelot was minded to seek them out after he had finished with Sir Tauleas. So ere he returned to the court of King Arthur he had first of all to go thitherward.

Now you are to know that the castle of Tintagel lay upon the way that he was to take upon that adventure, and so it was that he brought Sir Tristram to the castle of Tintagel, where King Mark of Cornwall was then holding court. For Sir Launcelot was minded to leave Sir Tristram there whilst he went upon that adventure aforetold of.

And Sir Launcelot was received in Tintagel with very great honor and acclaim, for it was the first time he had ever been there. And King Mark

*Sir Tristram comes to Tintagel.* besought Sir Launcelot for to abide a while in Tintagel; but Sir Launcelot refused this hospitality, saying: "I have an adventure to do for the sake of my master, King Arthur, and I may not abide here at this present. But I pray you to grant me a favor, and it is this: that you cherish this poor madman whom I found in the forest, and that you keep him here, treating him kindly until I shall return from the quest I am upon. For I have great love for this poor fellow and I would not have any harm befall him whilst I am away."

Then King Mark said: "I am sorry you will not remain with us, but as to this thing it shall be done as you desire, for we will cherish and care for this man while you are away." So said King Mark, speaking with great cheerfulness and courtesy; for neither he nor any of his court at that time wist who Sir Tristram was.

So Sir Launcelot went upon his way, and King Mark gave orders that Sir Tristram should be well-clothed and fed, and it was done as he commanded.

Thus it was that Sir Tristram was brought back to the castle of Tintagel again. And now it shall be told how it befell with him thereat.

ir Tristram leaps into y̆ Sea

# Chapter Third

*How Sir Tristram was discovered at Tintagel and of what befell thereby.*

NOW during the time that Sir Tristram abode thus unknown at the court of Tintagel, he was allowed to wander thereabouts whither-oever he chose, and no one hindered him either in going or in coming. For none in all that place suspected who he was, but everyone thought that he was only a poor gentle madman of the forest; so he was allowed to wander at will as his fancy led him.

And Sir Tristram's memory never awoke; but though it awoke not, yet it stirred within him. For though he could not remember what this place was whereunto he had come, yet it was very strangely familiar to him, so that whithersoever he went, he felt that those places were not altogether strange to him. And in some of those places he felt great pleasure and in other places somewhat of pain, yet he knew not why he should have the one feeling or the other.

*How Sir Tristram dwelt at Tintagel.*

Now of all those places whereunto he wandered, Sir Tristram found most pleasure in the pleasance of the castle where was a fair garden and fruit trees; for it was there that he and the Lady Belle Isoult had walked together aforetime ere his affliction had befallen him, and he remembered this place better than any other, and took more pleasure in it. Now one day Sir Tristram came wandering thus into that pleasance and, the day being warm, he sat under the shade of an appletree beside a marble fountain of water; and the appletree above his head was all full of red and golden fruit. So Sir Tristram sat there, striving to remember how it was that he had once aforetime beheld that fountain and that garden and that apple-tree beneath which he sat.

So whilst he sat there pondering in that wise, there came the Lady Belle Isoult into the garden of that pleasance and her lady, the dame Bragwaine, was with her, and the hound, hight Houdaine, which Sir Tristram had sent to her by Gouvernail, walked beside her on the other side. Then Belle Isoult perceived that there was a man sitting under the appletree, and she said to dame Bragwaine: "Who is yonder man who hath dared to come hither into our privy garden?" Unto this, dame Bragwaine replied: "That, lady, is the gentle madman of the forest whom Sir Launcelot brought hither two days ago."

Then the Lady Belle Isoult said, "Let us go nearer and see what manner of man he is"; and so they went forward toward where Sir Tristram sat, and the dog Houdaine went with them.

Then Sir Tristram was aware that someone was nigh; and therewith he turned his face and beheld the Lady Isoult for the first time since he had gone mad in the forest; and the lady was looking at him, but knew him not.

Then of a sudden, because of his great love for Belle Isoult, the memory of Sir Tristram came all back to him in the instant, and upon that instant he knew who he was and all that had befallen him, and how he had been brought there as a madman out of the forest. But though he knew her in that wise, yet, as has been said, she knew not him.

Then Sir Tristram was all overwhelmed with shame that he should be thus found by that dear lady; wherefore he turned away his face and bowed his head so that she might not remember him, for he perceived that as yet she did not know him who he was.

Now at that moment the dog, Houdaine, was aware of the savor of Sir Tristram; wherefore he leaped away from the Lady Belle Isoult and ran to Sir Tristram and smelt very eagerly of him. And with that he knew his master.

Then the two ladies who looked beheld Houdaine fall down at the feet of Sir Tristram and grovel there with joy. And they beheld that he licked Sir Tristram's feet and his hands, and that he leaped upon Sir Tristram and licked his neck and face, and at that they were greatly astonished.

*Houdaine knoweth Sir Tristram.*

Then of a sudden a thought came to dame Bragwaine, and she catched the Lady Isoult by the arm and she said: "Lady, know you not who yonder madman is?" But the Lady Belle Isoult said: "Nay, I know not who he is. Who is he, Bragwaine?" And Bragwaine said: "Certes, that is Sir Tristram, and no one else in all the world."

Therewith, at those words, the scales suddenly fell from Lady Belle Isoult's eyes and she knew him. Then, for a little space, she stood as though turned into stone; then she emitted a great loud cry of joy and ran to Sir Tristram where he sat, and flung herself down upon the ground at the feet of Sir Tristram and embraced him about the knees. And she cried out in a voice of great passion: "Tristram! Tristram! Is it thou? They told me thou wert dead, and lo!

*Belle Isoult knows Sir Tristram.*

thou art come to life again!'' And with that she fell to weeping with such fury of passion that it was as though the soul of her were struggling to escape from her body.

Then Sir Tristram got to his feet in great haste and agitation and he said: "Lady! Lady! This must not be—arise, and stay your passion or else it will be our ruin. For behold, I am alone and unarmed in this castle, and there are several herein who seek my life. So if it be discovered who I am, both thou and I are lost."

Then, perceiving how that Belle Isoult was in a way distracted and out of her mind with joy and grief and love, he turned him unto Bragwaine and said to her: "Take thy lady hence and by and by I will find means whereby I may come to speech with her in private. Meanwhile it is death both for her and for me if she remain here to betray me unto the others of this castle."

So Bragwaine and Sir Tristram lifted up the Lady Belle Isoult, and Bragwaine led her thence out of that place; for I believe that Belle Isoult knew not whither she went but walked like one walking half in a swoon.

Now it chanced at that time that Sir Andred was in a balcony overlooking that pleasance, and, hearing the sound of voices and the sound of a disturbance that was suppressed, he looked out and beheld all that passed. Then he also wist who was that *Sir Andred knoweth Sir Tristram.* madman whom Sir Launcelot had fetched to that place out of the forest, and that he was Sir Tristram.

Therewith he was filled with a great rage and fury and was likewise overwhelmed with great fear lest, if Sir Tristram should escape from that castle with his life, he would reclaim those possessions that he, Sir Andred, had seized upon.

So therewith he withdrew himself from that balcony very softly, into the apartment behind. And he sat down in that apartment for a little while as though not knowing rightly what to do. But after a little while he arose and went to King Mark; and King Mark looked up and beheld him and said, "What news do you bring, Messire?" Thereunto Sir Andred made reply: "Lord, know you who that madman is whom Sir Launcelot hath fetched hither?" King Mark said, "Nay, I know not who he is." But with that he fell to trembling throughout his entire body, for he began to bethink him who that madman was. "Lord," said Sir Andred, "it is Sir Tristram, and meseems Sir Launcelot was aware who it was, and that he was plotting treason when he fetched him hither." *Sir Andred betrays Sir Tristram to King Mark.*

At that King Mark smote his hands together and he cried in a terrible voice, "I know it! I know it!" And then he said: "Blind! Blind! How was it that I knew him not?" Then after a little he fell to laughing and he said to Sir Andred: "Lo! God hath assuredly delivered that traitor, Sir Tristram, into mine hands so that I may punish him for his treasons. For, behold! he is here in our midst and he is altogether unarmed. Go, Messire, with all haste, gather together such force as may be needful, and seize upon him and bind him so that he may do no further harm to any man. Then let justice be executed upon him so soon as it is possible to do so." And Sir Andred said: "Lord, it shall be done according to your demands and upon the instant."

Therewith Sir Andred went forth from where the King was, and he armed himself in complete armor, and he gathered together a number of knights and esquires and he led them to that place where he knew Sir Tristram would be; and there he found Sir Tristram sitting sunk in thought. And when Sir Tristram beheld those armed men come in thus upon him, he arose to defend himself. But then Sir Andred cried out in a loud voice: "Seize him ere he can strike and bind him fast, for he is unarmed and may do you no harm!"

With that a dozen or more of those who were with Sir Andred flung themselves upon Sir Tristram, shouting and roaring like wild beasts. And

*The castle folk seize Sir Tristram.* they bore him to the earth by numbers, and after a while, by dint of great effort, they held him and bound his hands together by the wrists. Then they lifted up Sir Tristram and stood him upon his feet, and lo! his bosom heaved with his struggles, and his eyes were all shot with blood and his lips afroth with the fury of his fighting; and his clothes were torn in that struggle so that his body was half naked. And they held him there, a knight in armor with a naked sword standing upon his right hand and another armed knight with a naked sword standing upon his left hand.

Then Sir Andred came and stood in front of Sir Tristram and taunted him, saying: "Ha, Tristram, how is it with thee now?" Lo! thou camest like a spy into this place, and now thou art taken with all thy treason upon thee. So thou shalt die no knightly death, but, in a little while, thou shalt be hanged like a thief."

Then he came close to Sir Tristram, and he laughed and said: "Tristram where is now the glory of thy strength that one time overcame all thine enemies? Lo! thou art helpless to strike a single blow in defence of thine honor." And therewith Sir Andred lifted his hand and smote Sir Tristram upon the face with the palm thereof.

At that blow the rage of Sir Tristram so flamed up in him that his eyes burned as with pure green fire. And in an instant, so quickly that no man wist what he did, he turned with amazing suddenness *Sir Tristram* upon that knight who stood at his left hand, and he lifted up *slays Sir Andred.* both hands that were bound, and he smote that knight such a blow upon the face that the knight fell down upon the ground and his sword fell out of his hand. Then Sir Tristram snatched the sword and, turning with astonishing quickness, he smote the knight upon his right hand such a buffet that he instantly fell down upon his knees and then rolled over upon the ground in a swoon. Then Sir Tristram turned upon Sir Andred, and lifting high the sword with both hands tied, he smote him so terrible a blow that the blade cut through his epulier and half through his body as far as the paps. At that great terrible blow the breath fled out of Sir Andred with a deep groan, and he fell down upon the ground and immediately died.

Now all this had happened so suddenly that they who beheld it were altogether amazed and stood staring as though bewitched by some spell. But when they beheld Sir Tristram turn upon them and make at them with that streaming sword lifted on high, the terror of his fury so seized upon them that they everywhere broke from before him and fled, yelling, and with the fear of death clutching them in the vitals. And Sir Tristram chased them out of that place and into the courtyard of the castle, and some he smote down and others escaped; but all who could do so scattered away before him like chaff before the wind.

Then, when they were gone, Sir Tristram stood panting and glaring about him like a lion at bay. Then he set the point of his sword upon the pavement of the court and the pommel thereof he set against his breast, and he drew the bonds that held his wrists across the edge of the sword so that they were cut and he was free.

But Sir Tristram wist that in a little the whole castle would be aroused against him, and that he would certainly be overwhelmed by dint of numbers, wherefore he looked about him for some place of refuge; *Sir Tristram* and he beheld that the door of the chapel which opened upon *defends the* the courtyard stood ajar. So he ran into the chapel and shut *chapel.* to that door and another door and locked and bolted them both, and set a heavy bar of wood across both of them so that for a while he was safe.

But yet he was only safe for a little while, for about the time of early nightfall, which came not long thereafter, a great party of several score of King Mark's people came against the chapel where he was. And when they found that the doors were locked and barred, they brought rams for to batter in the chief door of the chapel.

Then Sir Tristram beheld how parlous was his case, and that he must in a little while die if he did not immediately do something to save himself. So with that he ran to a window of the chapel and opened it and looked out thence.   And lo! below him and far beneath was the sea, and the rocks of the shore upon which the castle was built; and the sea and the rocks lay twelve fathoms beneath him.

But Sir Tristram said, "Better death there than here;" and therewith
*Sir Tristram* finding that the door was now falling in beneath the rams, he
*leaps into the* leaped out from the window-ledge, and thence he dived down
*sea.* into the sea; and no one saw that terrible leap that he made.

So he sank down deep into the sea, but met no rocks, so that he presently came up again safe and sound.   Then, looking about him, he perceived in the twilight a cave in the rocks, and thither he swam with the intent to find shelter for a little.

Now when they who had come against him had broken into the chapel they all ran in in one great crowd, for they expected to find Sir Tristram and to do battle with him.   But lo! Sir Tristram was not there, but only the empty walls.   Then at first they were greatly astonished, and knew not what to think.   And some who came cried out: "Is that man then a spirit that he can melt away into thin air?"   But after a little, one of them perceived where the window of the chapel stood open, and therewith several of them ran thereunto and looked out, and they wist that Sir Tristram had leaped out thence into the sea.

Then they said to one another: "Either that knight is now dead, or else he will perish when the tide rises and covers the rocks; so to-night we will do no more with this business; but to-morrow we will go and find his body where it lies among the rocks of the shore."   So thereupon they shut the window and went their ways.

Now Gouvernail was not at that time at Tintagel, nor did he return thereunto until all this affair was over and done.   But when he came there, there were many voices to tell him what had befallen, and to all of them Gouvernail listened without saying anything.

But afterward Gouvernail went and sought out a certain knight hight Sir Santraille de Lushon, who, next to himself, was the most faithful friend to Sir Tristram at that place.   To him Gouvernail said: "Messire, I do not think that Sir Tristram is dead, for he hath always been a most wonderful swimmer and diver.   But if he be alive, and we do not save him, he will assuredly perish when the tide comes up and covers over those rocks amongst which he may now be hidden."

So Gouvernail and Sir Santraille went to that chapel unknown to anyone, and they went to that window whence Sir Tristram had leaped, and they opened the window, and leaned out and called upon Sir Tristram in low voices: "Sir Tristram, if thou art alive, arise and answer us, for we are friends!"

Then after a while Sir Tristram recognized Gouvernail's voice and answered them: "I am alive; but save me, or I perish in a little while." Then Gouvernail said: "Lord, are you hurt, or are you whole?" Sir Tristram replied, "I am strong and well in body, but the tide rises fast." Gouvernail said, "Messire, can you wait a little?" Sir Tristram said, "Ay; for a little, but not for too long."

Then Gouvernail and Sir Santraille withdrew from where they were and they made all haste, and they got together a great number of sheets and napkins, and tied these together and made a rope, and lowered *Gouvernail and* the rope down to the rocks where Sir Tristram was. Then *Sir Santraille* Sir Tristram climbed up the rope of linen and so reached the *rescue Sir Tris-* chapel in safety. And at that time it was nigh to midnight *tram.* and very dark.

But when Sir Tristram stood with them in the chapel, he gave them hardly any greeting, but said at once: "Messires, how doth it fare with the Lady Belle Isoult?" For he thought of her the first of all and above all things else.

To this Sir Santraille made reply: "Sir, the lady hath been shut into a tower, and the door thereof hath been locked upon her, and she is a close prisoner."

Then Sir Tristram said: "How many knights are there in the place who are my friends, and who will stand with me to break out hence?" To this Gouvernail said: "Lord, there are twelve besides ourselves, and that makes fourteen in all who are with thee in this quarrel unto life or death."

Sir Tristram said: "Provide me presently with arms and armor and bring those twelve hither armed at all points. But first let them saddle horses for themselves and for us, and for the Lady Belle Isoult and for her waiting-woman, Dame Bragwaine. When this is done, we will depart from this place unto some other place of refuge, and I do not think there will be any in the castle will dare stop or stay us after we are armed."

So it was done as Sir Tristram commanded, and when all those were gathered together, and their horses ready, Sir Tristram and *Sir Tristram* several of the knights of his party went openly to that tower *arms himself.* where the Lady Belle Isoult was prisoner. And they burst open the doors and went in with torches, and found Belle Isoult and her attendant in the upper part of the castle.

But when Belle Isoult beheld the face of Sir Tristram, she said: "Is it thou, my love; and art thou still alive, and art thou come to me?" Sir Tristram said: "Yea, I am still alive nor will I die, God willing, until I have first brought thee out of this wicked castle and into some place of safety. And never again will I entrust thee unto King Mark's hands; for I have great fear that if he have thee in his hands he will work vengeance upon thee so as to strike at my heart through thee. So, dear love, I come to take thee away from this place; and never again right or wrong, shalt thou be without the shelter of my arm."

Then the Lady Belle Isoult smiled very wonderfully upon Sir Tristram so that her face appeared to shine with a great illumination of love. And she said: "Tristram, I will go with thee whithersoever thou wilt. Yea, I would go with thee even to the grave, for I believe that I should be happy even there, so that thou wert lying beside me."

Then Sir Tristram groaned in spirit and he said: "Isoult, what have I done, that I should always bring unhappiness upon thee?" But the Lady Belle Isoult spake very steadily, saying: "Never unhappiness, Tristram, but always happiness; for I have thy love for aye, and thou hast mine in the same measure, and in that is happiness, even in tears and sorrow, and never unhappiness."

With that Sir Tristram kissed Belle Isoult upon the forehead, and then he lifted her up and carried her in his arms down the stairs of the tower and sat her upon her horse. And Bragwaine followed after, and Gouvernail lifted her up upon her horse.

Now all they of that castle were amazed beyond measure to find all *Sir Tristram* those knights armed and prepared for battle so suddenly in *taketh Belle* their midst. And most of all were they filled with terror to *Isoult away from* find Sir Tristram at the head of these knights. Wherefore *Tintagel.* when Sir Tristram made demand that they should open the portcullis of the castle and let fall the drawbridge, the porters thereof dared not refuse him, but did as he said.

So Sir Tristram and his knights rode forth with the Lady Belle Isoult and Bragwaine and no one stayed them. And they rode into the forest, betaking their way toward a certain castle of Sir Tristram's, which they reached in the clear dawning of the daytime.

And so Sir Tristram brought the Lady Belle Isoult away from Tintagel and into safety.

———————

King Mark broods mischief

# Chapter Fourth

*How Sir Tristram and the Lady Belle Isoult returned to Cornwall and how they ended their days together*

And now remaineth to be told the rest of these adventures of Sir Tristram as briefly as may be.

For indeed I thought not, when I began this history, to tell you as much concerning him as I have done. But as I have entered into this history I have come so strongly to perceive how noble and true and loyal was the knighthood of Sir Tristram, that I could not forbear telling you of many things that I had not purposed to speak of.

Yet, as I have said before this, there are a great many adventures that I have not spoken of in this book. For I have told only those things that were necessary for to make you understand how it fared with him in his life.

So now shall be told those last things that concerned him.

Now two days after those things aforesaid had come to pass, Sir Launcelot returned unto Tintagel from that quest which he had been upon, and so soon as he came thither he made inquiry of King Mark con- *Sir Launcelot* cerning the welfare of that madman of the forest whom he had *reproves King* left in the care of King Mark. But when he heard that that *Mark.* madman was Sir Tristram, he was astonished beyond all measure; but when he heard how Sir Tristram had been served by King Mark and by the people of the castle under the lead of Sir Andred, he was filled with a great and violent indignation. So he arose and stood before King Mark and said: "Lord King, I have heard much ill said of thee and shameful things concerning thy unknightliness in several courts of chivalry where I have been; and now I know that those things were true; for I have heard from the lips of many people here, how thou didst betray Sir Tristram into bringing the Lady Belle Isoult unto thee; and I have heard from many how thou dost ever do ill and wickedly by him, seeking to take from him both his honor and his life. And yet Sir Tristram hath always been thy true and faithful knight, and hath served thee in all ways thou hast demanded of him. I know that thou hast jealousy for Sir Tristram in thy heart and that thou hast ever imputed wickedness and sin unto him. Yet all the world knoweth that Sir Tristram is a true knight and altogether innocent of any evil. For all the evil which thou hast imputed to him hath no existence saving only in thine own evil heart. Now I give thee and all thy people to know that had ill befallen Sir Tristram at your hands I should have held you accountable therefor and should have punished you in such a way that you would not soon have forgotten it. But of that there is no need, for Sir Tristram himself hath punished you in full measure without any aid from me. So now I will go away from this place and will never come hither again; nor will I acknowledge you should I meet you in court or in field."

So saying, Sir Launcelot turned and went away from that place very proudly and haughtily, leaving them all abashed at his rebuke.

So that day Sir Launcelot went forward through the forest until he reached that castle whereunto Sir Tristram had taken the *Sir Launcelot findeth Sir Tristram and Belle Isoult in the forest.* Lady Belle Isoult, and there he was received by Sir Tristram with all joy and honor. And Sir Launcelot abided at that place for two days, with great pleasure to himself and to Sir Tristram and to Belle Isoult.

At the end of that time Sir Launcelot said to Sir Tristram: "Messire, it is not well that you and this dear lady should abide here so nigh to Tintagel. For, certes, King Mark will some time work some grievous ill upon you. So I beseech you to come with me unto my castle of Joyous Gard. There this lady shall reign queen paramount and we shall be her very faithful servants to do her pleasure in all ways. That castle is a very beautiful place, and there she may dwell in peace and safety and tranquillity all the days of her life if she chooses to do so."

Now that saying of Sir Launcelot's seemed good to Sir Tristram and to *They depart for Joyous Gard.* Belle Isoult; wherefore in three days all they and their court made ready to depart. And they did depart from that castle in the forest unto Joyous Gard, where they were received with great honor and rejoicing.

So the Lady Belle Isoult abided for three years at Joyous Gard, dwelling there as queen paramount in all truth and innocence of life; and Sir Launcelot and Sir Tristram were her champions and all their courts were her servants. And during those three years there were many famous joustings held at Joyous Gard, and several bel-adventures were performed both by Sir Launcelot and Sir Tristram in her honor.

And indeed I believe that this was the happiest time of all the Lady Belle Isoult's life, for she lived there in peace and love and tranquillity and she suffered neither grief nor misfortune in all that time.

Then one day there came King Arthur to Joyous Gard, and he was received with such joy and celebration as that place had never before beheld. A great feast was set in his honor, and after the feast King Arthur and Sir Tristram and Belle Isoult withdrew to one side and sat together in converse. *King Arthur comes to Joyous Gard.*

Then after a while King Arthur said, "Lady, may I ask you a question?" And at that Lady Belle Isoult lifted up her eyes and looked very strangely upon the King, and after a while she said, "Ask thy question, Lord King, and I will answer it if I can." "Lady," said King Arthur, "answer me this question: is it better to dwell in honor with sadness or in dishonor with joy?"

Then Belle Isoult began to pant with great agitation, and by and by she said, "Lord, why ask you me that?" King Arthur said: "Because, lady, I think your heart hath sometimes asked you the selfsame question." Then the Lady Belle Isoult clasped her hands together and cried out: "Yea, yea, my heart hath often asked me that question, but I would not answer it." King Arthur said: "Neither shalt thou answer me, for I am but a weak and erring man as thou art a woman. But answer thou that question to God, dear lady, and then thou shalt answer it in truth."

Therewith King Arthur fell to talking of other things with Sir Tristram, but the lady could not join them in talk, but sat thenceforth in silence, finding it hard to breathe because of the oppression of tears that lay upon her bosom.

And Belle Isoult said no more concerning that question that King Arthur had asked. But three days after that time she came to Sir Tristram and said: "Dear lord, I have bethought me much of what King Arthur said, and this hath come of it, that I must return again unto Cornwall."

Then Sir Tristram turned away his face so that she might not see it, and he said, "Methought it would come to that." And then in a little he went away from that place, leaving her standing there.

So it came about that peace was made betwixt Sir Tristram and King Mark, and Belle Isoult and King Mark, and King Arthur was the peacemaker.

Thereafter Sir Tristram and his court and the Lady Belle Isoult returned unto Cornwall, and there they dwelt for some time in seeming peace. But in that time the Lady Belle Isoult would never see King Mark nor exchange a word with him, but lived entirely apart from him and in her own life in a part of the castle; and at *Belle Isoult scorns King Mark.* that King Mark was struck with such bitterness of despair that he was like to a demon in torment. For he saw, as it were, a treasure very

near and yet afar, for he could not come unto it.   And the more he suffered
that torment, the more he hated Sir Tristram, for in his suffering it appeared
to him that Sir Tristram was the cause of that suffering.

So it came about that King Mark set spies to watch Sir Tristram, for
in his evil heart he suspected Sir Tristram of treason, and he hoped that his
spies might discover Sir Tristram in some act for which he might be punished.   So those spies watched Sir Tristram both night and day, but they
could find nothing that he did that was amiss.

Now one day Belle Isoult felt such a longing for Sir Tristram that she
could not refrain from sending a note to him beseeching him for to come
to her so that they might see one another again ; and though Sir Tristram
misdoubted what he did, yet he went as she desired,  even if it should mean
the peril of death to him.

Then came those spies to King Mark and told him that Sir Tristram was
gone to the bower of the Lady Belle Isoult, and that she had bidden him
to come thither.

At that the vitals of King Mark were twisted with such an agony of
hatred and despair that he bent him double and cried out, "Woe!   Woe!
I suffer torments!"

Therewith he arose and went very quickly to that part of the castle
where the Lady Belle Isoult inhabited ; and he went very softly up by a
*King Mark*
*spies upon*
*Sir Tristram*
*and Isoult.*
back way and through a passage to where was a door with
curtains hanging before it ; and when he had come there he
parted the curtains and peeped within.   And he beheld that
the Lady Belle Isoult and Sir Tristram sat at a game of chess,
and he beheld that they played not at the game but that they sat talking
together very sadly ; and he beheld that Dame Bragwaine sat in a deep
window to one side—for Belle Isoult did not wish it to be said that she and
Sir Tristram sat alone.

All this King Mark saw and trembled with a torment of jealousy.   So
by and by he left that place and went very quietly back into that passage-
way whence he had come.   And when he had come there he perceived a
great glaive upon a pole two ells long.   This he took into his hand and
returned unto that curtained doorway again.

Then being in all ways prepared he parted the curtains silently and
stepped very quickly and without noise into the room.   And the back
of Sir Tristram was toward him.

Then King Mark lifted the glaive on high and he struck ; and Sir Tristram
sank without a sound.

Yea, I believe that that good knight knew naught of what had happened until he awoke in Paradise to find himself in that realm of happiness and peace.

Then Belle Isoult arose, overturning the table of chessmen as she did so, but she made no outcry nor sound of any sort. But she stood looking down at Sir Tristram for a little space, and then she kneeled *Of the passing* down beside his body and touched the face thereof as though *of Tristam and* to make sure that it was dead. Therewith, as though being *Isoult.* assured, she fell down with her body upon his; and King Mark stood there looking down upon them.

All this had passed so quickly that Dame Bragwaine hardly knew what had befallen; but now, upon an instant, she suddenly fell to shrieking so piercingly that the whole castle rang with the sound thereof.

Now there were in the outer room several of the knights of the court of Sir Tristram who had come thither with him as witnesses that he performed no treason to the King. These, when Dame Bragwaine shrieked in that wise, came running into the room and therewith beheld what had happened. Then all they stood aghast at that sight.

But there was in the court of Sir Tristram a very young, gallant knight hight Sir Alexander. This knight came to where King Mark stood looking down upon his handiwork as though entranced with what he *Sir Alexander* had done. Then Sir Alexander said to King Mark, "Is this thy *slays King* work?" And King Mark raised his eyes very heavily and *Mark.* looked at Sir Alexander and he answered, "Ay!" Then Sir Alexander cried out, "Thou hast lived too long!" And therewith drawing his misericordia, he catched King Mark by the left wrist and lifted his arm. And Sir Alexander drave the dagger into the side of King Mark, and King Mark groaned and sank down upon the ground, and in a little while died where he lay.

Then those knights went to where the Lady Belle Isoult lay and lifted her up; but, lo! the soul had left her, and she was dead. For I believe that it was not possible for one of those loving souls to leave its body without the other quitting its body also, so that they might meet together in Paradise. For there never were two souls in all the history of chivalry that clave to one another so tenderly as did the souls of Tristram and Isoult.

So endeth this story of Sir Tristram, with only this to say, that they two were buried with the graves close together, and that it is said by many who have written of them that there grew a rose-tree up from Sir Tristram's grave, and down upon the grave of Belle Isoult; and it is said that this rose-tree was a miracle, for that upon his grave there grew red roses, and upon her grave there grew pure white roses. For her soul was white like to thrice-carded wool, and so his soul was red with all that was of courage or knightly pride.

And I pray that God may rest the souls of those two as I pray He may rest the souls of all of us who must some time go the way that those two and so many others have travelled before us. Amen.

# The Book of Sir Percival

*H*ERE *beginneth the story of Sir Percival of Gales, who was considered to be one of the three great knights of the Round Table at that time. For, if Sir Launcelot was the chiefest of all the knights who ever came unto King Arthur's court, then it is hard to say whether Sir Tristram of Lyonesse or Sir Percival of Gales was second unto him in renown.*

*And I pray that it shall be given unto all of ye to live as brave and honorable and pure a life as he did; and that you, upon your part, may claim a like glory and credit in the world in which you dwell by such noble behavior as he exhibited.*

ir Percival of Gales

# Prologue.

THE father of Sir Percival was that king hight Pellinore who fought so terrible a battle with King Arthur as has been told in the Book of King Arthur. For it was after that fight that King Arthur obtained his famous sword Excalibur, as was therein told.

Now, King Pellinore was one of those eleven kings who, in the beginning of King Arthur's reign, were in rebellion against King Arthur as hath been told in the book aforesaid, and he was one of the last of all those kings to yield when he was overcome. So King Arthur drove him from town to town and from place to place until, at last, he was driven away from the habitations of men and into the forests like to a wild beast.

Now, King Pellinore took with him into the wilderness his wife and his four sons; to wit, Lamorack and Aglaval and Dornar and *King Pellinore* Percival. Of these, Percival was but three years of age; the *fleeth to the wil-* others, excepting Dornar, being nigh to the estate of manhood. *derness.* Thereafter that noble family dwelt in the forest like hunted animals, and that was a very great hardship for the lady who had been queen; and, likewise, it was greatly to the peril of the young child, Percival.

Now, Percival was extraordinarily beautiful and his mother loved him above all her other sons. Wherefore she feared lest the young child should die of those hardships in the wilderness.

So one day King Pellinore said: "Dear love, I am now in no wise prepared for to defend thee and this little one. Wherefore, for a while, I shall put ye away from me so that ye may remain in secret hiding until such time as the child shall have grown in years and stature to the estate of manhood and may so defend himself.

"Now of all my one-time possessions I have only two left to me. One of these is a lonely castle in this forest (unto which I am now betaking my way), and the other is a solitary tower at a great distance from this, and in a very desolate part of the world where there are many mountains. Unto that place I shall send ye, for it will not be likely that mine enemies will ever find ye there.

"So my will is this: that if this child groweth in that lonely place to manhood, and if he be weak in body or timid in spirit, thou shalt make of him a clerk of holy orders. But if when he groweth, he shall prove to be strong and lusty of frame and high of spirit, and shall desire to undertake deeds of knighthood, thou then shalt not stay him from his desires, but shall let him go forth into the world as he shall have a mind to do.

"And if a time should come when he desireth to go thus into the world behold! here is a ring set with a very precious ruby; let him bring that ring to me or to any of our sons wheresoever he may find us, and by that ring we shall know that he is my son and their brother, and we will receive him with great gladness."

And King Pellinore's lady said, "It shall be done as thou dost ordain."

So it was that King Pellinore betook himself to that lonely castle where *Percival's mother* King Arthur found him and fought with him; and Percival's *taketh him to the* mother betook herself to that dwelling-place in the mountains *mountains.* of which King Pellinore had spoken—which was a single tower that reached up into the sky, like unto a finger of stone.

There she abided with Percival for sixteen years, and in all that time Percival knew naught of the world nor of what sort it was, but grew altogether wild and was entirely innocent like to a little child.

In the mean time, during those years, it happened very ill to the house of King Pellinore. For though King Arthur became reconciled to King Pellinore, yet there were in King Arthur's court many who were bitter enemies to that good, worthy knight. So it came about that first King Pellinore was slain by treachery, and then Sir Aglaval and Sir Dornar were slain in the same way, so that Sir Lamorack alone was left of all that noble family.

(And it was said that Sir Gawaine and his brothers were implicated in those murders—they being enemies unto King Pellinore—and great reproach hath always clung to them for the treacherous, unknightly way in which those noble knights of the house of Pellinore were slain.)

Now the news of those several deaths was brought to that lonely tower of the mountain wilderness and to Sir Percival's mother; and when she heard how her husband and two of her sons were dead she gave great outcry of grief, and smote her hands together and wept with *Percival's mother grieveth* great passion. And she cried out: "Mefeareth it will be the *for the death of her dear ones.* time of Lamorack next to be slain. As for Percival; never shall I be willing for him to go out into that cruel world of wicked murderers. For if he should perish also, my heart would surely break."

So she kept Percival always with her and in ignorance of all that concerned the world of knighthood. And though Percival waxed great of body and was beautiful and noble of countenance, yet he dwelt *How Percival* there among those mountains knowing no more of the world *dwelt in the* that lay beyond that place in which he dwelt than would a *mountains.* little innocent child. Nor did he ever see anyone from the outside world, saving only an old man who was a deaf-mute. And this old man came and went betwixt that tower where Percival and his mother dwelt and the outer world, and from the world he would come back with clothing and provisions loaded upon an old sumpter horse for Percival and his mother and their few attendants. Yet Percival marvelled many times whence those things came, but no one told him and so he lived in entire ignorance of the world.

And Percival's mother would not let him touch any weapon saving only a small Scot's spear which same is a sort of javelin. But with this Percival played every day of his life until he grew so cunning in handling it that he could pierce with it a bird upon the wing in the air.

Now it chanced upon a time when Percival was nineteen years of age that he stood upon a pinnacle of rock and looked down into a certain valley. And it was very early in the spring-time, so that the valley appeared, as it were, to be carpeted all with clear, thin green. There was a shining stream of water that ran down through the midst of the valley, and it was a very fair and peaceful place to behold.

So Percival stood and gazed into that low-land, and lo! a knight rode up through that valley, and the sun shone out from behind a cloud of rain and smote upon his armor so that it appeared to be all ablaze *Percival beholds* as with pure light, and Percival beheld that knight and wist *a knight-rider.* not what it was he saw. So, after the knight had gone away from the valley, he ran straightway to his mother, all filled with a great wonder, and he said: "Mother! Mother! I have beheld a very wonderful thing." She said, "What was it thou didst see?" Percival said: "I beheld somewhat that was like a man, and he rode upon a horse, and he shone very brightly and with exceeding splendor. Now, I prithee tell me what it was I saw?"

Then Percival's mother knew very well what it was he had seen, and she was greatly troubled at heart, for she wist that if Percival's knightly spirit should be awakened he would no longer be content to dwell in those peaceful solitudes. Wherefore she said to herself: "How is this? Is it to be that this one lamb also shall be taken away from me and nothing left to me of all my flock?" Then she said to Percival: "My son, that which thou didst behold was doubtless an angel." And Percival said, "I would that I too were an angel!" And at that speech the lady, his mother, sighed very deeply.

Now it chanced upon the next day after that that Percival and his mother went down into the forest that lay at the foot of the mountain whereon that tower stood, and they had intent to gather such early flowers of the spring-time as were then abloom. And whilst they were there, lo! there came five knights riding through the forest, and, the leaves being thin like to a mist of green, Percival perceived them a great way off. So he cried out in a loud voice: "Mother! Mother! Behold! Yonder is a whole company of angels such as I saw yesterday! Now I will go and give them greeting."

But his mother said: "How now! How now! Wouldst thou make address unto angels!" And Percival said: "Yea; for they appear to be both mild of face and gentle of mien." So he went forward for to greet those knights.

Now the foremost of that party of knights was Sir Ewaine, who was always both gentle and courteous to everybody. Wherefore, when Sir Ewaine saw Percival nigh at hand, he gave him greeting and said, "Fair youth, what is thy name?" Unto this Percival made reply: "My name is Percival." Sir Ewaine said: "That is a very good name, and thy face *Percival holds* likewise is so extraordinarily comely that I take thee to be of *discourse with* some very high lineage. Now tell me, I prithee, who is thy *five knights.* father?" To this Percival said, "I cannot tell thee what is my lineage, for I do not know," and at that Sir Ewaine marvelled a very great deal. Then, after a little while, he said: "I prithee tell me, didst thou see a knight pass this way to-day or yesterday?" And Percival said, "I know not what sort of a thing is a knight." Sir Ewaine said, "A knight is such a sort of man as I am."

Upon this Percival understood many things that he did not know before, and he willed with all his soul to know more than those. Wherefore he said: "If thou wilt answer several questions for me, I will gladly answer thine." Upon this Sir Ewaine smiled very cheerfully (for he liked Percival exceedingly), and he said: "Ask what thou wilt and I will answer thee in so far as I am able."

So Percival said, "I prithee tell me what is this thing?" And he laid his hand thereon. And Sir Ewaine said, "That is a saddle." And Percival said, "What is this thing?" And Sir Ewaine said, "That is a sword." And Percival said, "What is this thing?" And Sir Ewaine said, "That is a shield." And so Percival asked him concerning all things that appertained to the accoutrements of a knight, and Sir Ewaine answered all his questions. Then Percival said: "Now I will answer thy question. I saw a knight ride past this way yesterday, and he rode up yonder valley and to the westward."

Upon this Sir Ewaine gave gramercy to Percival and saluted him, and so did the other knights, and they rode their way.

After they had gone Percival returned to his mother, and he beheld that she sat exactly where he had left her, for she was in great travail of soul because she perceived that Percival would not now stay with her very much longer. And when Percival came to where she sat he said to her: "Mother, those were not angels, but very good, excellent knights." And upon this the lady, his mother, burst into a great passion of weeping, so that Percival stood before her all abashed, not knowing why she wept. So by and by he said, "Mother, why dost thou weep?" But she could not answer him for a while, and after a while she said, "Let us return homeward." And so they walked in silence.

Now when they had come to the tower where they dwelt, the lady turned of a sudden unto Percival and she said to him, "Percival, what is in thy heart?" And he said, "Mother, thou knowest very well what is there." She said, "Is it that thou wouldst be a knight also?" And he said, "Thou sayst it." And upon that she said, "Thou shalt have thy will; come with me."

So Percival's mother led him to the stable and to where was that poor pack-horse that brought provisions to that place, and she said: "This is a sorry horse but I have no other for thee. Now let us make a saddle for him." So Percival and his mother twisted sundry cloths and wisps of hay and made a sort of a saddle thereof. And Percival's mother brought him a scrip with bread and cheese for his refreshment and she hung it about his shoulder. And she brought him his javelin which he took in his hand. And then she gave him the ring of King Pellinore with that precious ruby jewel inset into it, and she said: "Take thou this, Percival, and put it upon thy finger, for it is a royal ring. Now when thou leavest me, go unto the court of King Arthur and make diligent inquiry for Sir Lamorack of Gales. And when thou hast found him, show him that ring, and he will see that thou art made a very worthy knight; for, Percival, Sir Lamorack is thy brother. One time thou hadst a father alive, and thou hadst two other brothers. But all they were slain by treachery of our enemies, and only thou and Lamorack are left; so look to it that thou guard thyself when thou art in the world and in the midst of those enemies; for if thou shouldst perish at their hands, I believe my heart would break."

Then she gave Percival advice concerning the duty of one who would make himself worthy of knighthood, and that advice was as follows:

*Percival's mother giveth him advice.* "In thy journeying thou art to observe these sundry things: When thou comest to a church or a shrine say a pater-noster unto the glory of God; and if thou hearest a cry of anyone in trouble, hasten to lend thine aid—especially if it be a woman or a child who hath need of it; and if thou meet a lady or a damosel, salute her in seemly fashion; and if thou have to do with a man, be both civil and courageous unto him; and if thou art an-hungered or athirst and findest food and wine, eat and drink enough to satisfy thee, but no more; and if thou findest a treasure or a jewel of price and canst obtain those things without injustice unto another, take that thing for thine own—but give that which thou hast with equal freedom unto others. So, by obeying these precepts, thou shalt become worthy to be a true knight and, haply, be also worthy of thy father, who was a true knight before thee."

And Percival said, "All these things will I remember and observe to do."

And Percival's mother said, "But thou wilt not forget me, Percival?" And he said: "Nay, mother; but when I have got me power and fame and wealth, then will I straightway return thitherward and take thee away from this place, and thou shalt be like to a Queen for all the glory that I shall bestow upon thee." Upon this the lady, his mother, both laughed *Percival de-* and wept; and Percival stooped and kissed her upon the lips. *parts from the* Then he turned and left her, and he rode away down the *mountain.* mountain and into the forest, and she stood and gazed after him as long as she could see him. And she was very lonely after he had gone.

So I have told you how it came that Percival went out into the world for to become a famous knight.

———

he Lady Yvette the Fair.

# Chapter First

*How Percival departed into the world and how he found a fair damsel in a pavilion; likewise how he came before Queen Guinevere and how he undertook his first adventure.*

NOW after Percival had ridden upon his way for a very long time, he came at last out of that part of the forest and unto a certain valley where were many osiers growing along beside a stream of water. So he gathered branches of the willow-trees and peeled them and wove them very cunningly into the likeness of armor *Percival maketh* such as he had seen those knights wear who had come into his *himself armor* forest. And when he had armed himself with wattled osiers *of willow twigs.* he said unto himself, "Now am I accoutred as well as they." Whereupon he rode upon his way with an heart enlarged with joy.

By and by he came out of the forest altogether and unto a considerable village where were many houses thatched with straw. And Percival said to himself: "Ha! how great is the world; I knew not that there were so many people in the world."

But when the folk of that place beheld what sort of a saddle was upon the back of the pack-horse; and when they beheld what sort of armor it was that Percival wore—all woven of osier twigs; and when *How Percival* they beheld how he was armed with a javelin and with no other *rode in the* weapon, they mocked and laughed at him and jeered him. *world.* But Percival understood not their mockery, whereupon he said: "Lo! how pleasant and how cheerful is the world. I knew not it was so merry a place." So he laughed and nodded and gave them greeting who mocked him in that manner. And some of them said, "That is a madman." And others said, "Nay, he is a silly fool." And when Percival heard these he said to himself: "I wonder whether there are other sorts of knights that I have not yet heard tell of?"

So he rode upon his way very happy, and whenever he met travellers, they would laugh at him; but he would laugh louder than they and give them greeting because of pure pleasure that the great world was so merry and kind.

Now in the declining of the afternoon, he came to a certain pleasant glade, and there he beheld a very noble and stately pavilion in among the trees. And that pavilion was all of yellow satin so that it shone like to gold in the light of the declining sun.

Then Percival said to himself: "Verily, this must be one of those churches concerning which my mother spake to me." So he descended from his horse and went to that pavilion and knelt down and said a pater-noster.

And when he had ended that prayer, he arose and went into the pavilion, and lo! he beheld there a wonderfully beautiful young damsel of sixteen *Percival enters the golden pavilion.* years of age who sat in the pavilion upon a carved bench and upon a cushion of cloth of gold, and who bent over a frame of embroidery, which she was busy weaving in threads of silver and gold. And the hair of that damosel was as black as ebony and her cheeks were like rose leaves for redness, and she wore a fillet of gold around her head, and she was clad in raiment of sky blue silk. And near by was a table spread with meats of divers sorts and likewise with several wines, both white and red. And all the goblets were of silver and all the pattens were of gold, and the table was spread with a napkin embroidered with threads of gold.

Now you are to know that the young lady who sat there was the Lady Yvette the Fair, the daughter of King Pecheur.

When Percival came to that pavilion the Lady Yvette looked up and beheld him with great astonishment, and she said to herself: "That must either be a madman or a foolish jester who comes hither clad all in armor of wattled willow twigs." So she said to him, "Sirrah, what dost thou here?" He said, "Lady, is this a church?" Upon that she was angered thinking that he had intended to make a jest and she said: "Begone, fool, for if my father, who is King Pecheur, cometh and findeth thee here, he will punish thee for this jest." But Percival replied, "Nay; I think he will not, lady."

Then the damosel looked at Percival more narrowly and she beheld how noble and beautiful was his countenance and she said to herself: "This is no fool nor a jester, but who he is or what he is I know not."

So she said to Percival, "Whence comest thou?" and he said, "From *Percival breaks bread in the golden pavilion.* the mountains and the wilderness." Then he said: "Lady, when I left my mother she told me that whenever I saw good food and drink and was an-hungered, I was to take what I needed. Now I will do so in this case." Whereupon he sat him down to that table and fell to with great appetite.

Then when that damosel beheld what he did she laughed in great measure and clapped her hands together in sport. And she said: "If my father and brothers should return and find thee at this, they would assuredly punish thee very sorely, and thou couldst not make thyself right with them." Percival said, "Why would they do that, lady?" And she said: "Because that is their food and drink, and because my father is a king and my brethren are his sons." Then Percival said, "Certes, they would be uncourteous to begrudge food to a hungry man"; and thereat the damsel laughed again.

Now when Percival had eaten and drunk his fill, he arose from where he sat. And he beheld that the damsel wore a very beautiful ring of carved gold set with a pearl of great price. So he said to her: "Lady, my mother told me that if I beheld a jewel or treasure and desired it for my own, I was to take it if I could do so without offence to anyone. Now I prithee give me that ring upon thy finger, for I desire it a very great deal." At this the maiden regarded Percival very strangely, and she beheld that he was comely beyond any man whom she had ever seen and that his countenance was very noble and exalted and yet exceedingly mild and gentle. So she said to him, speaking very gently, "Why should I give thee my ring?" Whereunto he made reply: "Because thou art the most beautiful lady whom mine eyes ever beheld and I find that I love thee more than I had thought possible to love anyone."

At that the damosel smiled upon him and said, "What is thy name?" And he said, "It is Percival." She said, "That is a good name; who is thy father?" Whereunto he said: "That I cannot tell thee for my mother hath bidden me tell his name to no one yet whiles." She said, "I think he must be some very noble and worthy knight," and Percival said, "He is all that, for he too was a king."

Then the damsel said, "Thou mayst have my ring," and she gave it to him. And when Percival had placed it upon his finger he said: "My mother also told me that I should give freely of what is mine own, *The damsel* wherefore I do give thee this ring of mine in exchange for thine, *giveth Percival* and I do beseech thee to wear it until I have proved myself *her ring.* worthy of thy kindness. For I hope to win a very famous knighthood and great praise and renown, all of which, if I so accomplish my desires, shall be to thy great glory. I would fain come to thee another time in that wise instead of as I am at this present."

At that the damsel said: "I know not what thou art or whence thou comest who should present thyself in such an extraordinary guise as thou art pleased to do, but, certes, thou must be of some very noble strain. Wherefore I do accept thee for my knight, and I believe that I shall some time have great glory through thee."

Then Percival said: "Lady, my mother said to me that if I met a damosel I was to salute her with all civility. Now have I thy leave to salute thee?"

*Percival salutes the damsel of the golden pavilion.* And she said, "Thou hast my leave." So Percival took her by the hand, and kissed her upon the lips (for that was the only manner in which he knew how to salute a woman) and, lo! her face grew all red like to fire. Thereupon Percival quitted that pavilion and mounted his horse and rode away. And it seemed to him that the world was assuredly a very beautiful and wonderful place for to live in.

Yet he knew not what the world was really like nor of what a sort it was nor how passing wide, else had he not been so certainly assured that he would win him credit therein, or that he could so easily find that young damsel again after he had thus parted from her.

That night Percival came to a part of the forest where were many huts of folk who made their living by gathering fagots. These people gave him harborage and shelter for the night, for they thought that he was some harmless madman who had wandered afar. And they told him many things he had never known before that time, so that it appeared to him that the world was still more wonderful than he had thought it to be at first.

So he abided there for the night, and when the next morning had come he arose and bathed himself and went his way; and, as he rode upon his poor starved horse, he brake his fast with the bread and cheese that his mother had put into his wallet, and he was very glad at heart and rejoiced exceedingly in the wonderfulness and the beauty of the world in which he found himself to be.

So Percival journeyed on into that forest, and he took such great delight in the beauty of the world in which he travelled that he was at times like

*How Percival travelled in the forest.* to shed tears of pure happiness because of the joy he felt in being alive. For that forest path he travelled led beneath the trees of the woodland; and the trees at that time were in their early tender leaf, so that they appeared to shed showers of golden light everywhere down upon the earth. And the birds of the woodland sang in every bush and thicket; and, anon, the wood pigeon cooed so softly that the heart of Percival yearned with great passion for he knew not what.

Thus he rode, somewhiles all in a maze of green, and somewhiles out thence into an open glade where the light was wide and bright; and other whiles he came to some forest stream where was a shallow pool of golden gravel, and where the water was so thin and clear that you might not tell where it ended and the pure air began. And therethrough he would drive his horse, splashing with great noise, whilst the little silvery fish would dart away upon all sides, hither and thither, like sparks of light before his coming.

So, because of the beauty of this forest land in its spring-time verdure and pleasantness, the heart of Percival was uplifted with so much joy and delight that he was like to weep for pure pleasure as aforesaid.

Now it chanced at that time that King Arthur and several of his court had come into that forest ahawking; but, the day being warm, the Queen had grown weary of the sport, so she had commanded her attendants to set up a pavilion for her whilst the King continued his hawking. And the pavilion was pitched in an open glade of the forest whereunto Percival came riding.

Then Percival perceived that pavilion set up among the trees, and likewise he saw that the pavilion was of rose colored silk. Also he perceived that not far from him was a young page very gayly and richly clad.

Now when the page beheld Percival and what a singular appearance he presented, he laughed beyond all measure, and Percival, not knowing that he laughed in mockery, laughed also and gave him a very cheerful greeting in return. Then Percival said to the page: "I prithee tell me, fair youth, whose is that pavilion yonder?" And the page said: "It belongeth to Queen Guinevere; for King Arthur is coming hither into the forest with his court." *Percival bespeaketh the Lady Guinevere's page.*

At this Percival was very glad, for he deemed that he should now find Sir Lamorack. So he said: "I pray thee tell me, is Sir Lamorack of Gales with the court of the King, for I come hither seeking that good worthy knight?"

Then the page laughed a very great deal, and said: "Who art thou to seek Sir Lamorack? Art thou then a jester?" And Percival said, "What sort of a thing is a jester?" And the page said, "Certes, thou art a silly fool." And Percival said, "What is a fool?"

Upon this the page fell alaughing as though he would never stint his mirth so that Percival began to wax angry for he said to himself: "These people laugh too much and their mirth maketh me weary." So, without more ado, he descended from his horse with intent to enter the Queen's pavilion and to make inquiry there for Sir Lamorack.

Now when that page saw what Percival had a mind to do, he thrust in to prevent him, saying, "Thou shalt not go in!" Upon that Percival said, "Ha! shall I not so?" And thereupon he smote the page such a buffet that the youth fell down without any motion, as though he had gone dead.

Then Percival straightway entered the Queen's pavilion.

And the first thing he saw was a very beautiful lady surrounded by a court of ladies. And the Queen was eating a mid-day repast whilst a page waited upon her for to serve her, bearing for her refreshment pure wine in a cup

*Percival beholdeth Queen Guinevere.* of entire gold. And he saw that a noble lord (and the lord was Sir Kay the Seneschal), stood in the midst of that beautiful rosy pavilion directing the Queen's repast; for Sir Kay of all the court had been left in charge of the Queen and her ladies.

Now when Percival entered the tent Sir Kay looked up, and when he perceived what sort of a figure was there, he frowned with great displeasure. "Ha!" he said, "what mad fool is this who cometh hitherward?"

Unto him Percival made reply: "Thou tall man, I prithee tell me, which of these ladies present here is the Queen?" Sir Kay said, "What wouldst thou have with the Queen?" To this Percival said: "I have come hither for to lay my case before King Arthur, and my case is this: I would fain obtain knighthood, and meseems that King Arthur may best help me thereunto."

When the Queen heard the words of Percival she laughed with great merriment. But Sir Kay was still very wroth, and he said: "Sirrah, thou

*Sir Kay chides Percival.* certainly art some silly fool who hath come hither dressed all in armor of willow twigs and without arms or equipment of any sort save only a little Scots spear. Now this is the Queen's court and thou art not fit to be here."

"Ha," said Percival, "it seems to me that thou art very foolish—thou tall man—to judge of me by my dress and equipment. For, even though I wear such poor apparel as this, yet I may easily be thy superior both in birth and station."

Then Sir Kay was exceedingly wroth and would have made a very bitter answer to Percival, but at that moment something of another sort befell.

*Sir Boinde-gardus enters the Queen's pavilion.* For, even as Percival ceased speaking, there suddenly entered the pavilion a certain very large and savage knight of an exceedingly terrible appearance; and his countenance was very furious with anger. And this knight was one Sir Boindegardus le Savage, who was held in terror by all that part of King Arthur's realm. For Sir Boindegardus was surnamed the Savage because he dwelt like a wild man in the forest in a lonely dismal castle of the woodland; and because that from this castle he would issue forth at times to rob and pillage the wayfarers who passed by along the forest byways. Many knights had gone against Sir Boindegardus, with intent either to slay him or else to make him prisoner; but some of these knights he had overcome, and from others he had escaped, so that he. was as yet free to work his evil will as he chose.

So now this savage knight entered that pavilion with his helmet upon his hip and his shield upon his shoulder, and all those ladies who were there were terrified at his coming, for they wist that he came in anger with intent of mischief.

As for Sir Kay (he being clad only in a silken tunic of green color and with scarlet hosen and velvet shoes, fit for the court of a lady) he was afraid, and he wist not how to bear himself in the presence of Sir Boindegardus. Then Sir Boindegardus said, "Where is King Arthur?" And Sir Kay made no reply because of fear. Then one of the Queen's damsels said, "He is hawking out beyond here in the outskirts of the forest." Then Sir Boindegardus said: "I am sorry for that, for I had thought to find him here at this time and to show challenge to him and his entire court, for I fear no one of them. But, as King Arthur is not here, I may, at least, affront his Queen."

With that he smote the elbow of the page who held the goblet for the Queen, and the wine was splashed all in the Queen's face and over her stomacher. *Sir Boindegardus affronts the Queen.*

Thereupon the Queen shrieked with terror, and one of her maidens ran to her aid and others came with napkins and wiped her face and her apparel and gave her words of cheer.

Then Sir Kay found courage to say: "Ha! thou art a churlish knight to so affront a lady."

With that Sir Boindegardus turned very fiercely upon him and said: "And thou likest not my behavior, thou mayst follow me hence into a meadow a little distance from this to the eastward where thou mayst avenge that affront upon my person if thou art minded to do so."

Then Sir Kay knew not what to reply for he wist that Sir Boindegardus was a very strong and terrible knight. Wherefore he said, "Thou seest that I am altogether without arms or armor." Upon that Sir Boindegardus laughed in great scorn, and therewith seized the golden goblet from the hands of the page and went out from the pavilion, and mounting his horse rode away bearing that precious chalice with him.

Then the Queen fell aweeping very sorely from fright and shame, and when young Percival beheld her tears, he could not abide the sight thereof. So he cried out aloud against Sir Kay, saying: "Thou tall man! that was very ill done of thee; for, certes, with or without *Percival berates Sir Kay.* armor thou shouldst have taken the quarrel of this lady upon thee. For my mother told me I should take upon me the defence of all such as needed defence, but she did not say that I was to wait for arms or armor to aid me to do what was right. Now, therefore, though I know little of arms or of knighthood, I will take this quarrel upon myself and will do what I may to avenge this lady's affront, if I have her leave to do so."

And Queen Guinevere said: "Thou hast my leave, since Sir Kay does not choose to assume my quarrel."

Now there was a certain very beautiful young damsel of the court of the Queen hight Yelande, surnamed the "Dumb Maiden," because she would *The damsel* hold no commerce with any knight of the court. For in all *praises Percival.* the year she had been at the court of the King, she had spoken no word to any man, nor had she smiled upon any. This damsel perceiving how comely and noble was the countenance of Percival, came to him and took him by the hand and smiled upon him very kindly. And she said to him: "Fair youth, thou hast a large and noble heart, and I feel very well assured that thou art of a sort altogether different from what thine appearance would lead one to suppose. Now I do affirm that if thou art able to carry this adventure through with thy life, thou wilt some time become one of the greatest knights in all of the world. For never did I hear tell of one who, without arm or armor, would take up a quarrel with a well-approved knight clad in full array. But indeed thy heart is as brave as thy face is comely, and I believe that thou art as noble as thy speech and manner is gentle."

Then Sir Kay was very angry with that damsel and he said: "Truly, thou art ill taught to remain for all this year in the court of King Arthur *Sir Kay strikes* amid the perfect flower of chivalry and yet not to have given *the damsel.* to one of those noble and honorable knights a single word or a smile such as thou hast bestowed upon this boor." So saying, he lifted his hand and smote that damsel a box on the ear so that she screamed out aloud with pain and terror.

Upon this Percival came very close to Sir Kay and he said: "Thou discourteous tall man; now I tell thee, except that there are so many ladies here present, and one of these a Queen, I would have to do with thee in such a manner as I do not believe would be at all to thy liking. Now, first of all I shall follow yonder uncivil knight and endeavor to avenge this noble Queen for the affront he hath put upon her, and when I have done with him, then will I hope for the time to come in which I shall have to do with thee for laying hands upon this beautiful young lady who was so kind to me just now. For, in the fulness of time, I will repay the foul blow thou gavest her, and that twenty-fold."

Thereupon Percival straightway went out from that pavilion and mounted upon his sorry horse and rode away in the direction that Sir Boindegardus had taken with the golden goblet.

Now after a long time, he came to another level meadow of grass, and there he beheld Sir Boindegardus riding before him in great state with the golden goblet hanging to the horn of his saddle. And Sir *Percival* Boindegardus wore his helmet and carried his spear in his *follows Sir* right hand and his shield upon his other arm, and he was in *Boindegardus.* all ways prepared for an encounter at arms. And when he perceived Percival come riding out of the forest in pursuit of him, he drew rein and turned. And when Percival had come nigh enough Sir Boindegardus said, "Whence comest thou, fool?" Percival replied, "I come from Queen Guinevere, her pavilion." Then Sir Boindegardus said, "Does that knight who was there follow me hitherward?" Unto which Percival made reply: "Nay, but I have followed thee with intent to punish thee for the affront which thou didst put upon Queen Guinevere."

Then Sir Boindegardus was very wroth and he said: "Thou fool; I have a very good intention for to slay thee." Therewith he raised his spear and smote Percival with it upon the back of the neck so terrible a blow that he was flung violently down from off his horse. Upon this Percival was so angry that the sky all became like scarlet before his eyes. Wherefore, when he had recovered from the blow he ran unto Sir Boindegardus and catched the spear in his hands and wrestled with such terrible strength that he plucked it away from Sir Boindegardus. And having thus made himself master of that spear, he brake it across his knee and flung it away.

Then Sir Boindegardus was in furious rage, wherefore he drew his bright, shining sword with intent to slay Percival. But when Percival saw what he would be at, he catched up his javelin and, running to a *Percival* little distance, he turned and threw it at Sir Boindegardus *slays Sir* with so cunning an aim that the point of the javelin entered *Boindegardus.* the ocularium of the helmet of Sir Boindegardus and pierced through the eye and the brain and came out of the back of the head. Then Sir Boindegardus pitched down from off his horse all into a heap upon the ground, and Percival ran to him and stooped over him and perceived that he was dead. Then Percival said: "Well, it would seem that I have put an end to a terribly discourteous knight to ladies."

Now a little after Percival had quitted the pavilion of Queen Guinevere, King Arthur and eleven noble knights of the court returned *King Arthur* thither from hawking, and amongst those knights was Sir *sends Sir* Launcelot of the Lake and Sir Lamorack of Gales. Then those *Launcelot and* *Sir Lamorack* who were of the Queen's court told King Arthur what had be- *in quest of* fallen, and thereat the King felt great displeasure toward Sir *Percival.* Kay. And he said: "Kay, not only hast thou been very discourteous in not assuming this quarrel of the Queen's, but I believe that thou, a well-approved knight, hast in thy fear of Sir Boindegardus been the cause of sending this youth upon an adventure in which he will be subject to such great danger that it may very well be that he shall hardly escape with his life.   Now I will that two of you knights shall follow after that youth for to rescue him if it be not too late; and those two shall be Sir Launcelot of the Lake and Sir Lamorack of Gales.   So make all haste, Messires, lest some misfortune shall befall this brave, innocent madman."

Thereupon those two knights mounted straightway upon their horses and rode away in that direction whither Percival had gone.

# Sir Percival & Sir Lamorack ride together

# Chapter Second

*How Sir Percival was made knight by King Arthur; how he rode forth with Sir Lamorack and how he left Sir Lamorack in quest of adventure upon his own account; likewise how a great knight taught him craft in arms.*

SO after a considerable time they came to that meadow-land where Percival had found Sir Boindegardus.

But when they came to that place they perceived a very strange sight. For they beheld one clad all in armor of wattled willow-twigs and that one dragged the body of an armed knight hither and thither upon the ground. So they two rode up to where that affair was toward, and when they had come nigh enough, Sir Launcelot said: "Ha, fair youth, thou art doing a very strange thing. What art thou about?" *How the two knights find Percival in the meadow.*

To him Percival said: "Sir, I would get those plates of armor off this knight, and I know not how to do it!"

Then Sir Launcelot laughed, and he said: "Let be for a little while, and I will show thee how to get the plates of armor off." And he said: "How came this knight by his death."

Percival said: "Sir, this knight hath greatly insulted Queen Guinevere (that beautiful lady), and when I followed him thither with intent to take her quarrel upon me, he struck me with his spear. And when I took his spear away from him, and brake it across my knee, he drew his sword and would have slain me, only that I slew him instead."

Then Sir Launcelot was filled with amazement, and he said: "Is not that knight Sir Boindegardus?" And Percival said: "Ay." Then Sir Launcelot said: "Fair youth, know that thou hast slain one of the strongest and most terrible knights in all the world. In this thou hast done a great service unto King Arthur, so if thou wilt come with us to the court of King Arthur, he will doubtless reward thee very bountifully for what thou hast done."

Then Percival looked up into the faces of Sir Launcelot and Sir Lamorack and he perceived that they were very noble. So he smiled upon them

and said: "Messires, I pray you tell me who you are and what is your degree."
Then Sir Launcelot smiled in return and said: "I am called Sir Launcelot
of the Lake, and this, my companion, is called Sir Lamorack of Gales."

Then Percival wist that he stood in the presence of his own brother,
and he looked into the countenance of Sir Lamorack and marvelled how
*Percival knoweth* noble and exalted it was. And he felt a great passion of love
*Sir Lamorack.* for Sir Lamorack, and a great joy in that love. But he did not
tell Sir Lamorack who he was, for he had learned several things since he had
come out into the world, and one was that he must not be too hasty in such
things. So he said to himself: "I will not as yet tell my brother who I am,
lest he shall be ashamed of me. But first I shall win me such credit that
he shall not be ashamed of me, and then I will acknowledge to him who
I am."

Then Sir Launcelot said: "I prithee, fair youth, tell me what is thy name
since I have told thee ours, for I find that I have great love for thee so that
I would fain know who thou art."

Then Percival said: "My name is Percival."

At that Sir Lamorack cried out: "I knew one whose name was Percival,
and he was mine own brother. And if he be alive he must now be just such
a youth as thou art."

Then Percival's heart yearned toward Sir Lamorack, so that he looked
up and smiled with great love into his face; yet he would not acknowledge
to Sir Lamorack who he was, but held his peace for that while.

Then Sir Launcelot said: "Now, fair youth, we will show you how to take
the armor off of this dead knight, and after we have done that, we shall
take you back to King Arthur, so that he may reward you for what you
have done in the way that he may deem best."

So with that Sir Launcelot and Sir Lamorack dismounted from their
horses, and they went to that dead knight and unlaced his armor and re-
*The two knights* moved the armor from his body. And when they had done
*arm Percival.* that they aided Percival to remove the armor of wattled osier
twigs and they cased him in the armor of Sir Boindegardus; and thereafter
they all three rode back to that pavilion where the King and Queen were
holding court.

But when King Arthur heard that Sir Boindegardus was dead he was
filled with great joy; and when he heard how it was that Percival had slain
him, he was amazed beyond measure; and he said to Percival: "Surely
God is with thee, fair youth, to help thee to perform such a worthy feat
of arms as this that thou hast done, for no knight yet hath been able to
perform that service." Then he said: "Tell me what it is that thou hast

most desire to have, and if it is in my power to give it to thee thou shalt have it."

Then Percival kneeled down before King Arthur, and he said: "Lord, that which I most desire of all things else is to be made knight. So if it is in thy power to do so, I pray thee to make me a knight-royal with thine own hands."

Then King Arthur smiled upon Percival very kindly, and he said: "Percival, it shall be as thou dost desire, and to-morrow I will make thee a knight."

So that night Percival watched his armor in the chapel of a hermit of the forest, and the armor that he watched was the armor that had belonged to Sir Boindegardus (for Percival besought King Arthur that *King Arthur* he might wear that armor for his own because it was what he *makes Percival* himself had won in battle). And when the next morning had *a knight-royal.* come, Sir Launcelot and Sir Lamorack brought Percival before King Arthur, and King Arthur made him a knight.

After that Sir Percival besought King Arthur that he would give him leave to depart from court so that he might do some worthy deed of arms that might win him worship; and King Arthur gave him that leave he asked for.

Then Sir Percival went to where Sir Kay was sitting, and he said: "Messire, I have not forgot that blow you gave that fair damsel yesterday when she spake so kindly to me. As yet I am too young a knight *Sir Percival* to handle you; but by and by the time will come when I shall *threatens Sir* return and repay you that blow tenfold and twentyfold what *Kay.* you gave!" And at these words Sir Kay was in no wise pleased, for he wist that Sir Percival would one day become a very strong and worthy knight.

Now all this while the heart of Sir Lamorack yearned very greatly toward Sir Percival, though Sir Lamorack knew not why that should be; so when Sir Percival had obtained permission to go errant, Sir Lamorack asked King Arthur for leave to ride forth so as to be with him; and King Arthur gave Sir Lamorack that leave.

Thus it befell that Sir Lamorack and Sir Percival rode forth together very lovingly and cheerfully. And as they rode upon their way Sir Lamorack told Sir Percival many things concerning the circum- *Sir Percival and* stances of knighthood, and to all that he said Sir Percival *Sir Lamorack* gave great heed. But Sir Lamorack knew not that he was *ride together.* riding with his own brother or that it was his own brother to whom he was teaching the mysteries of chivalry, and Sir Percival told him nothing thereof. But ever in his heart Sir Percival said to himself: "If God will give me enough of His grace, I will some day do full credit unto thy teaching, O my brother!"

Now, after Sir Percival and Sir Lamorack had travelled a great way, they came at last out of that forest and to an open country where was a well-tilled land and a wide, smooth river flowing down a level plain.

And in the centre of that plain was a town of considerable size, and a very large castle with several tall towers and many roofs and chimneys that stood overlooking the town.

That time they came thitherward the day was declining toward its close, so that all the sky toward the westward shone, like, as it were, to a flame of gold—exceedingly beautiful. And the highway upon which they entered was very broad and smooth, like to a floor for smoothness. And there were all sorts of folk passing along that highway; some afoot and some ahorseback. Also there was a river path beside the river where the horses dragged deep-laden barges down to the town and thence again; and these barges were all painted in bright colors, and the horses were bedight with gay harness and hung with tinkling bells.

All these things Sir Percival beheld with wonder for he had never seen their like before; wherefore he cried out with amazement, saying: "Saints of Glory! How great and wonderful is the world!"

Then Sir Lamorack looked upon him and smiled with great loving-kindness; and he said: "Ha, Percival! This is so small a part of the world that it is but a patch upon it."

Unto this Sir Percival made reply: "Dear Messire, I am so glad that I have come forth into the world that I am hardly able to know whether I am in a vision or am awake."

So, after a considerable while, they came to that town with its castle, and these stood close beside the river—and the town and the castle were hight Cardennan. And the town was of great consideration, being very well famed for its dyed woollen fabrics.

So Sir Percival and Sir Lamorack entered the town. And when Sir Percival beheld all the people in the streets, coming and going upon their *Sir Percival and* businesses; and when he beheld all the gay colors and apparels *Sir Lamorack* of fine fabrics that the people wore; and when he beheld the *come to* many booths filled with rich wares of divers sorts, he wist not *Cardennan.* what to think for the wonder that possessed him; wherefore he cried out aloud, as with great passion: "What marvel do I behold! I knew not that a city could be so great as this."

And again Sir Lamorack smiled very kindly upon him and said: "Sayst thou so? Now I tell thee that when one compares this place with Camelot (which is the King's city) it is as a star compared to the full moon in her glory." And at that Sir Percival knew not what to think for wonder.

So they went up the street of the town until they came to the castle of Cardennan and there requested admission. And when the name and the estate of Sir Lamorack were declared, the porter opened the gate with great joy and they entered. Then, by and by, the lord and the lady of the castle came down from a carved wooden gallery and bade them welcome by word of mouth. And after that sundry attendants immediately appeared and assisted Sir Percival and Sir Lamorack to dismount and took their horses to the stable, and sundry other attendants conducted them to certain apartments where they were eased of their armor and bathed in baths of tepid water and given soft raiment for to wear. After that the lord and the lady entertained them with a great feast, where harpers and singers made music, and where certain actors acted a mystery before them.

So these two knights and the lord and the lady of the castle ate together and discoursed very pleasantly for a while; but, when the evening was pretty well gone, Sir Lamorack bade good-night, and he and Sir Percival were conducted to a certain very noble apartment where beds of down, spread with flame-colored cloth, had been prepared for their repose. *How the two knights were welcomed by the lord and lady of the castle.*

Thus ended that day which was the first day of the knighthood of Sir Percival of Gales.

Now though Sir Percival had travelled very contentedly with Sir Lamorack for all that while, yet he had determined in his own mind that, as soon as possible, he would leave Sir Lamorack and depart upon his own quest. For he said to himself: "Lo! I am a very green knight as yet, and haply my brother may grow weary of my company and cease to love me. So I will leave him ere he have the chance to tire of me, and I will seek knighthood for myself. After that, if God wills it that I shall win worthy knighthood, then my brother will be glad enough to acknowledge me as his father's son."

So when the next morning had come, Sir Percival arose very softly all in the dawning, and he put on his armor without disturbing Sir Lamorack. Then he stooped and looked into Sir Lamorack's face and beheld that his brother was still enfolded in a deep sleep as in a soft mantle. And as Sir Percival gazed upon Sir Lamorack thus asleep, he loved him with such ardor that he could hardly bear the strength of his love. But he said to himself: "Sleep on, my brother, whilst I go away and leave thee. But when I have earned me great glory, then will I return unto thee and will lay all that I have achieved at thy feet, so that thou shalt be very glad to acknowledge me." So saying to himself, he went away from that place very softly, and Sir Lamorack slept so deeply that he wist not that Sir Percival was gone.

Thereafter Sir Percival went to the courtyard of the castle and he bade

*Sir Percival leaves Sir Lamorack.*

certain attendants to prepare his horse for him, and they did so. And he bade certain others for to arm him, and they did so. Thereupon he mounted his horse and left that castle and rode away.

Now after Sir Percival had left Sir Lamorack still sleeping in the castle as aforetold, he journeyed upon his way, taking great pleasure in all things that he beheld. So he travelled all that morning, and the day was very bright and warm, so that by and by he was an-hungered and athirst. So after a while he came to a certain road that appeared to him to be good for his purpose, so he took that way in great hopes that some adventure would befall him, or else that he would find food and drink.

Then after a while he heard a bell ringing, and after he had followed that bell for some distance, he came to where was the dwelling-place of a hermit and where was a small chapel by the wayside. And Sir Percival beheld that the hermit, who was an old man with a long white beard, rang the bell of that chapel.

So Sir Percival thought that here he might find food and drink; and so he rode forward to where the hermit was ringing the bell. But when Sir

*Sir Percival meets his fate at the forest chapel.*

Percival came still more nigh he perceived that behind the chapel and to one side there was a very noble knight upon horseback; and he perceived that the knight was clad all in white armor and that his horse (which was white as milk and of very noble strength and proportions) was furnished altogether with furniture of white.

This knight, when he perceived Sir Percival, immediately rode up to meet him and saluted Sir Percival very courteously. And the knight said: "Sir, will you not joust a fall with me ere you break your fast? For this is a very fair and level field of green grass and well fitted for such a friendly trial at arms if you have the time for it."

Unto this Sir Percival said: "Messire, I will gladly try a fall with you, though I must tell you that I am a very young green knight, having been knighted only yesterday by King Arthur himself. But though I am un-skilled in arms, yet it will pleasure me a great deal to accept so gentle and courteous a challenge as that which you give me."

So with that each knight turned his horse and each took such stand as appeared to him to be best. And when they were in all ways prepared, they drave their horses together with great speed, the one against *Sir Percival is* the other, meeting one another, shield against spear, in the very *overthrown by* midst of the course. In that encounter (which was the first *the white knight.* that he ever ran) Sir Percival bare himself very well and with great knightliness of endeavor; for he broke his spear upon the white knight into small pieces. But the spear of the white knight held so that Sir Percival was lifted out of his saddle and over the crupper of his horse, and fell upon the ground with great violence and a cloud of dust.

Then the white knight returned from his course and came up to where Sir Percival was. And he inquired of him very courteously: "Sir, art thou hurt?" Thereunto Sir Percival replied: "Nay, sir! I am not hurt, only somewhat shaken by my fall."

Then the white knight dismounted from his horse and came to where Sir Percival was. And he lifted up the umbril of his helmet, and Sir Percival perceived that that white knight was Sir Launcelot of the Lake.

And Sir Launcelot said: "Percival, I well knew who you were from the first, but I thought I would see of what mettle you are, and I have found that you are of very good mettle indeed. But you are to know that it is impossible for a young knight such as you, who knoweth naught of the use of knightly weapons, to have to do with a knight well-seasoned in arms as I am, and to have any hope of success in such an encounter. Wherefore you need to be taught the craft of using your weapons perfectly."

To this Sir Percival said: "Messire, tell me, how may I hope to acquire craft at arms such as may serve me in such a stead as this?"

Sir Launcelot said: "I myself will teach thee, imparting to thee such skill as I have at my command. Less than half a day's journey to the southward of this is my castle of Joyous Gard. Thither I was upon my way when I met thee here. Now thou shalt go with me unto Joyous Gard, and there thou shalt abide until thou art in all ways taught the use of arms so that thou mayst uphold that knighthood which I believe God hath endowed thee withal."

So after that Sir Launcelot and Sir Percival went to the dwelling-place of the hermit, and the hermit fed them with the best of that simple fare which he had at his command.

After that, they mounted horse again and rode away to Joyous Gard, and there Sir Percival abided for a year, training himself in *How Sir* all wise so as to prepare himself to uphold that knighthood *Percival dwelt at* which in him became so famous. For, during that year Sir *Joyous Gard.* Launcelot was his teacher in the art of arms. Likewise he instructed him in all the civilities and the customs of chivalry, so it befell that ere Sir Percival came forth from Joyous Gard again he was well acquainted with all the ways in which he should comport himself at any time, whether in field or in court.

So when Sir Percival came forth again from Joyous Gard, there was no knight, unless it was Sir Launcelot himself, who could surpass him in skill at arms; nay, not even his own brother, Sir Lamorack; nor was there anybody, even if one were Sir Gawaine or Sir Geraint, who surpassed him in civility of courtliness or nobility of demeanor.

And now I shall tell you of the great adventure that befell Sir Percival after Sir Launcelot had thus taught him at Joyous Gard.

ir Percival overcometh ye
Enchantress Vivien.

# Chapter Third

*How Sir Percival met two strange people in the forest, and how he
succored a knight who was in very great sorrow and dole.*

NOW after Sir Percival had left Joyous Gard he rode for several
days seeking adventure but meeting none.

Then one day he came to a very dark and wonderful forest
which appeared to be so silent and lonely and yet so full of beauty that
Sir Percival bethought him that this must surely be some forest of magic.
So he entered into that forest with intent to discover if he might find any
worthy adventure therein.

(And that forest was a forest of magic; for you are to know that it
was the Forest of Arroy, sometimes called the Forest of Ad- *Sir Percival en-*
venture, which was several times spoken of in the book of *ters the Forest*
King Arthur. For no one ever entered into that forest but *of Arroy.*
some singular adventure befell him.)

So Sir Percival rode through this wonderful woodland for a long time
very greatly wondering, for everywhere about him was perfect silence, with
not so much as a single note of a bird of the woodlands to lighten that stillness. Now, as Sir Percival rode through that silence, he presently became
aware of the sound of voices talking together, and shortly thereafter he perceived a knight with a lady riding amid the thin trees that grew there.
And the knight rode upon a great white horse, and the lady rode upon a
red roan palfrey.

These, when they beheld Sir Percival, waited for him, and as Sir Percival
drew nigh to them he perceived that they were of a very singular appearance. For both of them were clad altogether in green, and *Sir Percival*
both of them wore about their necks very wonderful collars of *meets two*
wrought gold inset with opal stones and emeralds. And the *strange people.*
face of each was like clear wax for whiteness; and the eyes of each were very
bright, like jewels set in ivory. And these two neither laughed nor frowned,
but only smiled continually. And that knight whom Sir Percival beheld
was Sir Pellias, and the lady was the Lady Nymue of the Lake.

Now when Sir Percival beheld these two, he wist that they were fay,
wherefore he dismounted very quickly, and kneeled down upon the ground

and set his palms together.   Then the Lady of the Lake smiled very kindly upon Sir Percival, and she said: "Sir Percival, arise, and tell me what you do in these parts?"

Then Sir Percival arose and he stood before that knight and lady, and he said: "Lady, I wist not how you know who I am, but I believe you are fay and know many things.   Touching my purpose in coming here, it is that I am in search of adventure.   So if you know of any that I may undertake for your sake, I pray you to tell me of it."

The lady said: "If so be thy desire is of that sort, I may, perchance be able to bring thee unto an adventure that is worthy for any knight to undertake.   Go a little distance from this upon the way thou art following and by and by thou wilt behold a bird whose feathers shall shine like to gold for brightness.   Follow that bird and it will bring thee to a place where thou shalt find a knight in sore need of thy aid."

And Percival said: "I will do as thou dost advise."

Then the lady said: "Wait a little, I have something for thee."   Therewith she took from her neck a small golden amulet pendant from a silken *The Lady of the* cord very fine and thin.   And she said: "Wear this for it will *Lake giveth Sir* protect thee from all evil enchantments."   Therewith saying, *Percival a charm.* she hung the amulet about the neck of Sir Percival, and Sir Percival gave her thanks beyond measure for it.

Then the knight and the lady saluted him and he saluted them, and they each went their separate ways.

So Sir Percival travelled that path for some distance as the lady had advised him to do, and by and by he beheld the bird of which she had spoken. *How Percival* And he saw that the plumage of the bird glistered as though *followed the* it was of gold so that he marvelled at it.   And as he drew *golden bird.* nigh the bird flew a little distance down the path and then lit upon the ground and he followed it.   And when he had come nigh to it again it flew a distance farther and still he followed it.   So it flew and he followed for a very great way until by and by the forest grew thin and Sir Percival beheld that there was an open country lying beyond the skirts thereof.   And when the bird had brought him thus far it suddenly flew back into the forest again whence it had come, chirping very keenly and shrilly as it flew.

So Percival came out of the forest into the open country, the like of *Sir Percival* which he had never before seen, for it was a very desolate *beholds a* barren waste of land.   And in the midst of this desolate plain *wonderful castle.* there stood a castle of a very wonderful appearance; for in some parts it was the color of ultramarine and in other parts it was of crimson; and the ultramarine and the crimson were embellished with very extraordinary devices painted in gold. So because of all those extraordinary colors, that castle shone like a bright rainbow against the sky, wherefore Sir Percival sat his horse for some while and marvelled very greatly thereat.

Then, by and by Sir Percival perceived that the road that led to the castle crossed a bridge of stone, and when he looked at the bridge he saw that midway upon it was a pillar of stone and that a knight clad all in full armor stood chained with iron chains to that stone pillar, and at that sight Sir Percival was very greatly astonished. So he rode very rapidly along that way and so to the bridge and upon the bridge to where the knight was. And when Sir Percival came thus upon the bridge he perceived that the knight who was bound with chains was very noble and haughty of appearance, but that he seemed to be in great pain and suffering because of his being thus bound to that pillar. For the captive knight made continual moan so that it moved the heart of Sir Percival to hear him.

So Sir Percival said: "Sir Knight, this is a sorrowful condition thou art in." And the knight said: "Yea, and I am sorrowful; for I have stood here now for three days and I am in great torment of mind and body."

Sir Percival said, "Maybe I can aid thee," and thereupon he got down from off his horse's back and approached the knight. And he drew his sword so that it flashed in the sun very brightly.

Upon this the knight said: "Messire, what would you be at?" And Sir Percival said: "I would cut the chains that bind thee."

To this the knight said: "How could you do that? For who could cut through chains of iron such as these?"

But Sir Percival said: "I will try what I may do."

Thereupon he lifted up his sword and smote so terribly powerful a blow that the like of it had hardly ever been seen before. For that blow cut through the iron chains and smote the hauberk of the knight so smart a buffet that he fell down to the ground altogether deprived of breath.

*Sir Percival sets free the captive knight.*

But when Sir Percival saw the knight fall down in that wise, he cried out: "Woe is me! Have I slain this good, gentle knight when I would but do him service?" Thereupon he lifted the knight up upon his knee and eased the armor about his throat. But the knight was not dead, and by and by the breath came back to him again, and he said: "By my faith, that was the most wonderful stroke that ever I beheld any man strike in all of my life."

Thereafter, when the knight had sufficiently recovered, Sir Percival helped him to stand upon his feet; and when he stood thus his strength presently came back to him again in great measure.

And the knight was athirst and craved very vehemently to drink. So Sir Percival helped him to descend a narrow path that led to a stream of water that flowed beneath the bridge; and there the knight stooped and slaked his thirst. And when he had drunk his fill, his strength came altogether back to him again, and he said: "Messire, I have to give thee all thanks that it is possible for me to do, for hadst thou not come unto mine aid, I would else have perished very miserably and at no very distant time from this."

Then Sir Percival said: "I beseech you, Messire, to tell me how you came into that sad plight in which I found you."

To this the knight said: "I will tell you; it was thus: Two days ago I came thitherward and past yonder castle, and with me were two excellent *The knight* esquires—for I am a knight of royal blood. Now as we went *telleth his story.* past that castle there came forth a lady clad all in red and so exceedingly beautiful that she entirely enchanted my heart. And with this lady there came a number of esquires and pages, all of them very beautiful of face, and all clad, as she was, in red. Now when this lady had come nigh to me she spoke me very fair and tempted me with kind words so that I thought I had never fallen upon anyone so courteous as she. But when she had come real close to me, she smote me of a sudden across the shoulders with an ebony staff that she carried in her hand, and at the same time she cried out certain words that I remember not. For immediately a great darkness like to a deep swoon fell upon me and I knew nothing. And when I awakened from that swoon lo! I found myself here, chained fast to this stone pillar. And hadst thou not come hither I would else certainly have died in my torment. And as to what hath become of my esquires, I know not; but as for that lady, methinks she can be none other than a certain enchantress, hight Vivien, who hath wrought such powerful spells upon Merlin as to have removed him from the eyes of all mankind."

Unto all this Sir Percival listened in great wonder, and when the knight had ended his tale he said: "What is thy name?" And the knight said: "My name is Percydes and I am the son of King Pecheur—so called because he is the king of all the fisher-folk who dwell upon the West coast. And now I prithee tell me also thy name and condition, for I find I love thee a very great deal."

And Sir Percival said: "My name is Percival, but I may not at this present tell thee my condition and of whom I am born; for that I must keep secret until I have won me good credit as a knight. But now I have somewhat to do, and that is to deal with this lady Vivien as she shall deserve."

Upon that Sir Percydes cried out: "Go not near to that sorceress, else she will do some great harm to thee with her potent spells as she did to me."

But Sir Percival said: "I have no fear of her."

So Sir Percival arose and crossed the bridge and went toward that wonderful enchanted castle; and Sir Percydes would have gone with him, but Sir Percival said: "Stay where thou art." And so Sir Percydes stayed and Sir Percival went forward alone.

Now as he drew nigh to the castle the gate thereof was opened, and there came forth thence an extraordinarily beautiful lady surrounded by a court of esquires and pages all very beautiful of countenance. And this lady and all of her court were clad in red so that they shone like to several flames of fire. And the lady's hair was as red as gold, and she wore gold ornaments about her neck *The Lady Vivien cometh forth to Sir Percival.* so that she glistered exceedingly and was very wonderful to behold. And her eyebrows were very black and fine and were joined in the middle like two fine lines drawn together with a pencil, and her eyes were narrow and black, shining like those of a snake.

Then when Sir Percival beheld this lady how singularly beautiful she was he was altogether enchanted so that he could not forbear to approach her. And, lo! she stood still and smiled upon him so that his heart stirred within his bosom like as though it pulled at the strings that held it. Then she said to Sir Percival, speaking in a very sweet and gentle voice: "Sir Knight, thou art welcome to this place. It would pleasure us very greatly if thou wouldst consider this castle as though it were thine own and would abide within it with me for a while." Therewith speaking she smiled again upon Sir Percival more cunningly than before and reached out her hand toward him.

Then Sir Percival came toward her with intent to take her hand, she smiling upon him all the while so that he could not do otherwise than as she willed.

Now in the other hand this lady held an ebony staff of about an ell in length, and when Sir Percival had come close enough to her, she lifted this staff of a sudden and smote him with it very violently across the shoulders, crying out at the same time, in a voice terribly piercing and shrill: "Be thou a stone!"

Then that charm that the Lady of the Lake had hung around the neck of Sir Percival stood him in good stead, for, excepting for it, he would that instant have been transformed into a stone. But the charm of the sorceress did not work upon him, being prevented by the greater charm of that golden amulet.

But Sir Percival knew very well what the sorceress Vivien had intended to do to him, and he was filled with a great rage of indignation against her because she had meant to transform him into a stone.   There-

*Sir Percival draweth sword upon the Lady Vivien.*

fore he cried out with a loud voice and seized the enchantress by her long golden hair, and drew her so violently forward that she fell down upon her knees.   Then he drew his shining sword with intent to sever her long neck, so slender and white like alabaster. But the lady shrieked with great vehemence of terror and besought him mercy.   And at that Sir Percival's heart grew soft for pity, for he bethought him that she was a woman and he beheld how smooth and beautiful was her neck, and how her skin was like white satin for smoothness.   So when he heard her voice—the voice of a woman beseeching mercy—his heart grew soft, and he could not find strength within him to strike that neck apart with his sword.

So he bade her to arise—though he still held her by the hair (all warm, it was, and as soft as silk and very fragrant) and the lady stood up, trembling before him.

Then Sir Percival said to her: "If thou wouldst have thy life I command thee to transform back to their own shape all those people whom thou hast bewitched as thou wouldst have bewitched me."

Then the lady said: "It shall be done."   Whereupon she smote her hands very violently together crying out: "All ye who have lost your proper shapes, return thereunto."

Then, lo! upon the instant, a great multitude of round stones that lay

*The Lady Vivien undoes her enchantment.*

scattered about became quick, like to eggs; and they moved and stirred as the life entered into them.   And they melted away and, behold! there arose up a great many knights and esquires and several ladies to the number of four score and eight in all.

And certain other stones became quickened in like manner, and as Percival looked, lo! there rose up the horses of those people, all caparisoned as though for travel.

Now when those people who had been thus bewitched beheld the Lady Vivien, how Sir Percival held her by the hair of her head, they made great outcry against her for vengeance, saying: "Slay her!   Slay her!"   And therewith several made at her as though to do as they said and to slay her. But Percival waved his sword before her and said: "Not so!   Not so!   For this lady is my prisoner and ye shall not harm her unless ye come at her through me."

Thereat they fell silent in a little while, and when he had thus stilled them, he turned to the Lady Vivien and said: "This is my command that I lay upon thee: that thou shalt go into the court of King Arthur and shalt confess thyself to him and that thou shalt fulfil whatever penance he may lay upon thee to perform because of thy transgressions. Now wilt thou do this for to save thy life?"

And the Lady Vivien made reply: "All shall be done according to thy command."

Therewith Sir Percival released his hold upon her and she was free.

Then, finding herself to be thus free, she stepped back a pace or two and looked into Sir Percival his face, and she laughed. And she said: "Thou fool, didst thou think that I would do so mad a thing as that which thou hast made me promise? For what mercy could I expect at the hands of King Arthur seeing that it was I who destroyed the Enchanter Merlin, who was the right adviser of King Arthur! Go to King Arthur thyself and deliver to him thine own messages."

So saying, in an instant, she vanished from the sight of all those who stood there. And with her vanished that castle of crimson and ultra- *The Lady* marine and gold—and nothing was left but the bare rocks and *Vivien escapes.* the barren plain.

Then when those who were there recovered from their astonishment, upon beholding that great castle so suddenly disappear, they turned to Sir Percival and gave him worship and thanks without measure, saying to him: "What shall we do in return for saving us from the enchantment of this sorceress?"

And Percival said: "Ye shall do this: ye shall go to the court of King Arthur and tell him how that young knight, Percival, whom he made a knight a year ago, hath liberated you from the enchantment of this sorceress. And you shall seek out Sir Kay and shall say to him that, by and by, I shall return and repay him in full measure, twenty times over, that blow which he gave to the damosel Yelande, the Dumb Maiden because of her kindness to me."

So said Sir Percival, and they said: "It shall be done as thou dost ordain."

Then Sir Percydes said: "Wilt thou not come to my castle and rest thyself there for the night? For thou must be aweary with all thy toil." And Sir Percival said, "I will go with thee." So Sir Percydes and *Sir Percydes* Sir Percival rode away together to the castle of Sir Percydes. *knoweth the*

Now while Sir Percival and Sir Percydes sat at supper in *ring that Perci-* the castle of Sir Percydes, Sir Percival chanced to lay his hand *val wears.* in love upon the sleeve of Sir Percydes's arm, and that moment Sir Percydes saw the ring upon Sir Percival's finger which the young damosel of the pavilion had given unto him in exchange for his ring. When Sir Percydes saw that ring he cried out in great astonishment, "Where didst thou get that ring?"

Sir Percival said, "I will tell thee"; and therewith he told Sir Percydes all that had befallen him when he first came down into the world from the wilderness where he had aforetime dwelt, and how he had entered the yellow pavilion and had discovered the damosel who was now his chosen lady. When Sir Percydes heard that story he laughed in great measure, and then he said: "But how wilt thou find that young damosel again when thou hast a mind for to go to her once more?" To the which Sir Percival made reply: "I know not how I shall find her, nevertheless, I shall assuredly do so. For though the world is much wider and greater than I had thought it to be when I first came down into it, yet I know that I shall find that lady when the fit time cometh for me to seek her."

Then Sir Percydes said: "Dear friend, when thou desireth to find that damosel to whom belongeth the ring, come thou to me and I will tell thee where thou mayst find her; yet I know not why thou dost not go and find her now."

Unto this Sir Percival made reply: "I do not seek her immediately because I am yet so young and so unknown to the world that I could not be of any credit to her should I find her; so first I will seek to obtain credit as a knight, and then I will seek her."

Sir Percydes said: "Well, Percival, I think thou hast great promise of a very wonderful knighthood. Nor do I think thou wilt have difficulty in finding plenty of adventures to undertake. For even to-day I know of an adventure, which if thou couldst perform it successfully, would bring thee such worship that there are very few knights in all the world who will have more worship than thou."

Then Sir Percival said: "I prithee, dear friend, tell me what is that adventure."

Then Sir Percydes told Sir Percival what that adventure was as followeth:

"Thou art to know," quoth he, "that somewhat more than a day's journey to the north of this there is a fair plain, very fertile and beautiful *Sir Percydes tell-* to the sight. In the midst of that plain is a small lake of water, *eth Sir Percival* and in that lake is an island, and upon the island is a tall *of Beaurepaire.* castle of very noble size and proportions. That castle is called Beaurepaire, and the lady of that castle is thought to be one of the most beautiful damosels in the world. And the name of the lady is Lady Blanchefleur.

"Now there is a very strong and powerful knight hight Sir Clamadius, otherwise known as the King of the Isles; and he is one of the most famous knights in the world. Sir Clamadius hath for a long while loved the Lady Blanchefleur with such a passion of love that I do not think that the like of that passion is to be found anywhere else in the world. But the Lady Blanchefleur hath no love for Sir Clamadius, but ever turneth away from him with a heart altogether cold of liking.

"But Sir Clamadius is a wonderfully proud and haughty King, wherefore he can ill brook being scorned by any lady. Wherefore he hath now come against the castle of Beaurepaire with an array of knights of his court, and at present layeth siege to that castle aforesaid.

"Now there is not at that castle any knight of sufficient worship to serve as champion thereof, wherefore all they of Beaurepaire stay within the castle walls and Sir Clamadius holds the meadows outside of the castle so that no one enters in or goeth out thereof.

"If thou couldst liberate the Lady Blanchefleur from the duress which Sir Clamadius places upon her, I believe thou wouldst have as great credit in courts of chivalry as it is possible to have. For, since Sir Tristram is gone, Sir Clamadius is believed by many to be the best knight in the world except Sir Launcelot of the Lake; unless it be that Sir Lamorack of Gales is a better knight than he."

Then Sir Percival said: "What thou tellest me gives me great pleasure, for it would be a very good adventure for any young knight to undertake. For if he should lose there would be no shame in losing, and if he should win there would be great glory in winning. So to-morrow I will enter upon that adventure, with intent to discover what fortune I may have therein."

So I have told you how Sir Percival performed his first adventures in the world of chivalry after he had perfected himself in the mysteries of knighthood under the teaching of Sir Launcelot of the Lake, and I have told you how he achieved that adventure with great credit to himself and with great glory to the order of knighthood to which he now truly belonged as a most worthy member.

That night he abided in the castle of Sir Percydes with great comfort and rest to his body, and when the next morning had come he arose, much refreshed and strengthened in spirit. And he descended to the hall where was set a fair and generous breakfast for his further refreshment, and thereat he and Sir Percydes sat themselves down and ate with hearty appetite, discoursing with great amity of spirit as aforetold.

After he had broken his fast he bade farewell to Sir Percydes and mounted his horse and rode away through the bright sunlight toward Beaurepaire and those further adventures that awaited him thereat.

And, as it was with Sir Percival in that first adventure, so may you meet with a like success when you ride forth upon your first undertakings after you have entered into the glory of your knighthood, with your life lying before you and a whole world whereinto ye may freely enter to do your devoirs to the glory of God and your own honor.

So now it shall be told how it fared with Sir Percival in that adventure of the Castle of Beaurepaire.

# The Demoiselle Blanchefleur

# Chapter Fourth

*How Sir Percival undertook the adventure of the castle of Beaure-
paire and how he fared therein after several excellent adventures.*

NOW the way that Sir Percival travelled led him by the outskirts
of the forest, so that somewhiles he would be in the woodland
and somewhiles he would be in the open country. And about
noontide he came to a certain cottage of a neatherd that stood all alone in
a very pleasant dale. That place a little brook came bickering
out from the forest and ran down into the dale and spread out *Sir Percival breaks his*
into a small lake, besides which daffadowndillys bloomed in *fast at a forest*
such abundance that it appeared as though all that meadow *cottage.*
land was scattered over with an incredible number of yellow stars that
had fallen down from out of the sky. And, because of the pleasantness of
this place, Sir Percival here dismounted from his horse and sat him down
upon a little couch of moss under the shadow of an oak tree that grew
nigh to the cottage, there to rest himself for a while with great pleasure.
And as he sat there there came a barelegged lass from the cottage and
brought him fresh milk to drink; and there came a good, comely housewife
and brought him bread and cheese made of cream; and Sir Percival ate
and drank with great appetite.

Now whilst Sir Percival sat there resting and refreshing himself in that
wise, there appeared of a sudden coming thitherward, a tall and noble
knight riding upon a piebald war-horse of Norway strain. So when Sir
Percival beheld that knight coming in that wise he quickly put on his helmet
and mounted his horse and made him ready for defence in case the knight
had a mind to assail him.

Meantime that knight came riding up with great haughtiness of bearing
to where Sir Percival was, and when he had come nigh enough *Sir Percival*
he bespake Sir Percival, saying: "Sir Knight, I pray you to *bespeaketh the*
tell me your name and whither you go, and upon what quest?" *strange knight.*

Unto this Sir Percival made reply: "Messire, I do not choose to tell you
my name, for I am a young knight, very new to adventure, and I know
not how I shall succeed in that quest which I have undertaken. So I
will wait to try the success of that adventure before I tell my name. But

though I may not tell my name I will tell you whither I go and upon what quest. I go for to find a certain castle called Beaurepaire, and I intend to endeavor to liberate the lady of that castle from the duress of a certain knight hight Sir Clamadius, who, I understand, holds her by siege within the walls thereof."

Now, when Sir Percival had ceased speaking, the strange knight said: "Sir, this is a very singular thing: for that adventure of which you speak is the very adventure upon which I myself am bound. Now, as you say, you are a very young knight unused to arms, and as I am in the same degree a knight well seasoned in deeds of arms, it is more fitting that I should undertake this quest than you. For you may know how very well I am used to the service of arms when I tell you that I have had to do in four and twenty battles of various sorts; some of them friendly and some of them otherwise; and that I have had to do in more than four times that many affairs-at-arms with single knights, nearly all of them of great prowess. So now it would seem fitting that you should withdraw you from this affair and let me first essay it. Then, if I fail in my undertaking, you shall assume that adventure."

"Messire," quoth Sir Percival, "I see that you are a knight of much greater experience than I; but, ne'ertheless, I cannot find it in my heart to forego this adventure. So what I have to propose is this: that you and I do combat here in this place, and that he who proveth himself to be the better of us twain shall carry out this undertaking that we are both set upon."

Unto this, that strange knight lent a very willing assent, saying: "Very well, Messire, it shall be as you ask."

So with that each knight turned his horse and rode a little piece away; and each took such stand as pleased him; and each dressed his spear and

*Sir Percival doeth battle with the strange knight.* shield and made him in all wise ready for the encounter. And when they had so prepared themselves, each knight shouted to his horse, and drave spur into its flank and rushed, the one against the other, with such terrible noise and violence that the sound thereof was echoed back from the woods like to a storm of thunder.

So they met in the midst of the course with such a vehement impact that it was terrible to behold. And in that encounter the spear of each knight was burst all into fragments; and the horse of each fell back upon his haunches and would have been overthrown had not each knight voided his saddle with a very wonderful skill and agility.

Then each knight drew sword and came the one against the other, as furiously as two rams at battle. So they fought for nigh the space of an hour, foining and striking, and tracing hither and tracing thither most furiously; and the noise of the blows they struck might have been heard several furlongs away.

During that battle Sir Percival received several sore wounds so that by and by a great passion of rage seized upon him. So he rushed the battle with might and main, and therewith struck so many furious blows that by and by that other knight held his shield very low for weariness. This Sir Percival perceived, and therewith *Sir Percival overcometh the strange knight.* he smote the other so furious a blow upon the head that the knight sank down upon his knees and could not arise. Then Sir Percival ran to him and catched him by the neck and flung him down violently upon the ground, crying out, "Yield or I slay thee!"

Then that knight besought mercy in a very weak voice, saying: "Sir Knight, I beseech thee, spare my life!"

Sir Percival said: "Well, I will spare thee, but tell me, what is thy name?" To this the other said: "I am Sir Lionel, and I am a knight of King Arthur's court and of the Round Table."

Now when Sir Percival heard this he cried out aloud, for he was very greatly grieved, and he said: "Alas, what have I done for to fight against thee in this wise! I am Sir Percival, whom thine own kinsman, Sir Launcelot of the Lake, hath trained in arms. But indeed, *Sir Percival giveth aid to Sir Lionel.* I did never think to use that art which he taught me against one so dear to his heart as thou art, Sir Lionel." So with that Sir Percival assisted Sir Lionel to arise to his feet, and Sir Lionel was so weak from that woeful battle that he could hardly stand.

Now that stream and lake of water above spoken of was near by, so Sir Percival brought Sir Lionel thither, holding him up as he walked; and there Sir Lionel refreshed himself. Then, when he was revived a little, he turned his eyes very languidly upon Sir Percival, and he said: "Percival, thou hast done to me this day what few knights have ever done before. So all the glory that ever I have won is now thy glory because of this battle. For thou hast overcome me in a fair quarrel and I have yielded myself unto thee, wherefore it is now thy right to command me to thy will."

Then Percival said: "Alas, dear Sir Knight! It is not meet that I should lay command upon such as thou art. But, if thou wilt do so, I beseech thee when thou art come to King Arthur's court that thou wilt tell the King that I, who am his young knight Percival, have borne myself not unbecomingly in my battle with thee. For this is the first battle, knight against knight, that I have undertaken in all of my life. And I beseech thee that thou wilt greet Sir Kay the Seneschal, from me, and that thou wilt say to him that by and by I shall meet him and repay him that buffet which he gave to the damsel Yelande, the Dumb Maiden, in the Queen's pavilion.''

Sir Lionel said: "It shall be as thou sayst, and I will do thy bidding. But, touching Sir Kay, I do not believe that he will take very much joy at thy message to him. For he will find small pleasure in the thought of the payment of that buffet that thou hast promised to give him.''

Now, as the day by this time was waxing late, Sir Percival abided that night at that neatherd's hut nigh to which this battle had been fought and there had his wounds bathed and dressed; and when the next morning had come he arose early, and saddled his horse, and rode forward upon his way. And as he rode he was very

*Sir Percival goeth forward upon his adventure.*

well pleased at the thought of that battle he had fought with Sir Lionel, for he wist that he had obtained great credit to himself in that encounter, and he was aware, now that he had made trial of his strength against such a one as Sir Lionel, he must be one of the greatest knights of the world. So his heart was uplifted with great joy and delight at that thought; that he was now a well-approved knight-champion, worthy of his knighthood. Therefore he rode away for all that day, greatly rejoicing in spirit at the thought of what he had done the day before.

About the first slant of the afternoon Sir Percival came at last out of the woodlands and into a wide-open plain, very fertile and well tilled, with fields of wheat and rye abounding on all sides. And he saw that in the midst of that plain there was a considerable lake, and that in the midst of that lake there was an island, and that upon the island there stood a fair noble castle, and he wist that that castle must be the castle of Beaurepaire. So he rode down into that valley with some speed.

Now after he had so ridden for a while, he was aware of a knight, very haughty of appearance and bearing, who rode before him upon the same way that he was going. And that knight was clad all in red armor, and he rode upon a horse so black that I believe there was not a single white hair

*Sir Percival perceives a red knight.*

upon him. And all the trappings and the furniture of that horse were of red, so that he presented a very noble appearance. So Sir Percival made haste to overtake that knight, and when he had come nigh he drew rein at a little distance. Thereupon that knight in red bespake Sir Percival very proudly, saying: "Sir Knight, whither ride you, and upon what mission?''

"Messire,'' quoth Percival, "I ride toward yonder castle, which I take to be the castle of Beaurepaire, and I come hither with intent to succor the Lady Blanchefleur of that castle from a knight, hight Sir Clamadius, who keeps her there a prisoner against her will, so that it behooves any good knight to attempt her rescue.''

Upon this the red knight spake very fiercely, saying: "Messire, what business is that of yours? I would have you know that I am a knight of King Clamadius', wherefore I am able to say to you that you shall go no further upon that quest. For I am Sir Engeneron of Grandregarde, and I am Seneschal unto King Clamadius, and I will not have it that thou shalt go any farther upon this way unless you ride over me to go upon it.''

"Messire," quoth Sir Percival, "I have no quarrel with you, but if you have a mind to force a quarrel upon me, I will not seek to withdraw myself from an encounter with you. So make yourself ready, and I will make myself ready, and then we shall soon see whether or not I am to pass upon this way.''

So therewith each knight turned his horse away to such a place as seemed to him to be fitting; and when they were in all wise prepared they rushed together with an amazing velocity and a noise like to thunder. *Sir Percival* So they met in the midst of the course. And in that encounter *doeth battle with* the spear of Sir Engeneron broke into many pieces, but the *Sir Engeneron.* spear of Sir Percival held, so that he flung Sir Engeneron entirely out of his saddle and over the crupper of his horse and down upon the ground so violently that Sir Engeneron lay there in a swoon.

Then Sir Percival dismounted from his horse with all speed, and he rushed the helmet of Sir Engeneron off of his head with intent to *Sir Engeneron* slay him. But with that Sir Engeneron awoke to his danger, *yields himself to* and therewith gat upon his knees and clasped Sir Percival *Sir Percival.* about the thighs, crying out: "Sir, I beseech you upon your knighthood to spare my life.''

"Well," said Sir Percival, "since you beseech that upon my knighthood I must needs do as you ask. But I will only do so upon two conditions. The first of these conditions is that you go to the court of King Arthur, and that you surrender yourself as captive to a damsel of that court who is known as the Lady Yelande the Dumb Maiden. And you are to tell that maiden that the young knight who slew Sir Boindegardus greets her and that he tells her that in a little while he will return to repay to Sir Kay that buffet he gave her. This is my first condition." And Sir Engeneron said: "I will perform that condition.''

"And my second condition," said Sir Percival, "is this: that you give me your armor for me to use upon this adventure which I have undertaken, and that you take my armor and deposit it with the hermit of a little chapel you shall after a while come to if you return upon the road which brought me hither. After a while I will return and reclaim my armor and will return your armor. This is my second condition."

And Sir Engeneron said: "That condition also I shall fulfil according to your command."

Then Sir Percival said: "Arise." And Sir Engeneron did so. And after that Sir Engeneron put off his armor, and Sir Percival put off his
*Sir Percival and* armor. And Sir Percival put on the armor of Sir Engeneron,
*Sir Engeneron* and Sir Engeneron packed the armor of Sir Percival upon his
*exchange armor.* horse and prepared to depart in obedience to those conditions of Sir Percival. So they parted company, Sir Percival riding upon his way to Beaurepaire, and Sir Engeneron betaking his way to find the chapel of that hermit of whom Sir Percival had spoken.

So it was that after two adventures, Sir Percival entered upon that undertaking which he had come to perform in behalf of the Lady Blanchefleur.

And now, if it please you to read what follows, you shall hear how it befell with Sir Percival at the castle of Beaurepaire.

After that adventure with Sir Engeneron, Sir Percival rode onward upon his way, and by and by he came to the lake whereon stood the castle and the town of Beaurepaire. And Sir Percival beheld that a long narrow bridge crossed over that part of the lake from the mainland to the island and the town. So Sir Percival rode very boldly forth upon that bridge and across it, and no one stayed him, for all of the knights of Sir Clamadius who beheld him said: "Yonder rides Sir Engeneron." Thus Sir Percival crossed the bridge and rode very boldly forward until he came to the gate of the castle, and those who beheld him said: "Sir Engeneron haply beareth a message to the castle." For no one wist that that knight was not Sir Engeneron, but all thought that it was he because of the armor which he wore.

So Sir Percival came close to the castle, and when he was come there he
*Sir Percival* called very loudly to those within, and by and by there ap-
*cometh to Beau-* peared the face of a woman at an upper window and the face
*repaire.* was very pale and woe-begone.

Then Sir Percival said to the woman at the window: "Bid them open the gate and let me in; for I come to bring you succor at this place."

To this the woman said: "I shall not bid them open the gate, for I know from your armor who you are, and that you are Sir Engeneron the Seneschal. And I know that you are one of our bitterest enemies; for you have already slain several of the knights of this castle, and now you seek by guile to enter into the castle itself."

Then Sir Percival said: "I am not Sir Engeneron, but one who hath over-thrown Sir Engeneron in battle. I have put on his armor with intent that I might come hither to help defend this place against Sir *Sir Percival* Clamadius." So said Sir Percival, and therewith he put up *entereth Beau-* the umbril of his helmet, saying: "Look, see; I am not Sir *repaire.* Engeneron." Then the woman at the window saw his face and that it was not the face of Sir Engeneron. And she saw that the face of Sir Percival was mild and gentle, wherefore she ran and told the people of the castle that a knight who was a friend stood without. Therewith they of the castle let fall the drawbridge and opened the gates, and Sir Percival entered into the castle.

Then there came several of the chief people of the castle, and they also were all pale and woe-begone from long fasting, as was the woman whom Sir Percival had first seen; for all were greatly wasted because of the toil and anxiety of that siege. These asked Sir Percival who he was and whence he came and how he came thither; and Sir Percival told them all that it was necessary for them to know. For he told them how he was a young knight trained under the care of Sir Launcelot; and he told them that he had come thither with the hope of serving the Lady Blanchefleur; and he told them what adventures had befallen him in the coming and how he had already overthrown Sir Lionel and Sir Engeneron to get there. Wherefore, from these things, they of the castle perceived that Sir Percival was a very strong, worthy knight, and they gave great joy that he should have come thither to their aid.

So he who was chief of those castle people summoned several attendants, and these came and some took the horse of Sir Percival and led it to the stables, and others relieved Sir Percival of his armor; and others took him to a bath of tepid water, where he bathed himself, and was dried on soft linen towels; and others brought soft garments of gray cloth and clad Sir Percival in them and afterward brought him down into a fair large chamber where there was a table spread as though ready for meat.

Now in a little after Sir Percival was come to that supper-hall the door thereof was opened and there entered several people. With these came a damsel of such extraordinary beauty and gracefulness of figure *Sir Percival* that Sir Percival stood amazed. For her face was fair beyond *beholds the Lady* words; red upon white, like rose-leaves upon cream; and her *Blanchefleur.* eyes were bright and glancing like those of a falcon, and her nose was thin and straight, and her lips were very red, like to coral for redness, and her hair was dark and abundant and like to silk for softness. She was clad all in a dress of black, shot with stars of gold, and the dress was lined with ermine and was trimmed with sable at the collar and the cuffs and the hem thereof.

So Sir Percival stood and gazed at that lady with a pleasure beyond words to express, and he wist that this must be the Lady Blanchefleur, for whose sake he had come thither.

And the Lady Blanchefleur looked upon Sir Percival with great kindness, for he appeared to her like to a hero for strength and beauty; wherefore she smiled upon Sir Percival very graciously and came forward and gave him her hand. And Sir Percival took her hand and set it to his lips; and lo! her hand was as soft as silk and very warm, rosy and fragrant, and the fingers thereof glistered with bright golden rings and with gems of divers colors.

Then that beautiful Lady Blanchefleur said: "Messire, this is a very knightly thing for you to do to come hither to this place. And you come in good time, for food groweth very scarce with us so that in a little while we must face starvation. For because of the watch that Sir Clamadius keepeth upon this place, no one can either enter in or go out. Yea, thou art the very first one who hath come hither since he has sat down before Beaurepaire."

Then presently she ceased smiling and her face clouded over; then bright tears began to drop from the Lady Blanchefleur's eyes; and then she said: *The Lady* "I fear me greatly that Sir Clamadius will at last seize upon *Blanchefleur* this castle, for he hath kept us here prisoner for a long while. *telleth her sor-* Yet though he seize the castle, he shall never seize that *rows to Sir* *Percival.* which the castle contains. For I keep by me a little casket of silver, and therein is a dagger, very sharp and fine. Therefore the day that Sir Clamadius enters into this castle, I shall thrust that dagger into my heart. For, though Sir Clamadius may seize upon my castle, he shall never possess my soul."

Then Sir Percival was very sorry for the tears he saw shining upon the Lady Blanchefleur's face, wherefore he said: "Lady, I have great hopes that this affair may never reach to that woful extremity thou speakest of." The Lady Blanchefleur said: "I hope not also." And therewith she wiped away her tears and smiled again. Then she said: "See, Sir Percival, the evening has come and it is time to sit at supper, now I beseech thee for to come to table with me, for though we have but little to eat here, yet I assure thee that thou art very welcome to the best that we have."

So therewith Lady Blanchefleur led Sir Percival to the table, and they sat down to such feast as could be had at that place of starvation. For what they had was little enough, being only such fish as they could catch from the lake, and a little bread—but not much—and a very little wine.

Then after they had eaten and drunk what they had, the Lady Blanchefleur took a golden harp into her hand and played thereon, *The Lady sings*
and sang in a voice so clear and high and beautiful that Percival *to Sir Percival.*
was altogether enchanted and bewitched thereat.

Thus it was that that evening passed with them very pleasantly and
cheerfully, so that it was the middle of the night ere Sir Percival withdrew
to that couch that had been prepared for his rest.

Now word was brought to Sir Clamadius that Sir Engeneron the Seneschal
had been overcome by another knight, wherefore Sir Clamadius wist that
that was the knight in Sir Engeneron's armor who had entered into the
castle. So Sir Clamadius said: "Certes, this must be a champion of no
small prowess who hath undertaken single-handed such a dangerous quest
as this, and hath thus entered into the castle, for they appear to make great
rejoicings at his coming. Now if he remaineth there it may very well be
that they will be encouraged to resist me a great while longer, and so all
that I have thus far accomplished shall have been in vain."

Now there was among the counsellors of Sir Clamadius an old knight
who was very cunning and far-sighted. He said to the King: "Sire, I think
we may be able to devise some plan whereby we may withdraw
this knight-champion out of the castle. My plan is this: *The old counsel-*
*lor giveth advice*
Let ten of your best knights make parade before that castle to- *to Sir Clama-*
morrow, and let them give challenge to those within the castle *dius.*
to come forth to battle. Then I believe that this knight will come forth
with the other knights from the castle to accept that challenge. Thereafter
let it be that our knights withdraw as though in retreat, and so lead this
knight and the knights of the castle into an ambushment. There let many
fall upon them at once and either slay them or make them prisoners. So
the castle shall be deprived of this new champion that hath come to it, and
therewith may be so disheartened that it will yield to thee."

This advice seemed very good to King Clamadius, wherefore, when the
next morning had come, he chose him ten knights from among the foremost
of all his knights, and he bade them give that challenge in that wise. These
did so, and therewith Sir Percival and nine other knights issued out from
the castle against them.

But it did not fare as Sir Clamadius had expected; for the attack of
Sir Percival and his knights was so fierce and sudden that *Sir Percival do-*
those ten knights could not withdraw so easily as they intended. *eth great battle.*
For, ere they were able to withdraw, Sir Percival had struck down six of
these knights with his own hand and the other four were made prisoners.

Thus Sir Percival and his knights did not come into that ambush that had been prepared for them.

Then those who were in ambush perceived that their plan had failed, wherefore they broke from cover with intent to do what they could. But Sir Percival and his knights beheld them coming, and so withdrew, defending themselves with great valor. So they came into the castle again in safety.

Thus it was that the plans of King Clamadius and his counsellor failed of effect, whereupon Sir Clamadius was very angry at that wise old knight. So that, when that counsellor came to him again and said: "Sir, I have another plan," King Clamadius cried out very fiercely: "Away with thy plans! They are all of no avail." Then Sir Clamadius said: "When to-morrow comes, I myself will undertake this affair. For I will go and give challenge to this knight, and so I shall hope to decide this quarrel man to man. For unless yonder knight be Sir Launcelot of the Lake or Sir Lamorack of Gales, I do not think he will be my peer in an encounter of man to man."

So when the next morning had come, Sir Clamadius armed himself at
*Sir Clamadius arms himself for battle.* all points and straightway betook himself to a fair, smooth meadow beneath the walls of the castle. And when he had come there he cried out: "Sir Red Knight, come forth and speak with me."

So after a while Sir Percival appeared at the top of the castle wall, and he said: "Messire, here I am; what is it you would have of me?"

Then Sir Clamadius said: "Messire, are you Sir Launcelot of the Lake?" And Sir Percival said: "Nay, I am not he." Sir Clamadius said: "Art thou then Sir Lamorack of Gales?" And Sir Percival said: "Nay, I am not he." Then Sir Clamadius said: "Who, then, art thou?" Sir Percival said: "I am not any great knight-champion such as those two of whom you speak, but am a young knight who have not fought more than twice or thrice in my life."

At that Sir Clamadius was very glad, for he feared that Sir Percival might be some famous knight well-seasoned in arms. Wherefore when he found that Sir Percival was only a young and untried knight, he thought it would be an easy matter to deal with him. So he said: "Messire, I challenge thee to come forth to battle with me man to man so that thou and I may settle this quarrel betwixt us, for it is a pity to shed more blood than is necessary in this quarrel. So if thou wilt come forth and overthrow me, then I will withdraw my people from this place; but if I overthrow thee, then this castle shall be yielded up to me with all that it contains."

To this Sir Percival said: "Sir Knight, I am very willing to fight with thee upon that issue. But first of all I must obtain the consent of the Lady Blanchefleur to stand her champion."

So Sir Percival went to the Lady Blanchefleur, and he said: "Lady, will you accept me as your champion to fight the issue of this quarrel man to man with Sir Clamadius?"

She said: "Percival, thou art very young to have to do with so old and well-seasoned a knight. Now I greatly fear for your life in such a battle as that."

To this Sir Percival said: "Lady, I know that I am young, but indeed I feel a very big spirit stir within me, so that if thou wilt trust me, I have belief that, with the grace of God, I shall win this battle."

Then the Lady Blanchefleur smiled upon Sir Percival and she said: "Percival, I will gladly entrust my life and safety into thy keeping, for I too have great dependence in thy knighthood."

So straightway Sir Percival armed himself, and when he was in all wise prepared he went forth to that battle with a heart very full of great courage and hope.

There he found Sir Clamadius still parading in that meadow beneath the walls, awaiting the coming of his opponent.

Meanwhile many folk came and stood upon the walls of the castle to behold that encounter, whilst each knight took such stand as appeared good to him. Then, when they were in all wise prepared, *Sir Percival and* each knight drave spurs into his horse and rushed himself *Sir Clamadius* against the other with most terrible and fierce violence. *do battle.* Therewith they met in the very midst of the course with an uproar like to thunder that echoed back from the flat walls of the castle.

In that encounter the spear of Sir Percival held, but the spear of Sir Clamadius was riven into splinters. And so, Sir Percival riding forward with furious violence, Sir Clamadius was overthrown, horse and man, with such violence that he lay there upon the ground as though he were dead.

Then all those upon the walls shouted aloud with a great noise of rejoicing, whilst those of the party of Sir Clamadius gave lamentation in the same degree.

But Sir Percival voided his saddle in haste, and ran to where Sir Clamadius lay. And Sir Percival rushed the helmet off from the head of *Sir Clamadius* Sir Clamadius, and he catched him by the hair of the head, and *yields himself.* he raised his sword on high with intent to finish the work he had begun. Therewith Sir Clamadius aroused himself unto his danger, and he cried in a very piercing voice: "Messire, I beseech thee of thy knighthood to spare my life!"

"Well," said Sir Percival, "since you ask me upon my knighthood, I cannot refuse you, for so I was taught by the noble knight, Sir Launcelot, to refuse no boon asked upon my knighthood that I was able to grant. But I will only spare your life upon one condition, and that is this: That you disarm yourself in all wise, and that you go without armor to the court of King Arthur. There you shall deliver yourself as a servant unto a damsel of King Arthur's court, hight Yelande, surnamed the Dumb Maiden. Her you are to tell that the youth who slew Sir Boindegardus hath sent you unto her as a servant. And you are to say to Sir Kay, the Seneschal of King Arthur, that the young knight Percival will in a little while come to repay that buffet he gave to the damoiselle Yelande aforesaid."

So said Sir Percival, and Sir Clamadius said: "It shall be done in all wise as you command, if so be you will spare my life." Then Sir Percival said: "Arise"; and Sir Clamadius arose; and Sir Percival said: "Go hence"; and therewith Sir Clamadius departed as Sir Percival commanded.

So that day Sir Clamadius withdrew from the castle of Beaurepaire with all his array of knights, and after that he went to the court of King Arthur and did in all respects as Sir Percival had commanded him to do.

So it was that Sir Percival fulfilled that quest, and set the Lady Blanchefleur free from duress; and may God grant that you also fulfil all your quests with as great honor and nobility as therein exhibited.

Sir Kay interrupts ye meditations of Sir Percival :.

# Chapter Fifth

*How Sir Percival repaid Sir Kay the buffet he one time gave Yelande the Dumb Maiden, and how, thereafter, he went forth to seek his own lady of love.*

NOW, after these adventures aforesaid, Sir Percival remained for a long while at Beaurepaire, and during that time he was the knight-champion to the Lady Blanchefleur. And the Lady Blanchefleur loved Sir Percival every day with a greater and greater passion, but Sir Percival showed no passion of love for her in return, and thereat Lady Blanchefleur was greatly troubled.

Now one day the Lady Blanchefleur and Sir Percival were walking together on a terrace; and it was then come to be the fall of the year, so that the leaves of the trees were showering all down about them like flakes of gold. And that day the Lady Blanchefleur loved Sir Percival so much that her heart was pierced with that love as though with a great agony. But Sir Percival wist not of that.

*Sir Percival and the Lady Blanchefleur walk together.*

Then the Lady Blanchefleur said: "Messire, I would that thou wouldst stay here always as our knight-champion."

"Lady," quoth Percival, "that may not be, for in a little while now I must leave you. For, though I shall be sad to go from such a friendly place as this is, yet I am an errant knight, and as I am errant I must fulfil many adventures besides the one I have accomplished here."

"Messire," said the Lady Blanchefleur, "if you will but remain here, this castle shall be yours and all that it contains."

At this Sir Percival was greatly astonished, wherefore he said: "Lady, how may that be? Lo! this castle is yours, and no one can take it away from you, nor can you give it to me for mine own."

Then the Lady Blanchefleur turned away her face and bowed her head, and said in a voice as though it were stifling her for to speak: "Percival, it needs not to take the castle from me; take thou me for thine own, and then the castle and all shall be thine."

At that Sir Percival stood for a space very still as though without breathing. Then by and by he said: "Lady, meseems that no knight could have greater honor paid to him than that which you pay to me. Yet

*Sir Percival de-* should I accept such a gift as you offer, then I would be doing
*nies the Lady* such dishonor to my knighthood that would make it altogether
*Blanchefleur.* unworthy of that high honor you pay it. For already I have made my vow to serve a lady, and if I should forswear that vow, I would be a dishonored and unworthy knight."

Then the Lady Blanchefleur cried out in a great voice of suffering: "Say no more, for I am ashamed."

Sir Percival said: "Nay, there is no shame to thee, but great honor to me." But the Lady Blanchefleur would not hear him, but brake away from him in great haste, and left him standing where he was.

So Sir Percival could stay no longer at that place; but as soon as might be, he took horse and rode away. Nor did he see Blanchefleur again after they had thus talked together upon that terrace as aforesaid.

And after Sir Percival had gone, the Lady Blanchefleur abandoned herself to great sorrow, for she wept a long while and a very great deal; nor would she, for a long while, take any joy in living or in the world in which she lived.

So Sir Percival performed that adventure of setting free the duress of the castle of Beaurepaire. And after that and ere the winter came, he

*Of the further* performed several other adventures of more or less fame.
*adventures of* And during that time, he overthrew eleven knights in various
*Sir Percival.* affairs at arms and in all those adventures he met with no mishap himself. And besides such encounters at arms, he performed several very worthy works; for he slew a wild boar that was a terror to all that dwelt nigh to the forest of Umber; and he also slew a very savage wolf that infested the moors of the Dart. Wherefore, because of these several adventures, the name of Sir Percival became very famous in all courts of chivalry, and many said: "Verily, this young knight must be the peer of Sir Launcelot of the Lake himself."

Now one day toward eventide (and it was a very cold winter day) Sir Percival came to the hut of a hermit in the forest of Usk; and he abode all night at that place.

Now when the morning had come he went out and stood in front of the hut, and he saw that during the night a soft snow had fallen so that all the earth was covered with white. And he saw that it likewise had happened that a hawk had struck a raven in front of the hermit's habitation, and that some of the raven's feathers and some of its blood lay upon the snow.

Now when Sir Percival beheld the blood and the black feathers upon that white snow, he said to himself: "Behold! that snow is not whiter than the brow and the neck of my lady; and that red is not redder than her lips; and that black is not blacker than her hair." Therewith the thought of that lady took great hold upon him and he sighed so deeply that he felt his heart lifted within him because of that sigh. So he stood and gazed upon that white and red and black, and he forgot all things else in the world than his lady-love.

*Sir Percival stands in meditation.*

Now it befell at that time that there came a party riding through those parts, and that party were Sir Gawaine and Sir Geraint and Sir Kay. And when they saw Sir Percival where he stood leaning against a tree and looking down upon the ground in deep meditation, Sir Kay said: "Who is yonder knight?" (For he wist not that that knight was Sir Percival.) And Sir Kay said further: "I will go and bespeak that knight and ask him who he is."

But Sir Gawaine perceived that Sir Percival was altogether sunk in deep thought, wherefore he said: "Nay, thou wilt do ill to disturb that knight; for either he hath some weighty matter upon his mind, or else he is bethinking him of his lady, and in either case it would be a pity to disturb him until he arouses himself."

But Sir Kay would not heed what Sir Gawaine said, but forthwith he went to where Sir Percival stood; and Sir Percival was altogether unaware of his coming, being so deeply sunk in his thoughts. Then Sir Kay said: "Sir Knight,"—but Sir Percival did not hear him. And Sir Kay said: "Sir Knight, who art thou?" But still Sir Percival did not reply. Then Sir Kay said: "Sir Knight, thou shalt answer me!" And therewith he catched Sir Percival by the arm and shook him very roughly.

*Sir Kay shakes the arm of Sir Percival.*

Then Sir Percival aroused himself, and he was filled with indignation that anyone should have laid rough hands upon his person. And Sir Percival did not recognize Sir Kay because he was still entangled in that network of thought, but he said very fiercely: "Ha, sirrah! wouldst thou lay hands upon me!" and therewith he raised his fist and smote Sir Kay so terrible a buffet beside the head that Sir Kay instantly fell down as though he were dead and lay without sense of motion upon the ground. Then Sir Percival perceived that there were two other knights standing not far off, and therewith his thoughts of other things came back to him again and he was aware of what he had done in his anger, and was very sorry and ashamed that he should have been so hasty as to have struck that blow.

*Sir Percival smites Sir Kay a buffet.*

Then Sir Gawaine came to Sir Percival and spake sternly to him saying: "Sir Knight, why didst thou strike my companion so unknightly a blow as that?"

*Sir Gawaine chides Sir Percival.*

To which Sir Percival said: "Messire, it grieves me sorely that I should have been so hasty, but I was bethinking me of my lady, and this knight disturbed my thoughts; wherefore I smote him in haste."

To this Sir Gawaine made reply: "Sir, I perceive that thou hadst great excuse for thy blow. Ne'theless, I am displeased that thou shouldst have struck that knight. Now I make demand of thee what is thy name and condition?"

And Sir Percival said: "My name is Percival, and I am a knight of King Arthur's making."

At that, when Sir Gawaine and Sir Geraint heard what Sir Percival said, they cried out in great amazement; and Sir Gawaine said: "Ha, Sir Percival! this is indeed well met, for my name is Gawaine and I am a

*Sir Gawaine and Sir Geraint rejoice over Sir Percival.*

nephew unto King Arthur and am of his court; and this knight is Sir Geraint, and he also is of King Arthur's court and of his Round Table. And we have been in search of thee for this long time for to bring thee unto King Arthur at Camelot. For thy renown is now spread all over this realm, so that they talk of thee in every court of chivalry."

And Sir Percival said: "That is good news to me, for I wist not that I had so soon won so much credit. But, touching the matter of returning unto King Arthur's court with you; unto that I crave leave to give my excuses. For, since you tell me that I now have so much credit of knighthood, it behooves me to go immediately unto my lady and to offer my services unto her. For when I parted from her I promised her that I would come to her as soon as I had won me sufficient credit of knighthood. As for this knight whom I have struck, I cannot be sorry for that buffet, even if it was given with my fist and not with my sword as I should have given it. For I have promised Sir Kay by several mouths that I would sometime repay him with just such a buffet as that which he struck the damosel Yelande. So now I have fulfilled my promise and have given him that buffet."

Then Sir Gawaine and Sir Geraint laughed, and Sir Gawaine said: "Well, Sir Percival, thou hast indeed fulfilled thy promise in very good measure. For I make my vow that no one could have been better served with his dessert than was Sir Kay."

Now by this time Sir Kay had recovered from that blow, so that he rose up very ruefully, looking about as though he wist not yet just where he was.

Then Sir Gawaine said to Sir Percival: "As to thy coming unto the court of the King, thou dost right to fulfil thy promise unto thy lady before undertaking any other obligation. For, even though the *Sir Percival* King himself bid thee come, yet is thy obligation to thy lady *will not return* superior to the command of the King. So now I bid thee go *to court.* in quest of thy lady in God's name; only see to it that thou comest to the King's court as soon as thou art able."

So it was that Sir Percival fulfilled the promise of that buffet unto Sir Kay.

And now you shall hear how he found the Lady Yvette the Fair.

Now after Sir Percival had parted from Sir Gawaine, and Sir Geraint and Sir Kay, he went his way in that direction he wist, and by and by, toward eventide, he came again to the castle of Sir Percydes. *Sir Percival* And Sir Percydes was at home and he welcomed Sir Percival *cometh to the* with great joy and congratulations. For the fame of Sir *castle of Sir* Percival was now abroad in all the world, so that Sir Percydes *Percydes.* welcomed him with great acclaim.

So Sir Percival sat down with Sir Percydes and they ate and drank together, and, for the time, Sir Percival said nothing of that which was upon his heart—for he was of a very continent nature and was in no wise hasty in his speech.

But after they had satisfied themselves with food and drink, then Sir Percival spake to Sir Percydes of that which was upon his mind, saying: "Dear friend, thou didst tell me that when I was ready for to come to thee with a certain intent thou wouldst tell me who is the lady whose ring I wear and where I shall find her. Now, I believe that I am a great deal more worthy for to be her knight than I was when I first saw thee; wherefore I am now come to beseech thee to redeem thy promise to me. Now tell me, I beg of thee, who is that lady and where does she dwell?"

Then Sir Percydes said: "Friend, I will declare to thee that which thou dost ask of me. Firstly, that lady is mine own sister, hight Yvette, and she is the daughter of King Pecheur. Secondly, thou shalt *Sir Percydes* find her at the castle of my father, which standeth upon the *declares himself* west coast of this land. Nor shalt thou have any difficulty *to Sir Percival.* in finding that castle, for thou mayst easily come to it by inquiring the way of those whom thou mayst meet in that region. But, indeed, it hath been two years since I have seen my father and my sister, and I know not how it is with them."

Then Sir Percival came to Sir Percydes and he put his arm about him and kissed him upon either cheek, and he said: "Should I obtain the kind regard of that lady, I know nothing that would more rejoice me than to know that thou art her brother. For, indeed, I entertain a great deal of love for thee."

At that Sir Percydes laughed for joy and he said: "Percival, wilt thou not tell me of what house thou art come?" Percival said: "I will tell thee what thou dost ask : my father is King Pellinore who was a very good, noble knight of the court of King Arthur and of his Round Table."

Then Sir Percydes cried out with great amazement, saying: "That is very marvellous! I would that I had known this before, for thy mother and my mother were sisters of one father and one mother. So we are cousins german."

Then Sir Percival said: "This is great joy to me!" And his heart was expanded with pleasure at finding that Sir Percydes was of his kindred and that he was no longer alone in that part of the world.

So Sir Percival abided for two days with Sir Percydes and then he betook his way to the westward in pursuance of that adventure. And he was *Sir Percival* upon the road three days, and upon the morning of the fourth *departs for the* day he came, through diligent inquiry, within sight of the *castle of King* castle of King Pecheur. This castle stood upon a high crag of *Pecheur.* rock from which it arose against the sky so that it looked to be a part of the crag. And it was a very noble and stately castle, having many tall towers and many buildings within the walls thereof. And a village of white houses of the fisher-folk gathered upon the rocks beneath the castle walls like chicks beneath the shadow of their mother's wings.

And, behold! Percival saw the great sea for the first time in all his life, and was filled with wonder at the huge waves that ran toward the shore and burst upon the rocks, all white like snow. And he was amazed at the multitude of sea fowl that flew about the rocks in such prodigious numbers that they darkened the sky. Likewise he was astonished at the fisher-boats that spread their white sails against the wind, and floated upon the water like swans, for he had never seen their like before. So he sat his horse upon a high rock nigh to the sea and gazed his fill upon those things that were so wonderful to him.

Then after a while Sir Percival went forward to the castle. And as he drew nigh to the castle he became aware that a very reverend man, whose hair and beard were as white as snow, sat upon a cushion of crimson velvet upon a rock that overlooked the sea. Two pages, richly clad in black and silver, stood behind him; and the old man gazed out across the sea, and Sir Percival saw that he neither spake nor moved. But when Sir

Percival came near to him the old man arose and went into the castle, and the two pages took up the two crimson velvet cushions and followed him.

But Percival rode up to the castle, and he saw that the gateway of the castle stood open, wherefore he rode into the courtyard of the castle. And when he had come into the courtyard, two attendants immediately appeared and took his horse and assisted him to dismount; but neither of these attendants said aught to him, but both were as silent as deaf-mutes.

Then Percival entered the hall and there he saw the old man whom he had before seen, and the old man sat in a great carved chair beside a fire of large logs of wood. And Sir Percival saw that the eyes of *Sir Percival* the old man were all red and that his cheeks were channeled *finds King* with weeping; and Percival was abashed at the sadness of *Pecheur.* his aspect. Ne'ertheless, he came to where the old man sat and saluted him with great reverence, and he said: "Art thou King Pecheur?" And the old man answered, "Aye, for I am both a fisher and a sinner" (for that word Pecheur meaneth both fisher and sinner).

Then Sir Percival said: "Sire, I bring thee greetings from thy son, Sir Percydes, who is a very dear friend to me. And likewise I bring thee greeting from myself: for I am Percival, King Pellinore's son, and thy Queen and my mother are sisters. And likewise I come to redeem a pledge, for, behold, here is the ring of thy daughter Yvette, unto whom I am pledged for her true knight. Wherefore, having now achieved a not dishonorable renown in the world of chivalry, I am come to beseech her kindness and to redeem my ring which she hath upon her finger and to give her back her ring again."

Then King Pecheur fell to weeping in great measure and he said: "Percival thy fame hath reached even to this remote place, for every one talketh of thee with great unction. But, touching my daughter Yvette, if thou wilt come with me I will bring thee to her."

So King Pecheur arose and went forth and Sir Percival followed him. And King Pecheur brought Sir Percival to a certain tower; and he brought him up a long and winding stair; and at the top of the stairway was a door. And King Pecheur opened the door and Sir Percival entered the apartment.

The windows of the apartment stood open, and a cold wind came in thereat from off the sea; and there stood a couch in the mid- *Sir Percival* dle of the room, and it was spread with black velvet; and the *findeth the* Lady Yvette lay reclined upon the couch, and, lo! her face *Lady Yvette.* was like to wax for whiteness, and she neither moved nor spake, but only lay there perfectly still; for she was dead.

Seven waxen candles burned at her head, and seven others at her feet, and the flames of the candles spread and wavered as the cold wind blew upon them. And the hair of her head (as black as those raven feathers that Sir Percival had beheld lying upon the snow) moved like threads of black silk as the wind blew in through the window—but the Lady Yvette moved not nor stirred, but lay like a statue of marble all clad in white.

Then at the first Sir Percival stood very still at the door-way as though he had of a sudden been turned into stone. Then he went forward and stood beside the couch and held his hands very tightly together and gazed at the Lady Yvette where she lay. So he stood for a long while, and he wist not why it was that he felt like as though he had been turned into a stone, without such grief at his heart as he had thought to feel thereat. (For indeed, his spirit was altogether broken though he knew it not.)

Then he spake unto that still figure, and he said: "Dear lady, is it thus I find thee after all this long endeavor of mine? Yet from Paradise, haply, *Of the grief of* thou mayst perceive all that I have accomplished in thy be- *Sir Percival.* half. So shalt thou be my lady always to the end of my life and I will have none other than thee. Wherefore I herewith give thee thy ring again and take mine own in its stead." Therewith, so speaking, he lifted that hand (all so cold like the snow) and took his ring from off her finger and put her ring back upon it again.

Then King Pecheur said, "Percival, hast thou no tears?"

And Percival said, "Nay, I have none." Therewith he turned and left that place, and King Pecheur went with him.

After that Sir Percival abided in that place for three days, and King Pecheur and his lady Queen and their two young sons who dwelt at that place made great pity over him, and wept a great deal. But Sir Percival said but little in reply and wept not at all.

And now I shall tell you of that wonderful vision that came unto Sir Percival at this place upon Christmas day.

For on the third day (which was Christmas day) it chanced that Sir Percival sat alone in the hall of the castle, and he meditated upon the great *Sir Percival be-* sorrow that lay upon him. And as he sat thus this very won- *holds the grail.* derful thing befell him: He suddenly beheld two youths enter that hall. And the faces of the two youths shone with exceeding brightness, and their hair shone like gold, and their raiment was very bright and glistering like to gold. One of these youths bare in his hand a spear of mighty size, and blood dropped from the point of the spear; and the other youth bare in his hand a chalice of pure gold, very wonderful to behold, and he held the chalice in a napkin of fine cambric linen.

Then, at first, Sir Percival thought that that which he beheld was a vision conjured up by the deep sorrow that filled his heart, and he was afeard. But the youth who bare the chalice spake in a voice extraordinarily high and clear. And he said: "Percival! Percival! be not afraid! This which thou here beholdest is the Sangreal, and that is the Spear of Sorrow. What then may thy sorrow be in the presence of these holy things that brought with them such great sorrow and affliction of soul that they have become entirely sanctified thereby! Thus, Percival, should thy sorrow so sanctify thy life and not make it bitter to thy taste. For so did this bitter cup become sanctified by the great sorrow that tasted of it."

Percival said: "Are these things real or are they a vision that I behold?"

He who bare the chalice said, "They are real." And he who bare the spear said, "They are real."

Then a great peace and comfort came to Sir Percival's heart and they never left him to the day of his death.

Then they who bare the Sangreal and the Spear went out of the hall, and Sir Percival kneeled there for a while after they had gone and prayed with great devotion and with much comfort and satisfaction.

And this was the first time that any of those knights that were of King Arthur's Round Table ever beheld that holy chalice, the which Sir Percival was one of three to achieve in after-years.

So when Sir Percival came forth from that hall, all those who beheld him were astonished at the great peace and calmness that appeared to emanate from him. But he told no one of that miraculous vision which he had just beheld, and, though it appeareth in the history of these things, yet it was not then made manifest.

Then Sir Percival said to King Pecheur, his uncle and to his aunt and to their sons: "Now, dear friends, the time hath come when I must leave you. For I must now presently go to the court of King Arthur in obedience to his commands and to acknowledge myself unto my brother, Sir Lamorack."

So that day Sir Percival set forth with intent to go to Camelot, where King Arthur was then holding court in great estate of pomp. *Sir Percival departs for court.* And Sir Percival reached Camelot upon the fourth day from that time and that was during the feasts of Christmas-tide.

Now King Arthur sat at those feasts and there were six score of very noble company seated with him. And the King's heart was greatly uplifted and expanded with mirth and good cheer. Then, while all were feasting with great concord, there suddenly came into that hall an herald-messenger; the whom, when King Arthur beheld him, he asked: "What message hast thou brought?" Upon this the messenger said: "Lord, there hath come one asking permission to enter here whom you will be very well pleased to see." The King said, "Who is it?" And the herald-messenger said, "He saith his name is Percival."

Upon this King Arthur arose from where he sat and all the others uprose with him and there was a great sound of loud voices; for the fame of Sir Percival had waxed very great since he had begun his adventures. So King Arthur and the others went down the hall for to meet Sir Percival.

Then the door opened and Sir Percival came into that place, and his face shone very bright with peace and good-will; and he was exceedingly comely.

King Arthur said, "Art thou Percival?" And Percival said, "I am he." Thereupon King Arthur took Sir Percival's head into his hands, *Sir Percival* and he kissed him upon the brow. And Sir Percival kissed *is received with* King Arthur's hand and he kissed the ring of royalty upon *joy.* the King's finger, and so he became a true knight in fealty unto King Arthur.

Then Sir Percival said: "Lord, have I thy leave to speak?" And King Arthur said, "Say on." Sir Percival said, "Where is Sir Lamorack?" And King Arthur said, "Yonder he is." Then Sir Percival perceived where Sir Lamorack stood among the others, and he went to Sir Lamorack and knelt down before him; and Sir Lamorack was very much astonished, and said: "Why dost thou kneel to me, Percival?" Then Sir Percival said, "Dost thou know this ring?"

Then Sir Lamorack knew his father's ring and he cried out in a loud voice: "That is my father's ring; how came ye by it?"

Percival said: "Our mother gave it to me, for I am thy brother."

Upon this Sir Lamorack cried out with great passion; and he flung his *Sir Percival de-* arms about Sir Percival, and he kissed him repeatedly upon *clares himself to* the face. And so ardent was the great love and the great *Sir Lamorack.* passion that moved him that all those who stood about could in no wise contain themselves, but wept at that which they beheld.

Then, after a while, King Arthur said: "Percival, come with me, for I have somewhat to show thee."

*Sir Percival is* So King Arthur and Sir Lamorack and Sir Percival and sev-*made Knight of* eral others went unto that pavilion which was the pavilion of *the Round Table.* the Round Table, and there King Arthur showed Sir Percival a seat which was immediately upon the right hand of the Seat Perilous.

And upon the back of that seat there was a name emblazoned in letters of gold; and the name was this:

## 𝕻𝕰𝕽𝕮𝕴𝖁𝕬𝕷 𝕺𝕱 𝕲𝕬𝕷𝕰𝕾

Then King Arthur said: "Behold, Sir Percival, this is thy seat, for four days ago that name appeared most miraculously, of a sudden, where thou seest it; wherefore that seat is thine."

Then Sir Percival was aware that that name had manifested itself at the time when the Sangreal had appeared unto him in the castle of King Pecheur, and he was moved with a great passion of love and longing for the Lady Yvette; so that, because of the strength of that passion, it took upon it the semblance of a terrible joy. And he said to himself: "If my lady could but have beheld these, how proud would she have been! But, doubtless, she now, looketh down from Paradise and beholdeth us and all that we do." Thereupon he lifted up his eyes as though to behold her, but she was not there, but only the roof of that pavilion.

But he held his peace and said naught to anyone of those thoughts that disturbed him.

With this I conclude for the present the adventures of Sir Percival with only this to say: that thereafter, as soon as might be, he and Sir Lamorack went up into the mountains where their mother dwelt and brought her down thence into the world, and that she was received at the court of King Arthur with great honor and high regard until, after a while, she entered into a nunnery and took the veil.

Likewise it is to be said that Sir Percival lived, as he had vowed to do, a virgin knight for all of his life; for he never paid court to any lady from that time, but ever held within the sanctuary of his mind the image of that dear lady who waited for him in Paradise until he should come unto her in such season as God should see fit.

But you must not think that this is all that there is to tell of that noble, gentle and worthy young knight whose history we have been considering. For after this he performed many glorious services to the great honor of his knighthood and achieved so many notable adventures that the world spoke of him as being second in worship only to Sir Launcelot of the Lake. Yea; there were many who doubted whether Sir Launcelot himself was really a greater knight than Sir Percival; and though I may admit that Sir Launcelot had the greater prowess, yet Sir Percival was, certes, the more pure in heart and transparent of soul of those two.

So, hereafter, if God so wills, I shall tell more of Sir Percival, for I shall have much to write concerning him when I have to tell of the achievement of the Sangreal which he beheld in that vision at the Castle of King Pecheur as aforetold.

So, for this time, no more of these adventures, but fare you well.

## CONCLUSION.

*T*HUS *endeth the particular history of those three worthy, noble, excellent knights-champion—Sir Launcelot of the Lake, Sir Tristram of Lyonesse, and Sir Percival of Gales.*

*And I do hope that you may have found pleasure in considering their lives and their works as I have done. For as I wrote of their behavior and pondered upon it, meseemed they offered a very high example that anyone might follow to his betterment who lives in this world where so much that is ill needs to be amended.*

*But though I have told so much, yet, as I have just said, there remain many other things to tell concerning Sir Launcelot and Sir Percival, which may well afford anyone pleasure to read. These I shall recount in another volume at another time, with such particularity as those histories may demand.*